Birkhäuser

Operator Theory: Advances and Applications

Volume 292

Founded in 1979 by Israel Gohberg

Yemon Choi • Matthew Daws • Gordon Blower
Editors

Operators, Semigroups, Algebras and Function Theory

Volume from IWOTA Lancaster 2021

Birkhäuser

Editors
Yemon Choi
Department of Mathematics and Statistics
Lancaster University
Lancaster, UK

Matthew Daws
University of Central Lancashire
Preston, UK

Gordon Blower
Department of Mathematics and Statistics
Lancaster University
Lancaster, UK

ISSN 0255-0156 ISSN 2296-4878 (electronic)
Operator Theory: Advances and Applications
ISBN 978-3-031-38019-8 ISBN 978-3-031-38020-4 (eBook)
https://doi.org/10.1007/978-3-031-38020-4

Mathematics Subject Classification: 47-XX, 46L05, 20L05

This book is published under the imprint Birkhäuser, www.birkhauser-science.com by the registered company Springer Nature Switzerland AG
The registered company address is: Gewerbestrasse 11, 6330 Cham, Switzerland

Paper in this product is recyclable.

Preface

During 2016, the IWOTA steering committee informally approached Lancaster with the suggestion that we host a meeting in August 2020, and we accepted at once. The proposal was formally approved in 2017 at Chemnitz, and the meeting was announced in 2018 in Shanghai. Several special session organizers contributed ideas for topics, and we invited the main speakers. Their early support was instrumental in securing EPSRC financial support for the meeting, the first in the UK since 2004. IWOTA 2019 was an outstanding success, with the largest ever gathering of IWOTA in the beautiful city of Lisbon.

All was set fair until the winter of 2020, when COVID-19 crossed continents. On 1 March 2020, the American Physical Society cancelled their main meeting at two days' notice amid growing health concerns. Inevitably, MTNS Cambridge 2020 and IWOTA Lancaster 2020 followed, hopeful that normality would resume the next year. As cycles of lockdown and reopening passed, the organizers contemplated hybrid or online events, as meetings adopted what Bill Helton aptly described as a 2D format. Meanwhile, our colleagues in Chapman California organized another IWOTA with a complementary agenda. Our session organizers, main speakers, participants and funders showed exemplary patience with our uncertain communications, and they persisted with the project. Indeed, additional main speakers offered lectures and more session organizers enhanced the scientific programme.

Finally, IWOTA Lancaster took place from 16 to 20 August 2021 as an online meeting. On Monday, 16th August, the Deputy Vice-Chancellor of Lancaster University Steve Bradley and the Vice President of IWOTA Igor Klep opened the meeting by welcoming over 270 participants from 37 countries to IWOTA. These included over 107 early career researchers, some of whom contributed to the special sessions, while sessions 7 and 11 were chaired by early career researchers. Participants actively engaged in Special Sessions from the comfort of their homes and offices, although sometimes at unsociable hours.

Special Sessions

The meeting featured 15 Special Sessions, on the following topics.

1. Operator spaces, quantized function theory and noncommutative L_p spaces.
   Organizers: David Blecher (Houston, USA), Christian Le Merdy (Besancon, France)
2. Operator algebras.
   Organizer: Joachim Zacharias (Glasgow, UK)
3. Operator algebras in quantum theory.
   Organizers: Jason Crann (Carleton University, Canada), Ivan Todorov (Queen's University Belfast, UK, and Delaware, USA).
4. Noncommutative probability and random matrices.
   Organizers: Jani Virtanen (Reading, UK), Kenneth Dykema (Texas A&M, USA)
5. Operator semigroups and functional calculus.
   Organizers: Markus Haase (Kiel, Germany), Yuri Tomilov (IMPAN, Poland)
6. Spectral theory and differential operators.
   Organizers: Ian Wood (Kent, UK), Malcolm Brown (Cardiff, UK), Andrii Khrabustovskyi (Graz University of Technology, Austria)
7. Operator theory and communications.
   Organizer: Lucinda Hadley (Lancaster, UK)
8. Hilbert space operator theory and complex geometry.
   Organizers: Tirthankar Bhattacharyya (IISc Bangalore, India), Lucasz Kosinski (Jagiellonian University, Poland)
9. Free algebraic geometry and free analysis.
   Organizers: Victor Vinnikov (Ben Gurion, Israel), Juric Volcic (Texas A&M, USA)
10. Emerging topics: from operator algebras to geometric rigidity.
    A special session of IWOTA to mark the retirement of Professor Stephen C. Power from Lancaster University.
    Organizers: Derek Kitson (Mary Immaculate College, Ireland), Rupert Levene (University College Dublin, Ireland)
11. Early Career Researchers.
    Organizers: Eleftherios Kastis (Lancaster, UK), Maria Eugenia Celorrio (Lancaster, UK).
12. Quantum groups and algebraic quantum field theory.
    Organizers: Uwe Franz (Besancon, France), Robin Hillier (Lancaster, UK)
13. Multivariable operator theory.
    Organizers: Greg Knese (Washington, USA), Michael Dritschel (Newcastle, UK)
14. Complex analysis and operator theory.
    Organizers: Kehe Zhu (Albany, USA), Jani Virtanen (Reading, UK)

15. Operator ideals and operators on Banach spaces.
    Organizers: Kevin Beanland (Washington and Lee University, USA), Tomasz Kania (Jagiellonian University, Poland, and Czech Academy of Sciences), Niels Jakob Laustsen (Lancaster, UK)

Main Lectures

The main talks were delivered synchronously via Zoom, and produced to a high standard by Waggle Events. Most of these are available to view in the IWOTA YouTube Channel.

The Israel Gohberg Lecture was founded to commemorate the scientific achievements of Israel Gohberg, who was a driving force behind IWOTA. The inaugural lecture was by Vern Paulsen (Waterloo, Canada).

Monday 16th August

Charles Batty: Bounded functional calculi for unbounded operators

Birgit Jacob: Controllability and Riesz bases of infinite-dimensional port-Hamiltonian systems

Victor Vinnikov: A Beurling-type theorem for indefinite Hardy spaces on finite bordered Riemann surfaces

Hakan Hedenmalm: Planar orthogonal polynomials and soft Riemann–Hilbert problems

Tuesday 17th August

Benoit Collins: On the norm convergence of multi-matrix random matrix models

Magdalena Musat: Factorizable quantum channels, non-closure of quantum correlations and the Connes embedding problem

Zhengwei Liu: Quantum Fourier analysis

Kenneth Dykema: Spectral decompositions in finite von Neumann algebras

Wednesday 18th August

John E. McCarthy: Complete Pick spaces—what are they, and why are they interesting?

Raul Curto: The Beurling–Lax–Halmos theorem for infinite multiplicity

Mihai Putinar: Finding the cloud of a 2D point distribution

Zinaida Lykova: The μ-synthesis interpolation problem and some associated domains

Thursday 19th August

Vern Paulsen: Cooperative games and entanglement

Christian Le Merdy: ℓ^1 bounded maps and S^1 bounded maps on noncommutative L^p spaces

Juric Volcic: Positive noncommutative rational functions

Fritz Gesztesy: The limiting absorption principle and continuity properties of the spectral shift function for massless Dirac type operators

Friday 20th August At the start of the final day, the main organizer of the 2022 meeting of IWOTA outlined the scientific agenda, including the main speakers and the Special Sessions.

Marek Ptak: Invitation to IWOTA Krakow 2022
Daniel Alpay: Discrete analytic functions and Schur analysis
Omar El Fallah: Singular values of Hankel operators on Bergman spaces
J. W. Helton: Closing of IWOTA Lancaster 2021

In the closing address, the President of IWOTA asked participants to consider hosting future meetings.

The Proceedings

This volume of workshop proceedings features survey articles and original research papers which cover some of the main themes of the meeting. The editors are grateful to the referees who evaluated these contributions and suggested improvements.

The UK Engineering and Physical Sciences Research Council supported the meeting via grant EP/T007524/1 IWOTA Lancaster UK 2021 (principal investigator Gordon Blower, co-investigators Stephen Power and Derek Kitson). The local administration was by Anna Barnett and Stephanie Kutschmann.

Lancaster, UK — Yemon Choi
Preston, UK — Matthew Daws
Lancaster, UK — Gordon Blower

Contents

Criteria for Eventual Domination of Operator Semigroups and Resolvents

Sahiba Arora and Jochen Glück

Abstract We consider two C_0-semigroups $(e^{tA})_{t\geq 0}$ and $(e^{tB})_{t\geq 0}$ on function spaces (or, more generally, on Banach lattices) and analyse eventual domination between them in the sense that $\left|e^{tA} f\right| \leq e^{tB} |f|$ for all sufficiently large times t. We characterise this behaviour and prove a number of theoretical results which complement earlier results given by Mugnolo and the second author in the special case where both semigroups are positive for large times. Moreover, we study the analogous question of whether the resolvent of B eventually dominates the resolvent of A close to the spectral bound of B. This is closely related to the so-called maximum and anti-maximum principles. In order to demonstrate how our results can be used, we include several applications to concrete differential operators. At the end of the paper, we demonstrate that eventual positivity of the resolvent of a semigroup generator is closely related to eventual positivity of the Cesàro means of the associated semigroup.

1 Introduction

1.1 Motivation and Context

Quite recently, the theory of positive C_0-semigroups on functions spaces—or more generally, Banach lattices or ordered Banach spaces—(see, for instance, the monographs [8, 28] and the survey paper [9]) was complemented by the study of

S. Arora (✉)
Institut für Analysis, Fakultät für Mathematik, Technische Universität Dresden, Dresden, Germany
e-mail: sahiba.arora@mailbox.tu-dresden.de

J. Glück
Fakultät für Mathematik und Naturwissenschaften, Bergische Universität Wuppertal, Wuppertal, Germany
e-mail: glueck@uni-wuppertal.de

Y. Choi et al. (eds.), *Operators, Semigroups, Algebras and Function Theory*, Operator Theory: Advances and Applications 292,
https://doi.org/10.1007/978-3-031-38020-4_1

the following more subtle type of behaviour: a C_0-semigroup on a Banach lattice is called *eventually positive* if, for each positive initial value, the orbit becomes and stays positive for large times. The basics of the theory are laid out in the papers [14, 15] and have later been complemented in various articles which adapt the theory, for instance, to the more subtle concept of *local eventual positivity* [4]. Recent results indicate that large fractions even of the more involved parts of the theory of positive semigroups can be adapted to the eventually positive case, though significant changes to the employed methods are sometimes needed to achieve this. We refer to [34] for an instance of a very recent spectral-theoretic result of this type.

Concrete examples of semigroups that satisfy eventual positivity or a variation thereof abound. To list only a few examples, we explicitly mention the semigroup generated by $-\Delta^2$ on $\mathbb{R}^d$ (see [20, 21], and also [19]), the semigroup generated by the Dirichlet-to-Neumann operator on the unit circle in $\mathbb{R}^2$ [11], certain classes of semigroups on metric graphs [10, Proposition 5.5], and semigroups generated by Laplace operators which are coupled by point interactions [26, Proposition 2]. This list is not exhaustive but provides an impression of the large stock of examples for eventually positive behaviour. It is also worthwhile to note that, in finite dimensions, a general theory of eventually positive semigroups was developed earlier than its infinite-dimensional analogue; see, in particular, [29].

Building upon the theory of eventual positivity, the closely related phenomenon of *eventual domination* between two C_0-semigroups was studied in [23], where also a large variety of examples were presented.

In this article, we complement the theory of both eventual positivity and eventual domination by providing a number of natural extensions of earlier results. In particular, we sharpen and refine some of the results on eventual domination of semigroups from [23], and also demonstrate that similar results can be shown for eventual domination of resolvents of operators. Finally, we demonstrate that the interplay between eventual positivity of semigroups and of resolvents that plays an essential role in [14, 15], can also be extended to Cesàro means of semigroups.

1.2 Notation and Terminology

Throughout the paper, we assume familiarity with the theory of real and complex Banach lattices for which we refer to the monographs [27, 32]. So, let E be a complex Banach lattice whose real part is denoted by $E_\mathbb{R}$. An operator $A : E \supseteq \operatorname{dom}(A) \to E$ is called *real* if $\operatorname{dom}(A)$ is spanned by $\operatorname{dom}(A) \cap E_\mathbb{R}$ and A maps real vectors to real vectors, i.e., it maps $\operatorname{dom}(A) \cap E_\mathbb{R}$ into $E_\mathbb{R}$. Note that if A is a real operator, then so are the resolvent operators $\mathcal{R}(\lambda, A)$ for each real number λ in the resolvent set of A. In particular, an operator $T : E \to E$ is real if $TE_\mathbb{R} \subseteq E_\mathbb{R}$. Also, a C_0-semigroup $(e^{tA})_{t\geq 0}$ is called real if each operator e^{tA} is real.

For two vectors u, v in $E_{\mathbb{R}}$, we write $u \succeq v$ (or $v \preceq u$) if there exists a constant $c > 0$ such that $u \geq cv$. With the above notation in mind, the *principal ideal* of E generated by a positive element $u \in E$ is given by

$$E_u := \{x \in E : |x| \preceq u\}$$

which is also a complex Banach lattice when equipped with the *gauge norm* defined by

$$\|x\|_u := \inf\{c > 0 : |x| \leq cu\} \quad \text{for each } x \in E_u.$$

We say that u is a *quasi-interior point* of E if E_u is dense in E. For example, if $C(K)$ denotes the space of continuous functions on a compact Hausdorff space K, then the set of quasi-interior points is precisely the set of all $f \in C(K)$ such that $f(x) > 0$ for all $x \in K$. In fact, if $u \in C(K)$ is a quasi-interior point, then $C(K)_u = C(K)$. On the other hand, if (Ω, μ) is a finite measure space and $p \in [1, \infty)$, then the set of quasi-interior points of $L^p(\Omega, \mu)$ is the set of all functions $f \in L^p(\Omega, \mu)$ such that $f(x) > 0$ for almost all $x \in \Omega$.

For a vector $u \in E$ and a functional $\varphi \in E'$ (where E' denotes the dual space of E), we use the notation $u \otimes \varphi$ to denote the rank-one operator

$$E \ni f \mapsto (u \otimes \varphi)f := \langle \varphi, f \rangle u \in E.$$

The functional φ is called *strictly positive* if $\langle \varphi, f \rangle > 0$ for all $0 \lneq f \in E$ or equivalently, if the kernel of φ contains no positive non-zero element. Let $T, S : E \to E$ be linear operators. We say that T is *positive* and write $T \geq 0$ if $Tf \geq 0$ for all $0 \leq f \in E$. Of course, every positive operator is real. Moreover, for real operators $T, S : E \to E$, the notation $T \geq S$ is used to denote $T - S \geq 0$. Analogously to the case of vectors, we write $T \succeq S$ (or $S \preceq T$) if there exists $c > 0$ such that $T \geq cS$.

The space of bounded linear operators between two complex Banach spaces E and F is denoted as usual by $\mathcal{L}(E, F)$. In addition, we use the shorthand $\mathcal{L}(E) := \mathcal{L}(E, E)$. The kernel of $T \in \mathcal{L}(E, F)$ will be denoted by $\ker T$, the range of T by $\operatorname{Rg} T$, and the rank of T by $\operatorname{rk} T$. Further notations will be introduced as and when required.

1.3 Organization of the Article

In Sect. 2, we present criteria for eventual domination of resolvents—in addition to proving sufficient criteria, we also prove a necessary condition. Thereafter, in Sect. 3, we continue the theory of eventual domination of C_0-semigroups from [23]. Applications of the results presented in Sect. 2 and 3 will be given in Sect. 4. Finally, in Sect. 5, we demonstrate that many results about the eventual positivity of semigroups and resolvents can also be adapted to Cesàro means.

2 Eventual Domination of Resolvents

In this section, we give conditions for the resolvent of an operator B to eventually dominate the resolvent of A close to the spectral bound of B. This is reminiscent of the eventual domination of semigroups studied in [23]. For generators of C_0-semigroups, a characterisation for domination of resolvents is given in [28, Proposition C-II-4.1]. The main results of this section are the sufficient condition for eventual domination in Theorem 2.1 and the necessary condition in Theorem 2.5.

Theorem 2.1 *Let A and B be closed, densely defined, and real operators on a complex Banach lattice E, let $u \in E$ be positive, and let $\lambda_0 \in \mathbb{R}$ be a spectral value of B and a pole of its resolvent. Moreover, assume the following:*

(a) *The domination conditions* $\operatorname{dom}(A), \operatorname{dom}(B) \subseteq E_u$ *are satisfied.*
(b) *The spectral projection P associated with the spectral value λ_0 of B satisfies $Pf \succeq u$ whenever $0 \lneq f \in E$.*

If $\lambda_0 \in \rho(A)$, then for each $0 \neq f \in E$, we have

$$\mathcal{R}(\lambda, B)\,|f| - |\mathcal{R}(\lambda, A)f| \succeq u \quad \text{and} \quad \mathcal{R}(\mu, B)\,|f| + |\mathcal{R}(\mu, A)f| \preceq -u$$

for all λ in an f-dependent right neighbourhood of λ_0 and for all μ in an f-dependent left neighbourhood of λ_0.

Observe that because the operator A is densely defined, the domination assumption $\operatorname{dom}(A) \subseteq E_u$ in particular implies that u is a quasi-interior point of E.

The assumption (b) in the above theorem plays a crucial role in the abstract theory of the so-called individual (anti-)maximum principles. To understand this, let λ_0 be a spectral value of a closed and real operator B on a complex Banach lattice E and let $u \in E$ be a quasi-interior point. In addition, assume that λ_0 is a pole of the resolvent and consider the inequality

$$(\lambda - \lambda_0)\mathcal{R}(\lambda, B)f \geq cu \tag{2.1}$$

for $0 \lneq f \in E$, a real number $\lambda \in \rho(B)$, and some $c > 0$. If there exists a constant $c > 0$ such that (2.1) holds for all λ in an f-dependent *right* neighbourhood of λ_0, then, in particular, we have that $\mathcal{R}(\lambda, B)f \succeq u$ for all λ in this *right* neighbourhood of λ_0. In this case, B is said to satisfy an *individual maximum principle at* λ_0. On the other hand, if there exists a constant $c > 0$ such that (2.1) holds for all λ in an f-dependent *left* neighbourhood of λ_0, then, in particular, we have that $\mathcal{R}(\lambda, B)f \preceq -u$ for all λ in this *left* neighbourhood of λ_0. This time, we say that B satisfies an *individual anti-maximum principle at* λ_0. For recent contributions to the abstract theory of (anti-)maximum principles, we refer to [5, 6, 12, 14, 15]. In particular, it was proved in [12, Theorem 4.1] that if B satisfies the individual maximum or the individual anti-maximum principle at λ_0, then the spectral projection P corresponding to λ_0 satisfies $Pf \succeq u$ for all $0 \lneq f \in E$. If dom $(B) \subseteq E_u$, then a converse is also true, as will be made clear below. Before moving on, we note that a characterisation of assumption (b) above in terms of further spectral properties can be found in [14, Corollary 3.3].

Because of the above remarks, before proving Theorem 2.1, we first show the following:

Proposition 2.2 *Let A and B be closed, densely defined, and real operators on a complex Banach lattice E and let $u \in E$ be positive. Let $\lambda_0 \in \mathbb{R}$ be a spectral value of B and a pole of the resolvent $\mathcal{R}(\cdot, B)$. Moreover, assume that the domination condition* dom $(A) \subseteq E_u$ *is satisfied and that $\lambda_0 \in \rho(A)$.*

(i) *If for each $0 \lneq f \in E$, there exists $c > 0$ such that* (2.1) *holds for all λ in an f-dependent* right *neighbourhood of λ_0, then*

$$\mathcal{R}(\mu, B)\,|f| - |\mathcal{R}(\mu, A)f| \succeq u$$

for each $0 \neq f \in E$ and for all μ in an f-dependent right neighbourhood of λ_0.

(ii) *If for each $0 \lneq f \in E$, there exists $c > 0$ such that* (2.1) *holds for all λ in an f-dependent* left *neighbourhood of λ_0, then*

$$\mathcal{R}(\mu, B)\,|f| + |\mathcal{R}(\mu, A)f| \preceq -u$$

for each $0 \neq f \in E$ and for all μ in an f-dependent left neighbourhood of λ_0.

The following lemma contains an observation used in the proof of Proposition 2.2 below. As these arguments appear frequently in the theory of (anti-)maximum principles, we choose to state them separately.

Lemma 2.3 *Let A be a closed linear operator on a complex Banach lattice E and let $u \geq 0$ be a quasi-interior point of E such that the domination condition* dom $(A) \subseteq E_u$ *is satisfied. Let $\lambda_0 \in \mathbb{C}$.*

(i) *If λ_0 lies in the resolvent set of A, then $(\lambda - \lambda_0)\mathcal{R}(\lambda, A) \to 0$ in $\mathcal{L}(E, E_u)$ as $\lambda \to \lambda_0$.*

(ii) *If λ_0 is a simple pole of the resolvent $\mathcal{R}(\cdot, A)$ and the corresponding spectral projection is denoted by P, then $(\lambda - \lambda_0)\mathcal{R}(\lambda, A) \to P$ in $\mathcal{L}(E, E_u)$ as $\lambda \to \lambda_0$.*

Proof We assume without loss of generality that $\lambda_0 = 0$.

(ii) We know from [14, Lemma 4.7(i)] that $\lambda\mathcal{R}(\lambda, A) \to P$ in $\mathcal{L}(E, \text{dom}\,(A))$ as $\lambda \to 0$ (where dom (A) is endowed with the graph norm). Moreover, it follows from the domination assumption and the closed graph theorem that dom (A) embeds continuously into E_u, which implies the assertion.

(i) Firstly, $0 \in \rho(A)$ implies that there is a neighbourhood of 0 which contains no spectral value of A and $\mathcal{R}(\lambda, A) \to \mathcal{R}(0, A)$ in $\mathcal{L}(E)$ as $\lambda \to 0$. Now,

$$A\mathcal{R}(\lambda, A) = \lambda\mathcal{R}(\lambda, A) - I \to -I = A\mathcal{R}(0, A)$$

in $\mathcal{L}(E)$ as $\lambda \to 0$. Thus $\mathcal{R}(\lambda, A) \to \mathcal{R}(0, A)$ in $\mathcal{L}(E, \operatorname{dom}(A))$ as $\lambda \to 0$. The result follows as in (ii) above. □

Proof of Proposition 2.2 By a change of signs, we only need to prove (i). To this end, assume without loss of generality that $\lambda_0 = 0$ and fix a non-zero vector $f \in E$. By assumption there exist a constant $c > 0$ and a number $\delta_1 > 0$ such that $\lambda\mathcal{R}(\lambda, B)\,|f| \geq cu$ whenever $\lambda \in (0, \delta_1)$. Furthermore, by Lemma 2.3(i), there exists $\delta_2 > 0$ such that $\lambda\,|\mathcal{R}(\lambda, A)f| \leq c/2u$ for all $\lambda \in (0, \delta_2)$. Letting $\delta = \min\{\delta_1, \delta_2\} > 0$, we obtain

$$\lambda\big(\mathcal{R}(\lambda, B)\,|f| - |\mathcal{R}(\lambda, A)f|\big) \geq cu - \frac{c}{2}u = \frac{c}{2}u$$

for all $\lambda \in (0, \delta)$, from which the result follows. □

To prove Theorem 2.1 using Proposition 2.2, we will show that assumption (b) implies that (2.1) holds in some neighbourhood of λ_0. Essentially, this has already been shown in [14, Theorem 4.4], but the independence of the constant c of λ was not explicitly mentioned there. We state this here in a separate and general lemma, for the sake of easier reference.

Lemma 2.4 *Let $u \geq 0$ be a quasi-interior point in a complex Banach lattice E and let $(T_\alpha)_{\alpha\in\mathcal{A}}$ be a net of bounded and real linear operators in $\mathcal{L}(E)$ whose ranges are contained in E_u. Assume that $(T_\alpha)_{\alpha\in\mathcal{A}}$ converges strongly in $\mathcal{L}(E, E_u)$ to an operator $T \in \mathcal{L}(E, E_u)$.*

Fix $0 \lneq f \in E$. If there exists $c > 0$ such that $Tf \geq cu$, then for each $\epsilon \in (0, c)$, there exists $\alpha_0 \in \mathcal{A}$ such that $T_\alpha f \geq \epsilon u$ for all $\alpha \geq \alpha_0$.

Proof Let $\epsilon \in (0, c)$. Because $(T_\alpha)_{\alpha\in\mathcal{A}}$ converges strongly to T in $\mathcal{L}(E, E_u)$, there exists $\alpha_0 \in \mathcal{A}$ such that

$$\|T_\alpha f - Tf\|_u \leq c - \epsilon$$

and in turn, $|T_\alpha f - Tf| \leq (c - \epsilon)u$ for all $\alpha \geq \alpha_0$. Since the operators T_α are all real, we can write

$$T_\alpha f \geq Tf - (c - \epsilon)u \geq cu - (c - \epsilon)u = \epsilon u$$

for all $\alpha \geq \alpha_0$. □

Proof of Theorem 2.1 Without loss of generality, let $\lambda_0 = 0$. Using the necessary condition for our assumption (b) given in [14, Proposition 3.1], we get that the pole order of the spectral value 0 of B is one. Thus, it follows from assumption (a) and Lemma 2.3(ii) that $\lambda\mathcal{R}(\lambda, B) \to P$ in $\mathcal{L}(E, E_u)$ as $\lambda \to 0$.

Fix $0 \lneq f \in E$. Using assumption (b) along with Lemma 2.4, we see that (2.1) holds for all λ in some (f-dependent) neighbourhood of λ_0. The conclusion now follows from Proposition 2.2(ii). □

We now prove a converse to Theorem 2.1 and Proposition 2.2 in the spirit of [23, Theorem 3.1]. The point of the result is that, under appropriate technical assumptions, we cannot have both eventual domination and eventual positivity of the resolvents of A and B on the right of an eigenvalue λ_0 of B if λ_0 is also an eigenvalue of A (except in the trivial case where $A = B$).

As a consequence, we are able to disprove the eventual domination of resolvents for odd order operators in Sect. 4.5.

Theorem 2.5 *Let A and B be closed, densely defined, and real operators on a complex Banach lattice E and let $u \geq 0$ be a quasi-interior point of E. Let $\lambda_0 \in \mathbb{R}$ be a spectral value of B and a pole of the resolvent $\mathcal{R}(\,\cdot\,, B)$. Moreover, assume the following:*

(a) *For each $0 \leq f \in E$, we have $0 \leq \mathcal{R}(\lambda, A)f \leq \mathcal{R}(\lambda, B)f$ for all λ in an f-dependent right neighbourhood of λ_0.*
(b) *For each $0 \lneq f \in E$, we have $\mathcal{R}(\lambda, B)f \succeq u$ for all λ in an f-dependent right neighbourhood of λ_0.*

If λ_0 is, in addition, a spectral value of A and a pole of the resolvent $\mathcal{R}(\,\cdot\,, A)$, then $A = B$.

Proof There is no loss of generality in assuming $\lambda_0 = 0$. Denote by P the spectral projection associated with the spectral value 0 of B. The maximum principle assumption in (b) and [12, Theorem 4.1] together imply that $Pf \succeq u$ whenever $0 \lneq f \in E$. We can thus apply [14, Corollary 3.3] to obtain that $\ker B$ is spanned by a vector $v \succeq u$, the eigenspace $\ker B'$ contains a strictly positive functional ψ, and 0 is also a simple pole of the resolvent $\mathcal{R}(\,\cdot\,, B)$. The latter implies that $\operatorname{Rg} P = \ker B$ and $\operatorname{Rg} P' = \ker B'$. We may rescale ψ to have norm 1. Furthermore, we rescale v such that $\langle \psi\,,\, v\rangle = 1$, and thus obtain $P = v \otimes \psi$.

Next, let Q denote the spectral projection of A associated with the pole 0. As 0 is a simple pole of $\mathcal{R}(\,\cdot\,, B)$, we conclude from the eventual domination assumption in (a) that 0 is also a simple pole of $\mathcal{R}(\,\cdot\,, A)$. So, $\lambda\mathcal{R}(\lambda, A) \to Q$ in $\mathcal{L}(E)$ as $\lambda \to 0$. Moreover, the eventual positivity assumption on $\mathcal{R}(\,\cdot\,, A)$ in (a) implies that $Q \geq 0$ and the eventual domination in assumption (a) implies that $P \geq Q \geq 0$. We can thus infer from [17, Proposition 2.1.3] that $1 = \operatorname{rk} P \geq \operatorname{rk} Q > 0$, so $\operatorname{rk} Q = 1$. Thus there exists $w \in W$ and $\phi \in E'$ such that $Q = w \otimes \phi$; here, we may choose ϕ to have norm 1. Moreover, due to the positivity of Q, we may choose both w and ϕ to be positive. Note that $\langle \phi\,,\, w\rangle = 1$, because Q is a projection.

Next, we prove that $\psi = \phi$. To this end, observe that $\phi = Q'\phi \leq P'\phi = \langle \phi\,,\, v\rangle\, \psi$. Since

$$\langle \langle \phi\,,\, v\rangle\, \psi - \phi\,,\, v\rangle = 0$$

and v is a quasi-interior point of E, it follows that $\langle \varphi\,,\, v\rangle\, \psi = \phi$; see [32, Theorem II.6.3]. As ψ and ϕ are both positive functionals of norm 1, this implies that $\psi = \phi$.

Now consider a vector $0 \lneq f \in E$ and let $\delta > 0$ be such that $\mathcal{R}(\lambda, B)f \geq \mathcal{R}(\lambda, A)f$ for all $\lambda \in (0, \delta)$. Then, for $\lambda \in (0, \delta)$, the computation

$$\begin{aligned}\langle \psi\,,\, \mathcal{R}(\lambda, B)f - \mathcal{R}(\lambda, A)f\rangle &= \big\langle \mathcal{R}(\lambda, B')\psi\,,\, f\big\rangle - \big\langle \mathcal{R}(\lambda, A')\phi\,,\, f\big\rangle \\ &= \frac{1}{\lambda}[\langle \psi\,,\, f\rangle - \langle \phi\,,\, f\rangle] = 0\end{aligned}$$

and the strict positivity of ψ imply that $\mathcal{R}(\lambda, B)f = \mathcal{R}(\lambda, A)f$ for all $\lambda \in (0, \delta)$. Since the resolvent is analytic, the identity theorem for analytic functions yields that the same equality holds for all λ in a non-empty f-independent set $U \subseteq \mathbb{C}$. Hence, $\mathcal{R}(\,\cdot\,, A) = \mathcal{R}(\,\cdot\,, B)$ for all $\lambda \in U$, and we thus conclude that $A = B$, as desired. □

3 Eventual Domination of Semigroups

We now consider the case of C_0-semigroups rather than resolvents. Eventual domination for C_0-semigroups was studied in [23]; in this section, we provide a number of refinements and improvements to the results given there. Our first result, Theorem 3.1, extracts sufficient conditions for eventual domination from [23, Theorem 3.1], optimises the assumptions of the aforementioned theorem, and further improves the conclusion. In particular, in contrast to [23, Theorem 3.1], we do not need the dominated semigroup to be eventually positive, and we also consider non-positive initial values f (the latter is made possible by the inclusion of the modulus in the eventual domination estimate).

To state the theorem, we recall that for a closed operator A on a Banach space E, its *spectral bound* is defined as

$$s(A) := \sup\{\operatorname{Re}\lambda : \lambda \in \sigma(A)\} \in [-\infty, \infty].$$

If A generates a C_0-semigroup, then $s(A) < \infty$.

For two C_0-semigroups $(e^{tA})_{t\geq 0}$ and $(e^{tB})_{t\geq 0}$ on a complex Banach lattice E we say that $(e^{tA})_{t\geq 0}$ *individually eventually dominates* $(e^{tB})_{t\geq 0}$ if for each $f \in E$ there exists a time $t_0 \geq 0$ such that

$$\left|e^{tA} f\right| \leq e^{tB}\,|f|$$

for all $t \geq t_0$.

Theorem 3.1 *Let $(e^{tA})_{t\geq 0}$ and $(e^{tB})_{t\geq 0}$ be real C_0-semigroups on a complex Banach lattice E and let $u \geq 0$ be a quasi-interior point of E. In addition, assume the following:*

(a) *There exist $t_1, t_2 \geq 0$ such that $e^{t_1 A}E, e^{t_2 B}E \subseteq E_u$.*

(b) *The rescaled semigroup* $(e^{t(B-s(B))})_{t\geq 0}$ *converges strongly as* $t \to \infty$ *to an operator* P *that satisfies* $Pf \succeq u$ *whenever* $0 \lneq f \in E$.
(c) *The rescaled semigroup* $(e^{t(A-s(B))})_{t\geq 0}$ *converges strongly to* 0 *as* $t \to \infty$.

Then for each non-zero $f \in E$*, there exists a time* $\tau \geq 0$ *such that* $e^{tB}|f| - |e^{tA}f| \succeq u$ *for all* $t \geq \tau$*. In particular,* $(e^{tB})_{t\geq 0}$ *individually eventually dominates* $(e^{tA})_{t\geq 0}$.

The assumption (b) in the previous theorem plays an important role in the theory of individual eventual positivity. Let $(e^{tB})_{t\geq 0}$ be a real C_0-semigroup on a complex Banach lattice E and let $0 \leq u \in E$ be a quasi-interior point. If for each $0 \lneq f \in E$, there exists a time $t_0 \geq 0$ such that $e^{tB}f \succeq u$ for all $t \geq t_0$, then the semigroup $(e^{tB})_{t\geq 0}$ is said to be *individually eventually strongly positive* (with respect to u). The theory of eventually positive semigroups was developed in several recent articles. In particular, it was shown in [12, Theorem 5.1] that if the spectral bound $s(B)$ is a spectral value of B and a pole of the resolvent $\mathcal{R}(\,\cdot\,, B)$, then individual eventual strong positivity of $(e^{tB})_{t\geq 0}$ with respect to u implies that the spectral projection P associated with $s(B)$ satisfies $Pf \succeq u$ for all $0 \lneq f \in E$.

We prove the following result before proving Theorem 3.1:

Proposition 3.2 *Let* $(e^{tA})_{t\geq 0}$ *and* $(e^{tB})_{t\geq 0}$ *be real* C_0*-semigroups on a complex Banach lattice* E *and let* $u \geq 0$ *be a quasi-interior point of* E*. In addition, assume the following:*

(a) *There exists* $t_1 \geq 0$ *such that* $e^{t_1 A}E \subseteq E_u$.
(b) *For each* $0 \lneq f \in E$*, there exists a time* $t_0 \geq 0$ *and a constant* $c > 0$ *such that* $e^{tB}f \geq cu$ *for all* $t \geq t_0$.
(c) *The semigroup* $(e^{tA})_{t\geq 0}$ *converges to* 0 *strongly as* $t \to \infty$.

Then for each non-zero $f \in E$*, there exists a time* $\tau \geq 0$ *such that* $e^{tB}|f| - |e^{tA}f| \succeq u$ *for all* $t \geq \tau$*. In particular,* $(e^{tB})_{t\geq 0}$ *individually eventually dominates* $(e^{tA})_{t\geq 0}$.

Proof First of all, note that by the closed graph theorem we have $e^{t_1 A} \in \mathcal{L}(E, E_u)$. Therefore $e^{(t+t_1)A} = e^{t_1 A}e^{tA} \to 0$ in $\mathcal{L}(E, E_u)$ as $t \to \infty$. Fix a non-zero vector $f \in E$. Then there exists $t_2 \geq 0$ such that

$$\left\| e^{(t+t_1)A} f \right\|_u \leq \frac{c}{2}$$

for all $t \geq t_2$, where c is the constant from assumption (b) for the vector $|f|$. Hence $|e^{tA}f| \leq \frac{c}{2}u$ for all $t \geq t_1 + t_2$. Let $\tau = \max\{t_0, t_1 + t_2\}$. Then

$$e^{tB}|f| - \left|e^{tA}f\right| \geq cu - \frac{c}{2}u = \frac{c}{2}u$$

for all $t \geq \tau$, as required. □

To prove Theorem 3.1, we need to know that the strong convergence $e^{t(B-s(B))} \to P$ as $t \to \infty$ implies the assumption (b) in Proposition 3.2 for the semigroup generated by $B - s(B)$. This was essentially shown in [14, Theorem 5.2] but the independence of the constant c of time t was not mentioned explicitly there; it does, however, follow from Lemma 2.4. The details are as follows.

Proof of Theorem 3.1 Replacing B by $B - s(B)$ and A by $A - s(B)$, we may assume that $s(B) = 0$. We only need to prove that assumption (b) in Proposition 3.2 is satisfied. To this end, fix $0 \lneq f \in E$ and let $c > 0$ be such that $Pf \geq cu$. Since $e^{t_2 B} E \subseteq E_u$, the operator $e^{t_2 B}$ is bounded from E to E_u by the closed graph theorem. We therefore obtain that

$$e^{tB} = e^{t_2 B} e^{(t-t_2)B} \to P \text{ in } \mathcal{L}(E, E_u)$$

as $t \to \infty$. Combining this with assumption (b) and Lemma 2.4, we obtain that for each $0 \lneq f \in E$, there exist a constant $c > 0$ and time $t_0 \geq 0$ such that $e^{tB} f \geq cu$ for all $t \geq t_0$. Hence, all assumptions of Proposition 3.2 are satisfied, so the result follows. □

After obtaining sufficient conditions for individual domination of semigroups, we now turn to the uniform case. For C_0-semigroups $(e^{tA})_{t\geq 0}$ and $(e^{tB})_{t\geq 0}$ on a complex Banach lattice E we say that $(e^{tA})_{t\geq 0}$ *uniformly eventually dominates* $(e^{tB})_{t\geq 0}$ if there exists a time $t_0 \geq 0$ such that

$$\left|e^{tA} f\right| \leq e^{tB} \, |f|$$

for all $t \geq t_0$ and all $f \in E$. Before giving a sufficient condition for uniform eventual domination, we show by means of an example that the notions of individual and uniform eventual domination are not equivalent.

Example 3.3 Consider the Banach lattice $E = C[0, 1]$ and the strictly positive functionals $\phi_A, \phi_B : E \to \mathbb{C}$ given by

$$\langle \phi_A \,,\, f\rangle = \int_0^1 f(x)\, \mathrm{d}x \quad \text{and} \quad \langle \phi_B \,,\, f\rangle = 2\int_0^1 x f(x)\, \mathrm{d}x$$

for all $f \in E$. Let P_A and P_B be the rank-one projections on E given by $P_A = \mathbb{1} \otimes \phi_A$ and $P_B = \mathbb{1} \otimes \phi_B$; where $\mathbb{1} \in E$ denotes the constant one function.

Now consider the bounded linear operators

$$A = P_A - \frac{3}{2}\,\mathrm{id} \quad \text{and} \quad B = P_B - \mathrm{id}$$

on E. Then $s(B) = 0$ and $e^{tB} = P_B + e^{-t}(\mathrm{id} - P_B) \to P_B$ in operator norm as $t \to \infty$. Moreover, $s(A) = -\frac{1}{2}$, which implies—since A is a bounded operator—that $(e^{tA})_{t\geq 0}$ converges in operator norm to 0 as $t \to \infty$. Furthermore, because $E_{\mathbb{1}} = E$ and $P_B f \succeq \mathbb{1}$ whenever $0 \lneq f \in E$, all the assumptions of Theorem 3.1 are satisfied. Hence, $(e^{tB})_{t\geq 0}$ individually eventually dominates $(e^{tA})_{t\geq 0}$.

Next, note that P_B is the spectral projection of B associated with $s(B) = 0$. As $s(A) < 0$, Theorem 2.1 implies that for each $f \geq 0$, we have $\mathcal{R}(\lambda, B)f - \mathcal{R}(\lambda, A)f \succeq \mathbb{1}$ for all λ in an f-dependent right neighbourhood of 0.

In order to see that the eventual domination is not uniform, we define, for each $n \in \mathbb{N}$, a function $0 \leq f_n \in E$ in the following way: let $f_n(x) = 1 - nx$ for $x \in [0, 1/n)$ and $f_n(x) = 0$ for $x \in [1/n, 1]$. Then

$$P_A f_n = \frac{1}{2n}\,\mathbb{1} \qquad \text{and} \qquad P_B f_n = \frac{1}{3n^2}\,\mathbb{1}\,.$$

Using $f_n(1) = 0$, we obtain

$$\begin{aligned}(e^{tB} - e^{tA})f_n(1) &= \frac{1}{3n^2}(1 - e^{-t}) - \frac{e^{-t/2}}{2n}(1 - e^{-t}) \\ &= \left(\frac{1}{3n^2} - \frac{e^{-t/2}}{2n}\right)(1 - e^{-t})\end{aligned}$$

for all $t \geq 0$. Thus, for each fixed $t > 0$, we conclude for all sufficiently large n (namely, for $n > 2/3e^{t/2}$) that $(e^{tB} - e^{tA})f_n \not\geq 0$. Therefore, $(e^{tB})_{t\geq 0}$ does not uniformly eventually dominate $(e^{tA})_{t\geq 0}$.

Similarly, using the simple fact that the resolvent of every projection P is given by

$$\mathcal{R}(\lambda, P) = \frac{P + (\lambda - 1)\,\mathrm{id}}{\lambda(\lambda - 1)}$$

for all $\lambda \in \mathbb{C} \setminus \{0, 1\}$, we can compute that

$$(\mathcal{R}(\lambda, B) - \mathcal{R}(\lambda, A))f_n(1) = \frac{1}{3n^2}\left(\frac{1}{\lambda(\lambda + 1)}\right) - \frac{2}{n}\left(\frac{1}{(2\lambda + 3)(2\lambda + 1)}\right)$$

for each $\lambda > 0$. Hence, for each fixed $\lambda > 0$ we have

$$\mathcal{R}(\lambda, B)f_n(1) < \mathcal{R}(\lambda, A)f_n(1)$$

for all sufficiently large n.

Conditions for uniform eventual domination between semigroups were given in [23, Theorem 3.7]. However, only the case of self-adjoint semigroups on L^2-spaces was considered there. By adapting techniques from [13] we now give a sufficient condition for uniform eventual domination which also works for semigroups on more general spaces (and thus, in particular, without any self-adjointness assumption). As in Theorem 3.1, we do not need the dominated semigroup to be (eventually) positive, and by including a modulus on both sides of the eventual domination estimate, we are able to consider non-positive initial values as well.

Theorem 3.4 *Let* $(e^{tA})_{t\geq 0}$ *and* $(e^{tB})_{t\geq 0}$ *be real* C_0*-semigroups on a complex Banach lattice* E. *Let* $u \geq 0$ *be a quasi-interior point of* E *and* $\phi \in E'$ *be a strictly positive functional. In addition, assume the following:*

(a) *There exist times* $t_1, t_2 \geq 0$ *such that* $e^{t_1 A} E \subseteq E_u$ *and* $e^{t_2 A'} E' \subseteq (E')_\varphi$.
(b) *There exists a time* $t_3 \geq 0$ *and a constant* $c > 0$ *such that* $e^{tB} \succeq c(u \otimes \varphi)$ *for all* $t \geq t_3$.
(c) *The semigroup* $(e^{tA})_{t\geq 0}$ *converges to* 0 *in operator norm as* $t \to \infty$.

Then there exists a time $\tau \geq 0$ *such that* $e^{tB}\,|f| - \left|e^{tA} f\right| \succeq (u \otimes \varphi)\,|f|$ *for all* $t \geq \tau$ *and all* $f \in E$. *In particular,* $(e^{tB})_{t\geq 0}$ *uniformly eventually dominates* $(e^{tA})_{t\geq 0}$.

We make a few comments on the assumptions of the above theorem before proving the result.

Remarks 3.5

(a) *Smoothing assumptions.* Assumption (a) in Theorem 3.4 can be seen as an abstract version of an ultracontractivity property. The idea to use this assumption to obtain uniform eventual domination is directly borrowed from [13, Theorem 3.1], where similar smoothing assumptions were imposed to obtain uniform eventual positivity of semigroups.
(b) *Eventual positivity assumption.* Assumption (b) above is a uniform eventual strong positivity assumption and is in line with the individual eventual strong positivity assumption in Proposition 3.2 on the dominating semigroup. In fact, our assumption here is exactly the conclusion of [13, Theorem 3.1] (in the case where $s(B) = 0$), and is therefore satisfied whenever the hypothesis of [13, Theorem 3.1] is. In particular, assumption (b) can be replaced by the following two conditions:

   (i) There exist times $t_3, t_4 \geq 0$ such that $e^{t_3 B} E \subseteq E_u$ and $e^{t_4 B'} E' \subseteq (E')_\varphi$.
   (ii) The spectral bound $s(B)$ is a dominant spectral value of B and the corresponding eigenspace is generated by a vector $v \succeq u$. Moreover, the dual eigenspace $\ker(s(B) - B')$ contains a functional $\psi \succeq \varphi$.

   In this framework, it is worth mentioning an inaccuracy in the statement of the conclusion of [13, Theorem 3.1]: the estimate $e^{tA} \succeq \epsilon(u \otimes \varphi)$ for all $t \geq t_0$ in the theorem is clearly incorrect in cases where the semigroup converges to 0 as $t \to \infty$. This is due to the lack of a scaling factor: as is obvious from the proof, the correct conclusion there is $e^{t(A-s(A))} \succeq \epsilon(u \otimes \varphi)$ for all $t \geq t_0$.
(c) *Convergence in the operator norm.* In order to obtain individual eventual domination in Proposition 3.2, we assumed there that the dominated semigroup converges strongly to 0 as $t \to \infty$. Combining this with the assumption $e^{t_1 A} E \subseteq E_u$ for some $t_1 \geq 0$, we were able show that $e^{tA} f \preceq u$ for sufficiently large t, which was an important ingredient for the proof of individual eventual domination. In the same vein, in order to obtain uniform eventual domination, we require the dominated semigroup to converge to 0 in the operator norm,

and we will combine this with the smoothing assumption (a) in Theorem 3.4 to obtain a similar estimate.

We note that operator norm convergence of C_0-semigroups has been investigated in detail in [33].

(d) By a simple rescaling argument we can clearly replace the estimate in assumption (b) by the estimate $e^{t(B-s(B))} \geq c(u \otimes \varphi)$; in this case we also have to replace assumption (c) by the assumption that $(e^{t(A-s(B))})_{t\geq 0}$ converges to 0 in operator norm.

In the following proof not only do we need the principal ideal E_u, but also the Banach lattice E^φ which is, for a strictly positive functional $\varphi \in E'$, defined to be the completion of E with respect to the norm $\langle \varphi, |\cdot| \rangle$. This space plays an essential role in [13] and [5]; we also refer to these references for more details about E^φ.

Proof of Theorem 3.4 We proceed as in the proof of [13, Theorem 3.1]. For each $t \geq 0$ we deduce from assumption (a) together with [5, Proposition 2.5] that

$$e^{(t+t_1+t_2)A} = e^{t_1 A} e^{tA} e^{t_2 A}$$

extends to a bounded linear operator from E^φ to E_u for each time $t \geq 0$. Furthermore,

$$\left\| e^{(t+t_1+t_2)A} \right\|_{E^\varphi \to E_u} \leq \left\| e^{t_1 A} \right\|_{E \to E_u} \left\| e^{tA} \right\|_{E \to E} \left\| e^{t_2 A} \right\|_{E^\varphi \to E},$$

and the latter converges to 0 as $t \to \infty$. This gives the existence of $t_0 \geq 0$ such that $\| e^{(t+t_1+t_2)A} \|_{E^\varphi \to E_u} \leq c/2$ for all $t \geq t_0$; here the constant $c > 0$ is the same one as in assumption (b).

Fix $f \in E$. Then the preceding inequality yields

$$\left\| e^{tA} f \right\|_u \leq \frac{c}{2} \|f\|_{E^\varphi} = \frac{c}{2} \langle \varphi, |f| \rangle$$

for all $t \geq t_0 + t_1 + t_2$. We thus obtain $\left| e^{tA} f \right| \leq \frac{c}{2} \langle \varphi, |f| \rangle u = \frac{c}{2} (u \otimes \varphi) |f|$ for all $t \geq t_0 + t_1 + t_2$. Let $\tau := \max\{t_0 + t_1 + t_2, t_3\}$. The eventual positivity assumption (b) on $(e^{tB})_{t\geq 0}$ along with the above inequality gives

$$e^{tB} |f| - \left| e^{tA} f \right| \geq c(u \otimes \varphi) |f| - \frac{c}{2}(u \otimes \varphi) |f| = \frac{c}{2}(u \otimes \varphi) |f|$$

for all $t \geq \tau$. This proves the assertion. □

4 Applications

In this section, we demonstrate how our abstract results can be applied to a variety of differential operators. Several applications of eventual domination of semigroups were already given in [23, Section 4], but now we have more freedom as our dominated semigroup is not required to be eventually positive. Moreover, the theory in Sect. 2 enables us to prove eventual domination of the resolvents as well. Our intention is mainly to illustrate that eventual domination of semigroups and resolvents can happen in various situations, so we try to avoid technical difficulties by keeping the differential operators in our examples rather simple.

4.1 The Laplace Operator with Anti-Symmetric Boundary Conditions

Let us consider the realisation of the Laplace operator on $L^2(-1,1)$ with anti-symmetric boundary conditions given by

$$\begin{aligned}\operatorname{dom}\left(\Delta^{AS}\right) &= \{f \in H^2(-1,1) : f(-1) = -f(1),\ f'(-1) = -f'(1)\},\\ \Delta^{AS} f &= f''.\end{aligned}$$

This is a self-adjoint operator on $L^2(-1,1)$ and $-\Delta^{AS}$ is associated with the sesqui-linear form a given by $a(u,v) = \int_{-1}^{1} u\overline{v}\,\mathrm{d}x$ with form domain

$$\operatorname{dom}(a) = \{f \in H^1(-1,1) :\ f(-1) = -f(1)\}.$$

Since $H^2(-1,1)$ embeds compactly into $L^2(-1,1)$, it follows that Δ^{AS} has compact resolvent and hence all spectral values of Δ^{AS} are eigenvalues. Moreover, a straightforward computation shows that the eigenvalues of Δ^{AS} are precisely the numbers

$$-\left(k + \frac{1}{2}\right)^2 \pi^2 \quad \text{for integers } k \geq 0,$$

each of them with multiplicity 2; the eigenspace of the leading eigenvalue $s(\Delta^{AS}) = -\pi^2/4$ is spanned by the functions u_1, u_2 given by

$$u_1(x) = \cos\left(\frac{\pi}{2}x\right) \qquad \text{and} \qquad u_2(x) = \sin\left(\frac{\pi}{2}x\right)$$

for $x \in (-1, 1)$. Since Δ^{AS} is self-adjoint, the spectral projection P associated with $s(\Delta^{AS})$ is the orthogonal projection onto the eigenspace, given by

$$P = c\big(u_1 \otimes u_1 + u_2 \otimes u_2\big)$$

for an appropriate normalisation constant $c > 0$. In particular, the operator P is not positive since, for instance, $\big(P\,\mathbb{1}_{(0,1)}\big)\,(x)$ is strictly negative for x close to -1.

Due to the non-positivity of the projection P, the semigroup $(e^{t\Delta^{AS}})_{t\geq 0}$ is not positive and, more generally, not even eventually positive [13, Theorem 8.3] (in fact, this reference shows that the semigroup does not even have the weaker property of being *asymptotically positive*).

Now, as a second operator, we consider the Neumann Laplacian Δ^N on $L^2(-1, 1)$. It generates a positive semigroup, has spectral bound 0, and the corresponding eigenspace is spanned by the constant function $\mathbb{1}$. We have the following comparison result.

Theorem 4.1 *There exists a time $\tau \geq 0$ such that the estimate*

$$e^{t\Delta^N}\,|f| - \left|e^{t\Delta^{AS}} f\right| \succeq (\mathbb{1} \otimes \mathbb{1})\,|f|$$

holds for all $t \geq \tau$ and all $f \in L^2(-1, 1)$. On the other hand, the semigroup $(e^{t\Delta^N})_{t\geq 0}$ does not dominate $(e^{t\Delta^{AS}})_{t\geq 0}$ for all times, i.e., there exists a time $t > 0$ and a function $f \in L^2(-1, 1)$ such that

$$e^{t\Delta^N}\,|f| \not\geq \left|e^{t\Delta^{AS}} f\right|.$$

Proof The eventual domination claim follows from Theorem 3.4, applied to the vectors $u = \varphi = \mathbb{1}$. (There are various ways to see that assumption (b) in the theorem is satisfied for $B = \Delta^N$; within the framework of eventually positive semigroups, one can, for instance, use [13, Theorem 3.1] to see this.)

The fact that one does not have domination for all times can be easily checked by employing form methods: one can simply use the characterisation of domination given in [31, Theorem 2.21], since the form domain dom (a) introduced above is not an ideal (in the sense of [31, Definition 2.19]) of the form domain $H^1(-1, 1)$ of the sesqui-linear form associated with $-\Delta^N$. □

Remark 4.2 With essentially the same argument, one could show that the Laplace operator with periodic boundary conditions eventually dominates the Laplace operator with anti-symmetric boundary conditions. However, in this case, the domination even holds for all times, as was shown in [30, Example 2].

4.2 Laplace Operators with Non-Local Boundary Conditions

Our next application concerns an example that is used in several articles about eventually positive semigroups and (anti-)maximum principles—the Laplacian on an interval with certain non-local boundary conditions. Eventual positivity of the generated semigroup can be nicely demonstrated via two specific choices of non-local boundary conditions—one in which the boundary conditions are symmetric and the other subject to non-symmetric boundary conditions; see, for instance, [22, Section 11.7]. Here, we deal with the latter. We prove the following result.

Theorem 4.3 *Let $\beta_1, \beta_2 \in \mathbb{R}$. On $E := L^2(0, \pi)$, we consider the Laplace operators Δ_i with domains*

$$\operatorname{dom}(\Delta_i) = \{f \in H^2(0, \pi) : f'(0) = 0,\ f'(\pi) = \beta_i f(0)\}$$

for $i = 1, 2$. Let $\mathbb{1} \in E$ denote the constant function taking the value one and assume that $-\frac{1}{2} < \beta_1 < \beta_2 < 0$. Then there exists a time $\tau \geq 0$ such that

$$e^{t\Delta_2} |f| - \left|e^{t\Delta_1} f\right| \succeq (\mathbb{1} \otimes \mathbb{1}) |f|$$

for all $t \geq \tau$ and all non-zero $f \in E$.

Moreover, for each $0 \neq f \in E$, we have

$$\mathcal{R}(\lambda, \Delta_2) |f| - |\mathcal{R}(\lambda, \Delta_1) f| \succeq \mathbb{1} \quad \text{and} \quad \mathcal{R}(\mu, \Delta_2) |f| + |\mathcal{R}(\mu, \Delta_1) f| \preceq -\mathbb{1}$$

for all λ in an f-dependent right neighbourhood of $s(\Delta_2)$ and for all μ in an f-dependent left neighbourhood of $s(\Delta_2)$.

For a fixed $\beta \in \mathbb{R}$, the operator Δ_β introduced above was considered in [22, Theorem 11.7.4], where necessary and sufficient conditions were given for the corresponding semigroup to be positive and individually eventually positive. It can actually be shown using [13, Theorem 3.1] that the eventual positivity assertion there is uniform. In fact, with methods similar to [5, Proposition 6.3], one can even prove uniform (anti-)maximum principles at the spectral bound. We mention that the operator considered here is a slight modification of the simple thermostat models considered in [24, 25].

Proof of Theorem 4.3 Firstly, the semigroup is analytic and

$$\operatorname{dom}(\Delta_1) \subseteq H^1(0, \pi) \subseteq L^\infty(0, \pi) = E_{\mathbb{1}},$$

so,

$$e^{t\Delta_1} E \subseteq \cap_{n \in \mathbb{N}} \operatorname{dom}\left(\Delta_1^n\right) \subseteq E_u$$

for all $t > 0$. Also, because Δ_1 is real, the Hilbert space adjoint Δ_1^* coincides with the Banach space dual Δ_1' (one can find this argument in, for example, [13, page 46]). Moreover, the structure of Δ_1^* is same as that of Δ_1 with switched end points of the interval. It follows that assumption (a) of Theorem 3.4 is satisfied for $A := \Delta_1 - s(\Delta_2)$. Moreover, assumption (b) of Theorem 3.4 is also satisfied with $B := \Delta_2 - s(\Delta_2)$ (see Remark 3.5(b)) and $u = \varphi = \mathbb{1}$. Indeed, the proof is exactly the same as that of [13, Theorem 4.3] (the only difference is that one needs to use [22, Theorem 11.7.4] instead of [14, Theorem 6.10]).

Since A generates an analytic semigroup, the growth bound of the semigroup coincides with $s(A)$. So, if we are able to show that $s(A) < 0$, then assumption (c) of Theorem 3.4 also follows. To see that indeed $s(A) < 0$, we make use of the computations performed in [22, Theorem 11.7.4(c)]—which not only show that both $s(\Delta_1)$ and $s(\Delta_2)$ are negative, but also that the numbers $\mu_i = \sqrt{-s(\Delta_i)} > 0$ (for $i = 1, 2$) are the only numbers in $(0, 1/2)$ that satisfy the equations

$$\mu_1 \sin(\mu_1 \pi) = -\beta_1 \qquad \text{and} \qquad \mu_2 \sin(\mu_2 \pi) = -\beta_2,$$

respectively. Since the function

$$\mu \mapsto \mu \sin(\mu \pi)$$

is strictly increasing on $(0, \frac{1}{2})$ and $\beta_1 < \beta_2$, it follows that $\mu_1 > \mu_2$ and so $s(\Delta_1) < s(\Delta_2)$. Hence, we have shown that $s(A) < 0$.

Thus, we infer from Theorem 3.4, the existence of a time $\tau > 0$ such that

$$e^{t(\Delta_2 - s(\Delta_2))} |f| - \left| e^{t(\Delta_1 - s(\Delta_2))} f \right| \succeq (\mathbb{1} \otimes \mathbb{1}) |f|$$

for all $t \geq \tau$ and all non-zero $f \in E$, which proves the first conclusion.

Lastly, the inequality $s(\Delta_1) < s(\Delta_2)$ above also implies that $s(\Delta_2) \in \rho(\Delta_1)$. Moreover, the fact that the spectral projection P of Δ_2 associated with $s(\Delta_2)$ satisfies $Pf \succeq \mathbb{1}$ for all $0 \lneq f \in E$ follows from [22, Theorem 11.7.4] (use [14, Corollary 3.3]). The second assertion is thus a consequence of Theorem 2.1. □

4.3 *Dirichlet and Neumann Boundary Conditions: What Is (Eventually) in Between?*

The question of which semigroups are sandwiched between the Laplace operators with Dirichlet and Neumann boundary conditions dates back to Arendt and Warma [2]. Under locality and regularity assumptions, it was shown that the sandwiched semigroup is generated by a Laplace operator with general Robin boundary conditions. Later, it was shown by Akhlil [1] that the assumption of the locality of the form is redundant. More recently, Chill, Djida, and the first author revisited domination of semigroups generated by regular forms. The methods used there also show that the sandwiched semigroup must correspond to a local form; see [7, Section 4.1]. Here, we show that the semigroup generated by the operator considered in [2, Example 4.5] is eventually sandwiched between the Dirichlet and Neumann semigroups. Additionally, we prove eventual domination of the resolvents using our results in Sect. 2.

On $L^2(0, 1)$ we consider the Dirichlet Laplacian Δ^D with domain

$$\operatorname{dom}\left(\Delta^D\right) = \{f \in H^2(0,1) : f(0) = f(1) = 0\},$$

the Neumann Laplacian Δ^N with domain

$$\operatorname{dom}\left(\Delta^N\right) = \{f \in H^2(0,1) : f'(0) = f'(1) = 0\},$$

and a Laplace operator Δ^{NL} with non-local boundary conditions with domain

$$\operatorname{dom}\left(\Delta^{\mathrm{NL}}\right) = \{f \in H^2(0,1) : f'(0) = -f'(1) = f(0) + f(1)\}.$$

It is well-known that both Δ^D and Δ^N generate positive analytic C_0-semigroups on $L^2(0, 1)$. By contrast, while Δ^{NL} does generate an analytic semigroup, the generated semigroup is not positive but merely uniformly eventually positive [13, Theorem 4.2]. Moreover, uniform maximum and anti-maximum principles for the operator Δ^{NL} were recently proved in [5, Proposition 6.2].

Theorem 4.4 *For the Laplace operators considered above, we have the following domination of the resolvents.*

(i) *For each $0 \neq f \in L^2(0, 1)$, we have*

$$\mathcal{R}(\lambda, \Delta^{\mathrm{NL}})\,|f| - \left|\mathcal{R}(\lambda, \Delta^D) f\right| \succeq \mathbb{1}$$

and

$$\mathcal{R}(\mu, \Delta^{\mathrm{NL}})\,|f| + \left|\mathcal{R}(\mu, \Delta^D) f\right| \preceq -\mathbb{1}$$

for all λ in an f-dependent right neighbourhood of $s(\Delta^{\mathrm{NL}})$ and for all μ in an f-dependent left neighbourhood of $s(\Delta^{\mathrm{NL}})$.

(ii) *For each $0 \neq f \in L^2(0, 1)$, we have*

$$\mathcal{R}(\lambda, \Delta^N)\,|f| - \left|\mathcal{R}(\lambda, \Delta^{\mathrm{NL}}) f\right| \succeq \mathbb{1}$$

and

$$\mathcal{R}(\mu, \Delta^N)\,|f| + \left|\mathcal{R}(\mu, \Delta^{\mathrm{NL}})f\right| \preceq -\mathbb{1}$$

for all λ *in an* f*-dependent right neighbourhood of* 0 *and for all* μ *in an* f*-dependent left neighbourhood of* 0.

Proof We simultaneously verify the assumptions of Theorem 2.1 for both parts. To begin, note that the domain of each of the three operators lies inside $L^\infty(0, 1) = L^2(0, 1)_{\mathbb{1}}$. Also, each of the domains embeds into $H^1(0, 1) \hookrightarrow L^2(0, 1)$ and the latter embedding is compact. Therefore, each of the operators has compact resolvent (in particular, each spectral value is a pole of the corresponding resolvent).

Now, it was shown in [23, Theorem 4.4] that $s(\Delta^D) < s(\Delta^{\mathrm{NL}})$ and in [14, Lemma 6.9] that $s(\Delta^{\mathrm{NL}}) < 0$. Moreover, using the boundary conditions, one can check that $s(\Delta^N) = 0$. It follows that $s(\Delta^{\mathrm{NL}}) \in \rho(\Delta^D)$ and $0 \in \rho(\Delta^{\mathrm{NL}})$. In addition, assumption (b) in Theorem 2.1 for the operators Δ^N and Δ^{NL} was verified in [5, Propositions 6.1(b) and 6.2], respectively (more precisely, in [5, Propositions 6.1(b) and 6.2] a property was verified that is, according to [14, Corollary 3.3], equivalent to assumption (b) in Theorem 2.1).

With the above observations, the assertions readily follow from Theorem 2.1. □

As announced before, we are going to show that the semigroup generated by Δ^{NL} is eventually sandwiched between the Dirichlet and the Neumann semigroup. We point out that the eventual domination of the Dirichlet semigroup by the semigroup generated by Δ^{NL} is already known from [23, Theorem 4.4].

Theorem 4.5 *There exists a time* $t_0 \geq 0$ *such that* $e^{t\Delta^D} \leq e^{t\Delta^{\mathrm{NL}}} \leq e^{t\Delta^N}$ *for all* $t \geq t_0$.

Proof First of all, it was already proved in [23, Theorem 4.4] that there exists a time $t_1 \geq 0$ such that

$$e^{t\Delta^D} \leq e^{t\Delta^{\mathrm{NL}}}$$

for all $t \geq t_1$.

For the second estimate, we verify the assumptions of Theorem 3.4. Note that both the operators Δ^N and Δ^{NL} are self-adjoint. Moreover, the corresponding semigroups map $L^2(0, 1)$ into $H^2(0, 1) \subseteq L^\infty(0, 1) = L^2(0, 1)_{\mathbb{1}}$.

Next, note that for the Neumann Laplacian, the spectral bound $s(\Delta^N) = 0$ is a dominant spectral value and the corresponding eigenspace is spanned by $\mathbb{1}$. Therefore by Remark 3.5(b), we obtain that Δ^N satisfies the assumption (b) of Theorem 3.4. On the other hand, because Δ^{NL} generates an analytic semigroup, the growth bound of this semigroup satisfies $\omega_0(\Delta^{\mathrm{NL}}) = s(\Delta^{\mathrm{NL}})$. Moreover, it was shown in [14, Lemma 6.9] that $s(\Delta^{\mathrm{NL}}) < 0$. In particular, the semigroup generated by Δ^{NL} converges to 0 with respect to the operator norm as $t \to \infty$.

We now conclude from Theorem 3.4 that there exists $t_2 \geq 0$ such that

$$e^{t\Delta^D} \leq e^{t\Delta^{\mathrm{NL}}} \quad \text{and} \quad e^{s\Delta^{\mathrm{NL}}} \leq e^{s\Delta^N}$$

for all $t \geq t_1$ and for all $s \geq t_2$; the assertion is now immediate. □

4.4 Bi-Laplace Operator with Wentzell Boundary Conditions

Let $\Omega \subseteq \mathbb{R}^d$ be a bounded domain with Lipschitz boundary, where Ω is equipped with the Lebesgue measure and its boundary is equipped with the surface measure S. Fix functions $\alpha \in L^\infty(\Omega, \mathbb{R})$ and $\beta, \gamma_1, \gamma_2 \in L^\infty(\partial\Omega, \mathbb{R})$ such that there is a number $\eta > 0$ satisfying $\eta \leq \alpha$ and $\eta \leq \beta$ almost everywhere. Moreover, we assume that $\gamma_1, \gamma_2 \geq 0$.

Our state space is the Hilbert lattice

$$H := L^2(\Omega) \times L^2(\partial\Omega, \beta^{-1}dS)$$

equipped with the inner product

$$\langle u\,,\, v\rangle = \int_\Omega u_1\overline{v_1}\,\mathrm{d}x + \int_{\partial\Omega} u_2\overline{v_2}\beta^{-1}\,\mathrm{d}S$$

for $u = (u_1, u_2)$ and $v = (v_1, v_2)$ in H, and we consider the operators $-A_i$ associated with the forms

$$a_i((u_1, u_2), (v_1, v_2)) := \langle(\alpha\Delta u_1, \gamma_i u_2)\,,\,(\Delta v_1, v_2)\rangle$$

with form domains

$$\operatorname{dom}(a_i) := \left\{(u_1, u_2) \in H : u_1 \in \operatorname{dom}\left(\Delta^N\right), u_2 = \operatorname{tr} u_1\right\}$$

for $i = 1, 2$; here Δ^N denotes the Neumann Laplacian on Ω. It was shown in [16, Theorem 3.4] that the operators A_1 and A_2 are self-adjoint and generate analytic and contractive C_0-semigroups on H. In [16], it was also shown that the semigroups are not positive but eventually positive if $\gamma_i = 0$. Combining the results of the aforementioned reference with our current results, we are able to obtain eventual domination for both semigroups and resolvents:

Theorem 4.6 *Let $\gamma_2 = 0$ almost everywhere and let $\gamma_1 \geq 0$ be non-zero on a set of non-zero measure. In addition, set $\mathbb{1} := (\mathbb{1}_\Omega, \mathbb{1}_{\partial\Omega})$. Then there exists $\tau \geq 0$ such that*

$$e^{tA_2}\,|f| - \left|e^{tA_1}f\right| \succeq (\mathbb{1}\otimes\mathbb{1})\,|f|$$

for all $t \geq \tau$ and all $f \in E$.

Proof Fix $i \in \{1, 2\}$. Since the semigroups are analytic, we have

$$e^{tA_i} H \subseteq \cap_{n \geq 0} \operatorname{dom}\left(A_i^n\right)$$

for all $t > 0$. Moreover, it follows from [16, Theorem 5.4] that $\cap_{n \geq 0} \operatorname{dom}\left(A_i^n\right)$ embeds into $L^\infty(\Omega) \times L^\infty(\partial\Omega) = H_{\mathbb{1}}$. By virtue of [16, Lemma 6.4(i)], we have that $s(A_2) = 0$ and the corresponding eigenspace is spanned by $\mathbb{1}$. Since the operators A_1 and A_2 are self-adjoint, it follows by Remark 3.5(b) that assumptions (a) and (b) of Theorem 3.4 are satisfied. In fact, according to [16, Theorem 6.5(ii)], assumption (c) is also satisfied. Thus the assertion follows. □

Theorem 4.7 *Let $\gamma_2 = 0$ almost everywhere and let $\gamma_1 \geq 0$ be non-zero on a set of non-zero measure. In addition, set $\mathbb{1} := (\mathbb{1}_\Omega, \mathbb{1}_{\partial\Omega})$. If $d \leq 5$, then for each $f \in E$, we have*

$$\mathcal{R}(\lambda, A_2)\,|f| - |\mathcal{R}(\lambda, A_1)f| \succeq \mathbb{1} \quad \text{and} \quad \mathcal{R}(\mu, A_2)\,|f| + |\mathcal{R}(\mu, A_1)f| \preceq -\mathbb{1}$$

for all λ in an f-dependent right neighbourhood of 0 *and for all μ in an f-dependent left neighbourhood of* 0.

Proof Since $d \leq 5$, the proof of [16, Theorem 5.4] implies that

$$\operatorname{dom}(A_i) \subseteq L^\infty(\Omega) \times L^\infty(\partial\Omega) = H_{\mathbb{1}} \qquad (i = 1, 2).$$

Moreover, by [16, Lemma 6.4(i)], we have that $s(A_2) = 0$ with corresponding eigenspace being spanned by $\mathbb{1}$. Since the operator A_2 is self-adjoint, it follows that the spectral projection P of A_2 associated with 0 is a multiple of $\mathbb{1} \otimes \mathbb{1}$, and thus satisfies $Pf \succeq \mathbb{1}$ for all $0 \lneq f \in E$. Lastly, we know from [16, Lemma 6.4(ii)] that $0 \in \rho(A_1)$. In particular, all assumptions of Theorem 2.1 are satisfied which yields both the assertions. □

4.5 *Differential Operators of Odd Order*

We conclude this section by considering differential equations of odd order on an interval; these were recently considered in [5, Section 6.4]. It turns out, a little unexpectedly, that these operators satisfy a uniform maximum and anti-maximum principle. Furthermore, even though the first order operator generates a positive semigroup, if the order is strictly larger than one, then the corresponding semigroup is not even individually eventually positive. Here, we show that there is no eventual domination of resolvents.

Fix integers $m, \ell \in \mathbb{N}_0$. On the space $L^2(0, 1)$, consider the operators

$$\operatorname{dom}(A_k) := \left\{ f \in H^{2k+1}(0,1) : f^{(j)}(0) = f^{(j)}(1) \text{ for all } j = 0, 1, \ldots, 2k \right\}$$
$$A_k f := f^{(2k+1)}$$

for $k = m, \ell$.

Let $k \in \{m, \ell\}$. The number 0 is a spectral value of A_k and a simple pole of the resolvent $\mathcal{R}(\,\cdot\,, A_k)$ [5, Propositions 6.8 and 3.1(b)]. Also, according to [5, Theorem 6.9], we have the estimate $\mathcal{R}(\lambda, A_k) \succeq \mathbb{1} \otimes \mathbb{1}$ for all λ in a right neighbourhood of 0. Now if $m \neq \ell$, then $A_m \neq A_\ell$. Thus by Theorem 2.5, we have proved the following:

Theorem 4.8 *If for each $0 \leq f \in L^2(0, 1)$, we have $\mathcal{R}(\lambda, A_m) f \leq \mathcal{R}(\lambda, A_\ell) f$ for all λ in an f-dependent right neighbourhood of 0, then $m = \ell$.*

5 Some Remarks on Eventual Positivity of Cesàro Means

We close the paper with a few remarks on the eventual positivity of Cesàro means of C_0-semigroups. Since Cesàro means behave in several respects similarly to resolvents, it is reasonable to expect comparable behaviour concerning eventual positivity; the following results give a few first indications that this expectation is justified.

Let E be a complex Banach lattice and let $(e^{tA})_{t \geq 0}$ be a real C_0-semigroup on E. For a fixed quasi-interior point $u \in E$, we say that the Cesàro means of $(e^{tA})_{t \geq 0}$ are *individually eventually strongly positive with respect to u* if for each $0 \lneq f \in E$, there exists a time $t_0 \geq 0$ such that $C(r) f \succeq u$ for all $t \geq 0$; here the operators $C(r) \in \mathcal{L}(E)$ are defined by

$$C(r) f := \frac{1}{r} \int_0^r e^{sA} f \, ds \qquad \text{for all } f \in E$$

and are called the *Cesàro means of* $(e^{tA})_{t \geq 0}$. Some properties of the Cesàro means are given in [18, Lemma V.4.2]. Our first result here is the following criterion for the Cesàro means to be individually eventually positive.

Proposition 5.1 *Let $(e^{tA})_{t \geq 0}$ be a real C_0-semigroup on a complex Banach lattice E and let $u \geq 0$ be a quasi-interior point of E. Assume that $\sigma(A)$ is non-empty and consider the following assertions:*

(i) *The Cesàro means are individually eventually strongly positive with respect to u.*

(ii) *The rescaled semigroup* $(e^{t(A-s(A))})_{t\geq 0}$ *is mean ergodic and satisfies*

$$\lim_{t\to\infty}\frac{1}{t}\left\|e^{t(A-s(A))}\right\| = 0,$$

and the mean ergodic projection P *satisfies* $Pf \succeq u$ *for all* $0 \lneq f \in E$.

(iii) *The rescaled semigroup* $(e^{t(A-s(A))})_{t\geq 0}$ *is Abel ergodic and bounded, and the Abel ergodic projection* Q *satisfies* $Qf \succeq u$ *whenever* $0 \lneq f \in E$.

We always have that (iii) *implies* (ii). *If, in addition,* $\operatorname{dom}(A) \subseteq E_u$, *then* (ii) *implies* (i).

Before giving the proof, we briefly recall the notions used above. A C_0-semigroup $(e^{tA})_{t\geq 0}$ on a Banach space E is called *mean ergodic* if its Cesàro means $C(r)$ converge strongly as $r \to \infty$. If this is the case and, in addition $\left\|e^{t(A-s(A))}\right\|/t \to 0$ as $t \to \infty$, then the limit operator

$$E \ni f \mapsto Pf := \lim_{r\to\infty} C(r)f$$

is a projection and is called the corresponding *mean ergodic projection*. A brief overview of mean ergodicity of semigroups is given in [18, Section V.4]. Furthermore, the semigroup $(e^{tA})_{t\geq 0}$ is called *Abel ergodic* if there exists $\lambda_0 > 0$ such that $(0, \lambda_0) \subseteq \rho(A)$ and

$$\sup_{\lambda\in(0,\lambda_0)} \|\lambda\mathcal{R}(\lambda, A)\| < \infty$$

and the limit $\lim_{\lambda\downarrow 0}\lambda\mathcal{R}(\lambda, A)$ exists in the strong operator topology. In this case, the operator

$$E \ni f \mapsto Qf := \lim_{\lambda\downarrow 0}\lambda\mathcal{R}(\lambda, A)f$$

is also a projection and is called the corresponding *Abel ergodic projection*. We refer to [3, Section 4.3] for details about Abel ergodicity.

Proof of Proposition 5.1 We assume without loss of generality that $s(A) = 0$.

"(iii) $\Rightarrow$ (ii)" If the semigroup generated by A is Abel ergodic and bounded, then by [3, Proposition 4.3.4(b)] it is mean ergodic and the corresponding ergodic projections coincide.

We now assume that $\operatorname{dom}(A) \subseteq E_u$.

"(ii) $\Rightarrow$ (i)": Let $C(r) := \frac{1}{r}\int_0^r e^{sA}\,ds$ denote the Cesàro means of the semigroup. Since the semigroup is mean ergodic, we have $C(r) \to P$ strongly as $r \to \infty$ and $\operatorname{Rg} P = \ker A$. Therefore

$$AC(r) = \frac{A}{r}\int_0^r e^{sA}\,ds = \frac{e^{rA} - I}{r} \to 0 = AP$$

with respect to the strong operator topology in E as $r \to \infty$. As a result, $C(r)f \to Pf$ in $\operatorname{dom}(A)$ (endowed with the graph norm) as $r \to \infty$ for each $f \in E$. Since $\operatorname{dom}(A)$ embeds continuously into E_u (by the closed graph theorem) the convergence holds in E_u as well.

As the semigroup is real, so are the Cesàro means $C(r)$ and thus, assertion (i) follows from Lemma 2.4. □

Under certain technical assumptions, one can give a characterisation for eventual positivity of the Cesàro means which shows that this property is closely related to (anti-)maximum principles.

Theorem 5.2 *Let $(e^{tA})_{t\geq 0}$ be a real C_0-semigroup on a complex Banach lattice E and let $0 \leq u \in E$ be such that* $\operatorname{dom}(A) \subseteq E_u$. *Assume that the spectral bound $s(A)$ is a pole of the resolvent $\mathcal{R}(\,\cdot\,, A)$ and that the rescaled semigroup $(e^{t(A-s(A))})_{t\geq 0}$ is bounded. Then the following are equivalent.*

(i) *The Cesàro means are individually eventually strongly positive with respect to u.*
(ii) *The spectral projection P corresponding to $s(A)$ satisfies $Pf \succeq u$ whenever $0 \lneq f \in E$.*
(iii) *The individual strong maximum principle with respect to u holds, i.e., for every $0 \lneq f \in E$, we have*

$$\mathcal{R}(\lambda, A)f \succeq u$$

for all λ in an f-dependent right neighbourhood of λ_0.
(iv) *The individual strong anti-maximum principle with respect to u holds, i.e., for every $0 \lneq f \in E$, we have*

$$\mathcal{R}(\mu, A)f \preceq -u$$

for all μ in an f-dependent left neighbourhood of λ_0.

Proof By replacing A by $A - s(A)$, we may assume that $s(A) = 0$. The equivalence of (ii), (iii), and (iv) under the assumption $\operatorname{dom}(A) \subseteq E_u$ was proved in [14, Theorem 4.4]. On the other hand, boundedness of the semigroup implies that 0 is actually a simple pole of the resolvent $\mathcal{R}(\,\cdot\,, A)$. We infer that $(e^{tA})_{t\geq 0}$ is uniformly mean ergodic and the corresponding mean ergodic projection coincides with the spectral projection P associated with 0; see [18, Theorem V.4.10]. Since $\operatorname{dom}(A) \subseteq E_u$, the implication (ii) $\Rightarrow$ (i) now holds due to Proposition 5.1.

Lastly, if (i) holds, then so does (ii) by [14, Lemma 4.8(ii)]. □

It has recently been shown by the present authors [6, Theorem 1.2] that the condition $\operatorname{dom}(A) \subseteq E_u$ is, under mild assumptions on A, necessary for assertions (iii) and (iv) to hold simultaneously.

Clearly, the above results can only be considered as a starting point for a more thorough analysis, in particular since the methods used in this section are very closely related to the methods used in earlier articles about eventual positivity. In the future, it would, for instance, be desirable to better understand the relation between eventual positivity of the resolvent and the Cesàro means, without imposing the a priori assumption dom $(A) \subseteq E_u$. However, this seems to require more sophisticated analysis and probably different techniques.

We end this section with the observation that, by using Theorem 5.2, it can be shown for several semigroups considered in [5, 14, 15], that the Cesàro means are individually eventually positive.

Acknowledgments The authors are grateful to Delio Mugnolo for pointing out the reference [30].

Part of the work on this article was done during a pleasant visit of the first author to the second author at Universität Passau, Germany in 2021. The first named author was supported by Deutscher Akademischer Austauschdienst (Forschungsstipendium-Promotion in Deutschland).

References

1. K. Akhlil, Locality and domination of semigroups. Result. Math. **73**(2), 11 (2018). Id/No 59. https://doi.org/10.1007/s00025-018-0822-9
2. W. Arendt, M. Warma, Dirichlet and Neumann boundary conditions: What is in between? J. Evol. Equ. **3**(1), 119–135 (2003). Dedicated to Philippe Bénilan. https://doi.org/10.1007/s000280300005
3. W. Arendt, C.J.K. Batty, M. Hieber, F. Neubrander, *Vector-Valued Laplace Transforms and Cauchy Problems*, vol. 96, 2nd edn. (Birkhäuser, Basel, 2011). https://doi.org/10.1007/978-3-0348-0087-7
4. S. Arora, Locally eventually positive operator semigroups. J. Oper. Theory **88**(1), 203–242 (2022). https://doi.org/10.7900/jot.2021jan26.2316. Available online at https://arxiv.org/abs/2101.11386v3
5. S. Arora, J. Glück, An operator theoretic approach to uniform (anti-)maximum principles. J. Differ. Equ. **310**, 164–197 (2022). https://doi.org/10.1016/j.jde.2021.11.037
6. S. Arora, J. Glück, A characterization of the individual maximum and anti-maximum principle. Math. Z. (accepted for publication, 2022). Preprint. http://arxiv.org/abs/2203.05680v1
7. S. Arora, R. Chill, J.-D. Djida, Domination of semigroups generated by regular forms (2021). Preprint. https://arxiv.org/abs/2111.15489v2
8. A. Bátkai, M.K. Fijavž, A. Rhandi, *Positive Operator Semigroups: From Finite to Infinite Dimensions*, vol. 257 (Birkhäuser/Springer, Basel, 2017). https://doi.org/10.1007/978-3-319-42813-0
9. C.J.K. Batty, D.W. Robinson, Positive one-parameter semigroups on ordered Banach spaces. Acta Appl. Math. **2**, 221–296 (1984). https://doi.org/10.1007/BF02280855
10. S. Becker, F. Gregorio, D. Mugnolo, Schrödinger and polyharmonic operators on infinite graphs: parabolic well-posedness and p-independence of spectra. J. Math. Anal. Appl. **495**(2), 44 (2021). Id/No 124748. https://doi.org/10.1016/j.jmaa.2020.124748
11. D. Daners, Non-positivity of the semigroup generated by the Dirichlet-to-Neumann operator. Positivity **18**(2), 235–256 (2014). https://doi.org/10.1007/s11117-013-0243-7
12. D. Daners, J. Glück, The role of domination and smoothing conditions in the theory of eventually positive semigroups. Bull. Aust. Math. Soc. **96**(2), 286–298 (2017). https://doi.org/10.1017/S0004972717000260
13. D. Daners, J. Glück, A criterion for the uniform eventual positivity of operator semigroups. Integr. Equ. Oper. Theory **90**(4), 19 (2018). Id/No 46. https://doi.org/10.1007/s00020-018-2478-y

14. D. Daners, J. Glück, J.B. Kennedy, Eventually and asymptotically positive semigroups on Banach lattices. J. Differ. Equ. **261**(5), 2607–2649 (2016). https://doi.org/10.1016/j.jde.2016.05.007
15. D. Daners, J. Glück, J.B. Kennedy, Eventually positive semigroups of linear operators. J. Math. Anal. Appl. **433**(2), 1561–1593 (2016). https://doi.org/10.1016/j.jmaa.2015.08.050
16. R. Denk, M. Kunze, D. Ploß, The Bi-Laplacian with Wentzell boundary conditions on Lipschitz domains. Integr. Equ. Oper. Theory **93**(2), 26 (2021). Id/No 13. https://doi.org/10.1007/s00020-021-02624-w
17. E.Y. Emel'yanov, *Non-Spectral Asymptotic Analysis of One-Parameter Operator Semigroups*, vol. 173 (Birkhäuser, Basel, 2007). https://doi.org/10.1007/978-3-7643-8114-1
18. K.-J. Engel, R. Nagel, *One-Parameter Semigroups for Linear Evolution Equations*, vol. 194 (Springer, Berlin, 2000). https://doi.org/10.1007/b97696
19. L.C.F. Ferreira, V.A. Ferreira, Jr. On the eventual local positivity for polyharmonic heat equations. Proc. Am. Math. Soc. **147**(10), 4329–4341 (2019). https://doi.org/10.1090/proc/14565
20. A. Ferrero, F. Gazzola, H.-C. Grunau, Decay and local eventual positivity for biharmonic parabolic equations. Discrete Contin. Dyn. Syst. **21**(4), 1129–1157 (2008). https://doi.org/10.3934/dcds.2008.21.1129
21. F. Gazzola, H.-C. Grunau, Eventual local positivity for a biharmonic heat equation in $\mathbb{R}^n$. Discrete Contin. Dyn. Syst. Ser. S **1**(1), 83–87 (2008). https://doi.org/10.3934/dcdss.2008.1.83
22. J. Glück, Invariant Sets and Long Time Behaviour of Operator Semigroups. Ph.D. Thesis, Universität Ulm, 2016. https://doi.org/10.18725/OPARU-4238
23. J. Glück, D. Mugnolo, Eventual domination for linear evolution equations. Math. Z. **299**(3–4), 1421–1443 (2021). https://doi.org/10.1007/s00209-021-02721-x
24. P. Guidotti, S. Merino, Hopf bifurcation in a scalar reaction diffusion equation. J. Differ. Equ. **140**(1), 209–222 (1997). https://doi.org/10.1006/jdeq.1997.3307
25. P. Guidotti, S. Merino, Gradual loss of positivity and hidden invariant cones in a scalar heat equation. Differ. Integr. Equ. **13**(10–12), 1551–1568 (2000). https://bit.ly/1356061139.
26. A. Hussein, D. Mugnolo, Laplacians with point interactions – expected and unexpected spectral properties, in *Semigroups of Operators – Theory and Applications. Selected Papers Based on the Presentations at the Conference, SOTA 2018, Kazimierz Dolny, Poland, September 30 – October 5, 2018. In honour of Jan Kisyński's 85th birthday*(Springer, Cham, 2020), pp. 47–67. https://doi.org/10.1007/978-3-030-46079-2_3
27. P. Meyer-Nieberg, *Banach Lattices* (Springer-Verlag, Berlin/Heidelberg, 1991). https://doi.org/10.1007/978-3-642-76724-1
28. R.Nagel (ed.), *One-Parameter Semigroups of Positive Operators*. Lecture Notes in Mathematics, vol. 1184 (Springer, Cham, 1986). https://doi.org/10.1007/BFb0074922
29. D. Noutsos, M.J. Tsatsomeros, Reachability and holdability of nonnegative states. SIAM J. Matrix Anal. Appl. **30**(2), 700–712 (2008). https://doi.org/10.1137/070693850
30. E.M. Ouhabaz, Invariance of closed convex sets and domination criteria for semigroups. Potential Anal. **5**(6), 611–625 (1996). https://doi.org/10.1007/BF00275797
31. E.M. Ouhabaz, Analysis of Heat Equations on Domains, vol. 31 (Princeton University Press, Princeton, 2005). https://doi.org/10.1515/9781400826483
32. H.H. Schaefer, *Banach Lattices and Positive Operators*, vol. 215 (Springer, Cham, 1974). https://doi.org/10.1007/978-3-642-65970-6
33. H.R. Thieme, Balanced exponential growth of operator semigroups. *J. Math. Anal. Appl.* **223**(1), 30–49 (1998). https://doi.org/10.1006/jmaa.1998.5952
34. H. Vogt, Stability of uniformly eventually positive C_0-semigroups on L_p-spaces. Proc. Am. Math. Soc. (2022). Published electronically. https://doi.org/10.1090/proc/15926

Bounded Functional Calculi for Unbounded Operators

Charles Batty, Alexander Gomilko, and Yuri Tomilov

Abstract This article summarises the theory of several bounded functional calculi for unbounded operators that have recently been discovered. They extend the Hille-Phillips calculus for (negative) generators A of certain bounded C_0-semigroups, in particular for bounded semigroups on Hilbert spaces and bounded holomorphic semigroups on Banach spaces. They include functions outside the Hille-Phillips class, and they generally give sharper bounds for the norms of the resulting operators $f(A)$. The calculi are mostly based on appropriate reproducing formulas for the relevant classes of functions, and they rely on significant and interesting developments of function theory. They are compatible with standard functional calculi and they admit appropriate convergence lemmas and spectral mapping theorems. They can also be used to derive several well-known operator norm-estimates, provide generalisations of some of them, and extend the general theory of operator semigroups. Our aim is to help readers to make use of these calculi without having to understand the details of their construction.

1 Introduction

Functional calculus for operators A has a long history, originating with self-adjoint operators on Hilbert spaces, considered by Hilbert himself in the case of bounded operators, and by von Neumann for unbounded operators. In that context, $f(A)$

C. Batty (✉)
St. John's College, University of Oxford, Oxford, UK
e-mail: charles.batty@sjc.ox.ac.uk

A. Gomilko
Faculty of Mathematics and Computer Science, Nicolas Copernicus University, Toruń, Poland
e-mail: alex@gomilko.com

Y. Tomilov
Polish Academy of Sciences, Institute of Mathematics, Warsaw, Poland
e-mail: ytomilov@impan.pl

Y. Choi et al. (eds.), *Operators, Semigroups, Algebras and Function Theory*,
Operator Theory: Advances and Applications 292,
https://doi.org/10.1007/978-3-031-38020-4_2

can be defined as a bounded operator for every bounded measurable function f on the spectrum of A. For other operators on Hilbert spaces or Banach spaces, a functional calculus is a way of converting one operator (often unbounded) into a coherent collection of associated bounded operators. Many applications may require $f(A)$ to be defined for only a few functions $f(z)$, such as $\exp(-tz)$ for operator semigroups $(e^{-tA})_{t\geq 0}$, and $(z-1)^n(z+1)^{-n}$ for powers of cogenerators. Results concerning specific functions can be discussed individually as has often been the case, but constructing functional calculi covering many functions has advantages. A unified approach is likely to be efficient in the long run, and the treatment of a functional calculus is likely to be more attractive mathematically than considering many different functions individually. Moreover, including more functions f often improves the sharpness of estimates of the norm of $f(A)$, and may possibly widen the scope of applications in the future.

For operators which are not self-adjoint, a functional calculus usually has to be restricted to holomorphic functions f defined on an open set Ω containing the spectrum $\sigma(A)$ of A. If A is bounded, the Riesz-Dunford calculus can be defined by

$$f(A) = \frac{1}{2\pi i}\int_\gamma f(z)(z-A)^{-1}\,dz,$$

where γ is a contour within Ω around $\sigma(A)$. Extending this to unbounded operators runs into complications, as the contours would naturally have to go to infinity. Sometimes the contour γ may have to pass through a point in the boundary of Ω, and then convergence of the integral above may depend on further assumptions on f and A.

If A is an injective sectorial operator of angle $\theta \in [0,\pi)$, one can circumvent this problem by using the extended sectorial calculus as described in the book of Haase [21]. The prototype for functional calculi of this type was a paper by Bade on strip-type operators [5]. If $f \in H^\infty(\Sigma_\psi)$ for some $\psi \in (\theta,\pi)$, $f(A)$ can be defined as a closed operator on X. Many differential operators A have a bounded H^∞-calculus, in the sense that $f(A)$ is bounded for all $f \in H^\infty(\Sigma_\psi)$, but there are operators which do not have bounded H^∞-calculus (see [7, Remark 4.7], for example). If $-A$ is the generator of a bounded C_0-semigroup on a Banach space X, then A is sectorial of angle $\pi/2$ and $\sigma(A) \subseteq \overline{\mathbb{C}}_+$. In this case it may be more natural to consider a half-plane calculus as in [6].

For the sectorial (or half-plane) calculus, the functions f have to be defined on a larger sector than the sectorial angle of A (or a larger half-plane than $\mathbb{C}_+$), and this is sometimes an unnatural restriction. If A is sectorial of angle strictly less than $\pi/2$ (written as $A \in \mathrm{Sect}(\pi/2-)$), we may consider functions $f \in H^\infty(\mathbb{C}_+)$. Then $f(A)$ is defined in the sectorial calculus, but not necessarily bounded. Vitse [37] showed that $f(A)$ is a bounded operator for all f in the Banach algebra $\mathcal{B}$ of "analytic Besov functions" on $\mathbb{C}_+$ which is continuously embedded in $H^\infty(\mathbb{C}_+)$ (see Sect. 3 and Theorem 4.1). Thus every operator $A \in \mathrm{Sect}(\pi/2-)$ has a bounded $\mathcal{B}$-calculus. Subsequently Schwenninger [32] improved and extended Vitse's results, including an estimate of the norms of operators of the form $f(A)e^{-tA}$ for small $t > 0$ and $f \in H^\infty(\mathbb{C}_+)$.

An early example of a bounded functional calculus for an unbounded operator A is the Hille-Phillips calculus. This calculus defines $f(A)$ whenever f is the Laplace transform of a bounded Borel measure on $\mathbb{R}_+$ and $-A$ is the generator of a bounded C_0-semigroup on a Banach space. It was described in [24, Chapter XV] and it played a substantial role in developing semigroup theory.

The use of analytic Besov functions for functional calculus goes back to Peller [29] in the discrete case. He showed that if T is a power-bounded operator on a Hilbert space, then $\|p(T)\| \le C\|p\|_{\mathcal{B}_d}$ for all polynomials $p(z)$, where $\mathcal{B}_d$ is the analogue of $\mathcal{B}$ for the unit disc. Since the polynomials are dense in $\mathcal{B}_d$, it follows that T has a bounded functional calculus for $\mathcal{B}_d$. In the continuous-parameter case, the analogous class of analytic Besov functions on $\mathbb{C}_+$ is more complicated (for example, it is not separable), and it has received little attention in the literature. The earliest research which successfully considered functional calculus for negative generators of bounded C_0-semigroups on Hilbert space was carried out by White in his doctoral thesis [38], but it remained virtually unknown for many years. He proved estimates of the form $\|f(A)\| \le C\|f\|_{\mathcal{B}}$ for functions f in a subclass of the Hille-Phillips algebra. There was a further contribution by Haase [22] who (inspired by [37], and unaware of White's work) obtained a similar estimate for a larger class of functions.

Many other functional calcuii have been proposed and developed by numerous contributors. This article is not a comprehensive survey of the subject. Instead, we set out the basic properties of a small group of other functional calculi which have recently been developed and are related to the calculus in [37]. A bounded $\mathcal{B}$-calculus for a class of operators A, including the negative generators of all bounded C_0-semigroups on Hilbert spaces and all bounded holomorphic C_0-semigroups, is described in Sect. 3, based on our papers [8] and [9]. Section 4 gives an account of two classes of function spaces, $\mathcal{D}_s$ and $\mathcal{H}_\psi$, and the associated $\mathcal{D}$-calculus and $\mathcal{H}$-calculus for operators $A \in \operatorname{Sect}(\pi/2-)$, based on [10]. Spaces in both these classes contain $\mathcal{B}$, and they also included functions which are unbounded on $\mathbb{C}_+$ or only defined on proper subsectors of $\mathbb{C}_+$. Another bounded calculus for a Banach algebra $\mathcal{A}$ of functions which can be applied to negative generators of bounded C_0-semigroups on Hilbert spaces has been constructed by Arnold and Le Merdy [3] and it is briefly described in Sect. 5.

The aim of this article is to present the main results about these calculi in a coherent fashion which will enable other researchers to appreciate their value and to apply them when possible, without having to piece together information from several lengthy papers. We describe the definitions and main properties of the algebras and the calculi, how they relate to each other and to the sectorial and half-plane calculi, and how they relate to some problems. Proofs are omitted or briefly sketched. In general terms, the proofs in our papers depend mainly on complex analysis and functional analysis, with some aspects of harmonic analysis.

We present the functional calculi only for negative generators of bounded C_0-semigroups, but they can be adapted to unbounded semigroups. Assume that A is the negative generator of a C_0-semigroup $(T(t))_{t\geq 0}$ and $\|T(t)\| \leq Me^{\omega t}$, and X is a Hilbert space or $(T(t))_{t\geq 0}$ is a holomorphic semigroup. Then $f(A)$ can be defined to be $g(A+\omega)$, where $g(z) := f(z-\omega)$, provided that g belongs to $\mathcal{B}$, $\mathcal{D}_s$ or $\mathcal{H}_\psi$, as appropriate.

In the IWOTA conference, the first author of this article lectured mostly on the $\mathcal{B}$-calculus, with only a brief mention of the $\mathcal{D}$- and $\mathcal{H}$-calculi. The first author of [3] lectured on the $\mathcal{A}$-calculus.

Notation We will use the following notation:

1. $\mathbb{R}_+ := [0, \infty)$,
2. $\mathbb{C}_+ := \{z \in \mathbb{C} : \Re z > 0\}$, $\overline{\mathbb{C}}_+ = \{z \in \mathbb{C} : \Re z \geq 0\}$,
3. $\Sigma_\theta := \{z \in \mathbb{C} : z \neq 0, |\arg z| < \theta\}$ for $\theta \in (0, \pi]$, $\Sigma_0 := (0, \infty)$,
4. $L(X, Y)$ is the space of bounded linear operators from one Banach space X to another Y; we write $L(X)$ when $Y = X$.

2 Semigroup Generators and Hille-Phillips Calculus

In this paper A will be a closed operator on a Banach space X with a dense domain $D(A)$, and its spectrum $\sigma(A)$ will be contained in $\overline{\Sigma_\theta}$ for some $\theta \in [0, \pi)$. Except for material on the $\mathcal{H}$-calculus in Sect. 4, θ will be in $[0, \pi/2]$.

Let $\mathfrak{A}$ be a Banach algebra which is continuously included in $H^\infty(\Sigma_\theta)$ where $\theta \in (0, \pi]$, and assume that $\mathfrak{A}$ contains the functions $r_z(\lambda) := (\lambda + z)^{-1}$ for $\lambda \in \Sigma_\theta$ and $z \in \Sigma_{\pi-\theta}$. An $\mathfrak{A}$*-calculus* for A is a bounded algebra homomorphism $\Phi : \mathfrak{A} \to L(X)$ such that $\Phi(r_z) = (z + A)^{-1}$ for all $z \in \Sigma_{\pi-\theta}$.

An operator A on a Banach space X is *sectorial of angle* $\theta \in [0, \pi)$ if $\sigma(A) \subset \overline{\Sigma_\theta}$ and, for each $\psi \in (\theta, \pi)$,

$$M_\psi(A) := \sup_{z \in \Sigma_{\pi-\psi}} \|z(z+A)^{-1}\| < \infty. \tag{2.1}$$

We write Sect(θ) for the class of all sectorial operators of angle θ for $\theta \in [0, \pi)$ on Banach spaces, and Sect($\pi/2-$) $:= \bigcup_{\theta \in [0,\pi/2)}$ Sect(θ). Then $A \in$ Sect($\pi/2-$) if and only if $-A$ generates a (sectorially) bounded holomorphic C_0-semigroup on X.

We refer to [21] for the general theory of sectorial operators, and to [2, Section 3.7] for the theory of holomorphic semigroups.

Let A be an operator and assume that $\sigma(A) \subset \overline{\mathbb{C}}_+$ and

$$M_A := M_{\pi/2}(A) = \sup_{z \in \mathbb{C}_+} \|z(z+A)^{-1}\| < \infty. \tag{2.2}$$

It follows from Neumann series that $\sigma(A) \subset \Sigma_\theta \cup \{0\}$ and $M_\theta(A) \le 2M_A$, where

$$\theta := \arccos(1/(2M_A)) < \pi/2.$$

So $A \in \mathrm{Sect}(\theta) \subset \mathrm{Sect}(\pi/2-)$. Conversely, if $A \in \mathrm{Sect}(\theta)$ where $\theta \in [0, \pi/2)$, then (2.2) holds, with M_A equal to $M_{\pi/2}(A)$ in (2.1). Thus $-A$ generates a bounded holomorphic semigroup if and only if $\sigma(A) \subset \overline{\mathbb{C}}_+$ and (2.2) holds. In that case, M_A is a basic quantity associated with A, known as the *sectoriality constant* of A.

Let $M(\mathbb{R}_+)$ be the space of all bounded Borel measures on $\mathbb{R}_+$. Then $M(\mathbb{R}_+)$ is a Banach algebra under convolution of measures. For $\mu \in M(\mathbb{R}_+)$, let $\mathcal{L}\mu$ be the Laplace transform of μ, and let

$$\mathcal{LM} = \{\mathcal{L}\mu : \mu \in M(\mathbb{R}_+)\}.$$

This is a subalgebra of $H^\infty(\mathbb{C}_+)$, and the map $\mu \mapsto \mathcal{L}\mu$ is an injective algebra homomorphism from $M(\mathbb{R}_+)$ into $H^\infty(\mathbb{C}_+)$. Thus $\mathcal{LM}$ becomes a Banach algebra in the norm $\|\mathcal{L}\mu\|_{\mathrm{HP}} := \|\mu\|_{M(\mathbb{R}_+)}$.

The following functions are in $\mathcal{LM}$:

$$e_t(z) := e^{-tz},\ t \ge 0; \quad r_\lambda(z) := (\lambda + z)^{-1},\ \lambda \in \mathbb{C}_+; \quad \frac{z-1}{z+1}; \quad \frac{1-e^{-z}}{z}.$$

Assume that $-A$ is the generator of a bounded C_0-semigroup $(T(t))_{t\ge0}$ on a Banach space X. Then we define

$$K_A := \sup_{t\ge0} \|T(t)\|.$$

For $f = \mathcal{L}\mu \in \mathcal{LM}$, define

$$\Pi_A(f)x = \int_0^\infty T(t)x\, d\mu(t), \qquad x \in X. \tag{2.3}$$

Then Π_A is an $\mathcal{LM}$-functional calculus for A, with $T(t) = \Pi_A(e_t) = \exp(-tA)$ and $\|\Pi_A\| \le K_A$. Moreover it is the unique $\mathcal{LM}$-calculus for A, and it is known as the *Hille-Phillips calculus* or HP-*calculus*.

The Hille-Phillips calculus is compatible with the sectorial functional calculus in the sense that $\Pi_A(f) = f(A)$ whenever $f(A)$ is defined in the sectorial calculus [21, Section 3.3]. Moreover there is a spectral inclusion theorem [21, Theorem 2.7.4], and a convergence lemma for the HP-calculus [21, Proposition 5.1.4].

We now discuss two questions which can be addressed via the HP-calculus. Both questions are considered in [19] and we will return to them in later sections of this article.

Cayley Transform Question Assume that $-A$ generates a bounded C_0-semigroup on a Banach space X and let $V(A)$ be the cogenerator, so $V(A) = (A-1)(A+1)^{-1}$.

Let V be the Cayley transform, so $V(z) = (z-1)(z+1)^{-1}$. Then V is the Laplace transform of $\delta_0 - 2e^{-t}dt \in M(\mathbb{R}_+)$, so $V \in \mathcal{LM}$ and $\|V\|_{\mathrm{HP}} = 3$. The spectral radius of $V(A)$ is at most 1, and this raised a question about sharp upper bounds for the sequence $(\|V(A)^n\|)_{n\in\mathbb{N}}$ for large n.

For the general case, it was shown in [12] that $\|V^n\|_{\mathrm{HP}} = O(n^{1/2})$, so the Hille-Phillips calculus establishes that $\|V(A)^n\| = O(n^{1/2})$. This estimate is known to be optimal for arbitrary A [19, Section 9], so $\|V^n\|_{\mathrm{HP}} \asymp n^{1/2}$. We shall return to the Cayley transform question for operators on Hilbert space in Sect. 5 (Application 1) and in Sect. 5.

Inverse Generator Question Assume that $-A$ generates a bounded C_0-semigroup on X, and A has dense range in X. Then A is injective and the operator A^{-1} is densely defined. The question was raised in [15] whether $-A^{-1}$ also generates a C_0-semigroup. If $A \in \mathrm{Sect}(\pi/2-)$ then $A^{-1} \in \mathrm{Sect}(\pi/2-)$, so the answer is positive in this case. For $t > 0$ and $z \in \overline{\mathbb{C}}_+ \setminus \{0\}$, let $f_t(z) = e^{-t/z}$. If operators $f_t(A)$ can be defined as bounded operators in some functional calculus for A, then the question should have a positive answer. However f_t cannot be defined continuously at 0, so f_t is not in $\mathcal{LM}$. We shall return to this question in Sect. 3 (Application 2).

3 $\mathcal{B}$-Calculus

To motivate the definition of the $\mathcal{B}$-calculus, we start by reconsidering the definition of the HP-calculus in Sect. 2. Let $-A$ be the generator of a bounded C_0-semigroup $(T(t))_{t\geq 0}$. Let $f = \mathcal{L}\mu$, where $\mu \in M(\mathbb{R}_+)$ and $\mu(\{0\}) = 0$. For $x \in X$, we may argue formally that

$$\begin{aligned} f(A)x &= \int_0^\infty T(t)x \, d\mu(t) \\ &= 4\int_0^\infty \int_0^\infty \alpha t^2 e^{-2\alpha t} T(t)x \, d\alpha \, d\mu(t) \\ &= \frac{2}{\pi}\int_0^\infty \int_0^\infty \alpha t e^{-\alpha t} \int_{\mathbb{R}} (\alpha - i\beta + A)^{-2} x \, e^{-i\beta t} \, d\beta \, d\alpha \, d\mu(t) \\ &= \frac{2}{\pi}\int_0^\infty \alpha \int_{\mathbb{R}} (\alpha - i\beta + A)^{-2} x \int_0^\infty t e^{-(\alpha+i\beta)t} \, d\mu(t) \, d\beta \, d\alpha \\ &= -\frac{2}{\pi}\int_0^\infty \alpha \int_{\mathbb{R}} (\alpha - i\beta + A)^{-2} x \, f'(\alpha + i\beta) \, d\beta \, d\alpha. \end{aligned}$$

Here we have assumed that the Fourier inversion formula can be applied to the function $t \mapsto \pi t e^{-\alpha t} T(t)x$ on $\mathbb{R}_+$ (extended to $\mathbb{R}$ by 0) and its Fourier transform $\beta \mapsto (\alpha + i\beta + A)^{-2}x$, and that Fubini's theorem is valid. The first assumption can be weakened by working in the weak operator topology. Then both assumptions can be satisfied by making assumptions on f and A which involve two spaces $\mathcal{B}$ and $\mathcal{E}$ of holomorphic functions on $\mathbb{C}_+$.

The Space $\mathcal{B}$ Let $\mathcal{B}$ be the space of all holomorphic functions on $\mathbb{C}_+$ such that

$$\|f\|_{\mathcal{B}_0} := \int_0^\infty \sup_{\beta\in\mathbb{R}} |f'(\alpha + i\beta)|\, d\alpha < \infty. \tag{3.1}$$

The literature on this space $\mathcal{B}$ was very limited before the appearance of [37] and [8], and the corresponding space of functions on the unit disc had received rather more attention. The spaces are sometimes described as spaces of analytic Besov functions. The origin for this terminology will be explained in Sect. 3 of this article.

We now set out some of the properties of individual functions in $\mathcal{B}$ (see [8, Proposition 2.2]).

Proposition 3.1 *Let $f \in \mathcal{B}$.*

1. $f(\infty) := \lim_{\Re z\to\infty} f(z)$ *exists in* $\mathbb{C}$.
2. f *is bounded, and* $\|f\|_\infty \le |f(\infty)| + \|f\|_{\mathcal{B}_0}$.
3. $f^b(\beta) := \lim_{\alpha\to 0+} f(\alpha + i\beta)$ *exists for all* $\beta \in \mathbb{R}$.
4. *The function f extends to a uniformly continuous function on* $\overline{\mathbb{C}_+}$.

We let $\mathcal{B}_0 := \{f \in \mathcal{B} : f(\infty) = 0\}$, so $\mathcal{B} = \mathcal{B}_0 \oplus \mathbb{C}$, where $\mathbb{C}$ represents the constant functions.

A norm on $\mathcal{B}$ is defined by

$$\|f\|_{\mathcal{B}} := \|f\|_\infty + \|f\|_{\mathcal{B}_0}.$$

The space $\mathcal{B}$ equipped with this norm has the following properties (see [8, Proposition 2.3, Lemma 2.14]).

Proposition 3.2

1. *The closed unit ball U of $\mathcal{B}$ is compact in the topology of uniform convergence on compact subsets of* $\mathbb{C}_+$.
2. *$\mathcal{B}$ is a Banach algebra, and $\mathcal{B}$ is not separable.*
3. *The norm $\|\cdot\|_{\mathcal{B}_0}$ is equivalent to $\|\cdot\|_{\mathcal{B}}$ on $\mathcal{B}_0$, and $\mathcal{B}_0$ is a closed ideal in the Banach algebra $\mathcal{B}$.*
4. *$\mathcal{LM} \subset \mathcal{B}$ with continuous inclusion; $\mathcal{LM}$ is not closed in $\mathcal{B}$.*
5. *$\mathcal{LM}$ is not dense in $\mathcal{B}$ in the norm-topology, but $\mathcal{LM}$ is dense in $\mathcal{B}$ in the topology of uniform convergence on compact subsets of* $\mathbb{C}_+$.

Examples 3.3

1. The function $e^{-1/z}$ is not in $\mathcal{B}$, as it is not uniformly continuous near 0. However $z(z+1)^{-1}e^{-1/z}$ is in $\mathcal{LM}$ (see the proof of [16, Theorem 3.3]).

2. The function $\operatorname{arccot} z$ is defined on $\mathbb{C}_+$ by

$$\operatorname{arccot} z = \frac{1}{2i} \log\left(\frac{z+i}{z-i}\right), \qquad z \in \mathbb{C}_+. \tag{3.2}$$

This function is not bounded on $\mathbb{C}_+$, so it is not in $\mathcal{B}$. The function $\exp(\operatorname{arccot} z)$ is bounded on $\mathbb{C}_+$, but it is not uniformly continuous near i, so $\exp(\operatorname{arccot} z)$ is not in $\mathcal{B}$ [10, Example 3.19].
3. The following two functions are in $\mathcal{B}$ but not in $\mathcal{LM}$:

$$e^{-z}(1+z)^i, \qquad e^{-2z} \int_0^1 \frac{\sinh(zt)}{t}\, dt.$$

Other classes of functions which are contained in $\mathcal{B}$ may be found in [8, Section 3]. We now give a description of all functions in $\mathcal{B}$.

Let $\operatorname{Hol}(\mathbb{C}_+)$ be the space of all holomorphic functions on $\mathbb{C}_+$, and $\mathcal{W}$ be the space of all (equivalence classes of) measurable functions g on $\mathbb{C}_+$ such that

$$\|g\|_{\mathcal{W}} := \int_0^\infty \operatorname*{ess\,sup}_{\beta\in\mathbb{R}} |g(\alpha+i\beta)|\, d\alpha < \infty, \tag{3.3}$$

with the norm given by (3.3). Then $\mathcal{W}$ is a Banach space, and

$$\mathcal{W} \cap \operatorname{Hol}(\mathbb{C}_+) = \{f' : f \in \mathcal{B}\}, \qquad \|f'\|_{\mathcal{W}} = \|f\|_{\mathcal{B}_0}. \tag{3.4}$$

For $g \in \mathcal{W}$, let

$$(Qg)(z) := -\frac{2}{\pi} \int_0^\infty \alpha \int_{\mathbb{R}} \frac{g(\alpha+i\beta)}{(\alpha - i\beta + z)^2}\, d\beta\, d\alpha. \tag{3.5}$$

This defines an operator Q which serves to obtain a reproducing formula for functions in $\mathcal{B}$ and to construct functions in $\mathcal{B}$ from arbitrary functions in $\mathcal{W}$, as in the following proposition. The first statement is proved in [8, Proposition 2.20], and the second and third statements are proved in [9, Proposition 3.1].

Proposition 3.4

1. Let $f \in \mathcal{B}$. Then $f' \in \mathcal{W}$ and

$$f(z) = f(\infty) + Q(f')(z) = f(\infty) - \frac{2}{\pi} \int_0^\infty \alpha \int_{\mathbb{R}} \frac{f'(\alpha+i\beta)}{(\alpha - i\beta + z)^2}\, d\beta\, d\alpha. \tag{3.6}$$

2. Let $g \in \mathcal{W}$. Then $Qg \in \mathcal{B}_0$.
3. The operator $Q \in L(\mathcal{W}, \mathcal{B}_0)$, and it maps $\mathcal{W} \cap \operatorname{Hol}(\mathbb{C}_+)$ bijectively onto $\mathcal{B}_0$.

The Banach algebra $\mathcal{B}$ satisfies the following Convergence Lemma [9, Lemma 8.1].

Lemma 3.5 *Let $(f_k)_{k\geq 1} \subset \mathcal{B}$ be such that $\sup_{k\geq 1} \|f_k\|_{\mathcal{B}} < \infty$. Assume that for every $z \in \mathbb{C}_+$ there exists*

$$f(z) := \lim_{k\to\infty} f_k(z) \in \mathbb{C},$$

and for every $r > 0$,

$$\lim_{\delta\to 0+} \int_0^\delta \sup_{|\beta|\leq r} |f_k'(\alpha + i\beta)| \, d\alpha = 0,$$

uniformly in k. Let $g \in \mathcal{B}$ with $\lim_{|z|\to\infty} g(z) = 0$. Then $f \in \mathcal{B}$, and

$$\lim_{k\to\infty} \|(f_k - f)g\|_{\mathcal{B}} = 0.$$

Shifts and Density Results for $\mathcal{B}$ In order to establish the existence of a $\mathcal{B}$-calculus for many operators, we will need a dense subalgebra of $\mathcal{B}_0$ with certain properties. To establish the density, we will make use of the shift operators $(T_{\mathcal{B}}(\tau))_{\tau\in\overline{\mathbb{C}}_+}$ on $\mathcal{B}$, which are defined by the formula

$$(T_{\mathcal{B}}(\tau)f)(z) := f(z+\tau), \qquad f \in \mathcal{B},\ \tau \in \overline{\mathbb{C}}_+,\ z \in \mathbb{C}_+.$$

The following properties are established in [8]: see Lemma 2.6 for statements (1)–(4), the proof of Proposition 2.10 for (5) and (6), and Proposition 4.6 for (7).

Proposition 3.6

1. *For each $f \in \mathcal{B}$,*

$$\|T_{\mathcal{B}}(\tau)f\|_{\mathcal{B}} \leq \|f\|_{\mathcal{B}}, \qquad \tau \in \overline{\mathbb{C}}_+.$$

   Moreover

$$\lim_{\tau\in\overline{\mathbb{C}}_+,\tau\to 0} \|T_{\mathcal{B}}(\tau)f - f\|_{\mathcal{B}} = 0.$$

2. *Let $-A_{\mathcal{B}}$ be the generator of the C_0-semigroup $(T_{\mathcal{B}}(t))_{t\geq 0}$ on $\mathcal{B}$. Then*

$$D(A_{\mathcal{B}}) = \{f \in \mathcal{B} : f' \in \mathcal{B}\}, \quad A_{\mathcal{B}}f = -f'.$$

3. *The generator of the C_0-group $(T_{\mathcal{B}}(-is))_{s\in\mathbb{R}}$ is $iA_{\mathcal{B}}$.*
4. *$\sigma(A_{\mathcal{B}}) = \mathbb{R}_+$.*

5. *For each* $f \in \mathcal{B}_0$,

$$\lim_{t\to\infty} \|T_{\mathcal{B}}(t)f\|_{\mathcal{B}_0} = \lim_{t\to\infty} \left\| f - A_{\mathcal{B}} \int_0^t T_{\mathcal{B}}(s)f\,ds \right\|_{\mathcal{B}_0} = 0.$$

6. *The closure of the range of* $A_{\mathcal{B}}$ *is* $\mathcal{B}_0$.
7. *The family* $(T_{\mathcal{B}}(\tau))_{\tau\in\mathbb{C}_+}$ *is a holomorphic* C_0*-semigroup of contractions.*

As a preliminary exercise, we consider the rational functions in $\mathcal{B}_0$. Let $\mathcal{R}_0(\mathbb{C}_+)$ be the linear span of $\{r_\lambda : \lambda \in \mathbb{C}_+\}$. It follows from the resolvent identity and the continuity of the map $\lambda \mapsto r_\lambda$ from $\mathbb{C}_+$ to $\mathcal{LM}$, that $\mathcal{R}_0(\mathbb{C}_+)$ is dense in the algebra of all rational functions which vanish at infinity and have no poles in $\overline{\mathbb{C}}_+$, with respect to the HP-norm and hence with respect to the $\mathcal{B}$-norm and the H^∞-norm.

Proposition 3.7 *The closure of* $\mathcal{R}_0(\mathbb{C}_+)$ *in* $\mathcal{B}$*, with respect to either the* $\mathcal{B}$*-norm or the* H^∞*-norm, is* $\mathcal{B} \cap C_0(\overline{\mathbb{C}}_+)$.

The case of the H^∞-norm in Proposition 3.7 is well-known (see [21, Appendix F, Proposition F.3], for example). The coincidence of the closures in the different norms is a special case of a more general result [9, Theorem 4.4] for subspaces of $\mathcal{B}$ which are invariant under the horizontal shifts $(T_{\mathcal{B}}(\tau))_{\tau\ge 0}$ considered above. As stated in Proposition 3.2, the Hille-Phillips algebra $\mathcal{LM}$ is neither closed nor dense in $\mathcal{B}$. We do not know of any explicit description of the closure of $\mathcal{LM}$ in $\mathcal{B}$, but the closures in the $\mathcal{B}$-norm and in the H^∞-norm coincide.

We now identify a dense subalgebra $\mathcal{G}$ of $\mathcal{B}_0$ consisting of (restrictions of) entire functions of exponential type. Various manipulations can be carried out initially for these functions, and then extended to all functions in $\mathcal{B}_0$ by density arguments. Thus the subalgebra $\mathcal{G}$ plays a very important role in the construction and theory of the $\mathcal{B}$-calculus.

It is well-known that the map $f \mapsto f^b$ is an isometric isomorphism from $H^\infty(\mathbb{C}_+)$ onto

$$H^\infty(\mathbb{R}) := \left\{ g \in L^\infty(\mathbb{R}) : \operatorname{supp}(\mathcal{F}^{-1}g) \subset \mathbb{R}_+ \right\}$$

where $\mathcal{F}$ is the (distributional) Fourier transform (see [23, Section II.1.5], for example). Let I be a closed subset of $\mathbb{R}_+$. The spectral subspace H_I^∞ of $H^\infty(\mathbb{C}_+)$ is defined to be

$$H_I^\infty = \left\{ f \in H^\infty(\mathbb{C}_+) : \operatorname{supp}\left(\mathcal{F}^{-1} f^b\right) \subset I \right\}.$$

Proposition 3.8 *Let* $I := [\epsilon, \sigma]$ *and* J *be non-empty compact intervals in* $(0, \infty)$*, and* $f \in H_I^\infty$, $g \in H_J^\infty$.

1. $f \in \mathcal{B}$ *and*

$$\|f\|_{\mathcal{B}} \le \|f\|_\infty \left(1 + 4\log\left(1 + \frac{\sigma}{\epsilon}\right)\right). \tag{3.7}$$

2. $fg \in H_{I+J}^\infty$.

See [8, Lemma 2.5] and [25, Lemma VI.4.7] for proofs. It follows that

$$\mathcal{G} := \bigcup \{H_I^\infty : I \subset (0,\infty),\, I \text{ compact}\} \tag{3.8}$$

is a subalgebra of $\mathcal{B}_0$. The following result is proved in [8, Proposition 2.10].

Theorem 3.9 *The closure of the subalgebra $\mathcal{G}$ in $\mathcal{B}$ is $\mathcal{B}_0$.*

We outline the proof of this statement, for which we will use Arveson's spectral theory for representations of locally compact abelian groups on Banach spaces (see [4], for example) in the special case of C_0-groups of isometries. For a bounded C_0-group $(T(t))_{t\in\mathbb{R}}$ on a Banach space X, let $\mathrm{sp}_T(x)$ be the Arveson spectrum of an element $x \in X$ with respect to T. For a closed subset I of $\mathbb{R}$, let $X_T(I) = \{x \in X : \mathrm{sp}_T(x) \subset I\}$.

We apply this theory to the C_0-group of shifts on $\mathrm{BUC}(\mathbb{R})$:

$$(S(t)g)(s) = g(s-t), \qquad s,t \in \mathbb{R},\ g \in \mathrm{BUC}(\mathbb{R}),$$

and to the C_0-group of vertical shifts:

$$G(t) := T_{\mathcal{B}}(-it), \qquad t \in \mathbb{R},$$

on $\mathcal{B}_0$, as in Proposition 3.6. Let $K : \mathcal{B} \to \mathrm{BUC}(\mathbb{R})$ be the isometric injection given by $Kf = f^b$. Then

$$S(t)K = KG(t), \qquad t \in \mathbb{R}.$$

For $f \in \mathcal{B}$, it follows from this and the definition of the Arveson spectrum that $\mathrm{sp}_G(f) = \mathrm{sp}_S(f) = \mathrm{supp}(\mathcal{F}^{-1} f^b)$. Hence $H_I^\infty = \mathcal{B}_G(I)$ for all compact subsets I of $(0,\infty)$.

The density of $\mathcal{G}$ in $\mathcal{B}_0$ follows from a combination of Proposition 3.6 and a further abstract result given in Proposition 3.10 below. It is a variant of the fact that the Arveson spectrum of a bounded C_0-group, with generator A, is $-i\sigma(A)$ [14, Theorem 8.19]. A direct proof is given in [8, Proposition 2.8].

Proposition 3.10 *Let $(T(t))_{t\ge 0}$ be a bounded C_0-group on a Banach space X, with generator A, and assume that the range of A is dense in X.*

1. *The set $\bigcup\{X_T(I) : I$ compact, $0 \notin I\}$ is dense in X.*
2. *If, in addition, $\sigma(A) \subset i[0,\infty)$, then $\bigcup\{X_T(I) : I$ compact, $I \subset (0,\infty)\}$ is dense in X.*

The Space $\mathcal{E}$ and Duality Let $\mathcal{E}$ be the space of all holomorphic functions g on $\mathbb{C}_+$ such that

$$\|g\|_{\mathcal{E}_0} := \sup_{\alpha>0} \alpha \int_{\mathbb{R}} |g'(\alpha + i\beta)| \, d\beta < \infty.$$

If $g \in \mathcal{E}$, then $g(\infty) := \lim_{\operatorname{Re} z \to \infty} g(z)$ exists in $\mathbb{C}$. A norm on $\mathcal{E}$ is defined by

$$\|g\|_{\mathcal{E}} := |g(\infty)| + \|g\|_{\mathcal{E}_0}.$$

Then $\mathcal{E}$ is a Banach space.

There is a (partial) duality between $\mathcal{E}$ and $\mathcal{B}$ given by

$$\langle g, f \rangle_{\mathcal{B}} := \int_0^\infty \alpha \int_{\mathbb{R}} g'(\alpha - i\beta) f'(\alpha + i\beta) \, d\beta \, d\alpha, \qquad g \in \mathcal{E}, \quad f \in \mathcal{B}. \tag{3.9}$$

It follows from the definitions of $\mathcal{B}$ and $\mathcal{E}$ that the integral exists and

$$|\langle g, f \rangle_{\mathcal{B}}| \le \|g\|_{\mathcal{E}_0} \|f\|_{\mathcal{B}_0}. \tag{3.10}$$

For $z \in \mathbb{C}_+$, the function $r_z \in \mathcal{E}$ and the formula (3.6) for $f \in \mathcal{B}$ can now be written as

$$f(z) = f(\infty) + \langle r_z, f \rangle_{\mathcal{B}}.$$

The duality is only partial in the sense that $\mathcal{B}$ and $\mathcal{E}$ are not the dual or predual of each other with respect to this duality, and the constant functions in each space are annihilated in the duality. The duality appeared in [37, p.266] (in slightly different form), and it was noted there that Green's formula on $\mathbb{C}_+$ transforms (3.9) into

$$\langle g, f \rangle_{\mathcal{B}} = \frac{1}{4} \int_{\mathbb{R}} g(-i\beta) f(i\beta) \, d\beta \tag{3.11}$$

for "good functions". For sample results about the interpretation of "good functions", see [36, Lemma 17] and [10, Proposition 2.2]. Informally, suppose that $f(\infty) = 0$ (without essential loss of generality), and that (3.11) holds for $g = r_z$, where $z \in \mathbb{C}_+$. Then (3.6) becomes

$$f(z) = \frac{1}{2\pi} \int_{\mathbb{R}} \frac{f(i\beta)}{z - i\beta} \, d\beta,$$

which is a Cauchy formula for $f(z)$. However the integral here is not necessarily absolutely convergent, and the formula is not valid for some $f \in \mathcal{B}$.

The duality plays a crucial role in the construction of the $\mathcal{B}$-calculus, via the following approximation lemma. Theorem 3.9 shows that, in order to prove the lemma, it suffices to consider the case when $f \in \mathcal{G}$, and then an inequality of Bohr's type is applicable. See [8, Lemma 3.19] for details of the proof.

Lemma 3.11 *Let $\mu \in M(\mathbb{R}_+)$ satisfy*

$$\mu(\mathbb{R}_+) = 1, \qquad \int_{\mathbb{R}_+} t\, d|\mu|(t) < \infty.$$

Let $m = \mathcal{L}\mu \in \mathcal{LM}$, and $f \in \mathcal{B}$. For $\delta > 0$, let

$$f_\delta(z) := m(\delta z) f(z), \qquad z \in \mathbb{C}_+.$$

Then, for all $g \in \mathcal{E}$,

$$\lim_{\delta \to 0+} \langle g, f_\delta \rangle_{\mathcal{B}} = \langle g, f \rangle_{\mathcal{B}}.$$

Definition of the $\mathcal{B}$-Calculus We can now proceed towards a definition of a $\mathcal{B}$-calculus by replacing z by an operator A in Proposition 3.4. To make this effective for as many operators A as possible, we will consider the integral in (3.6) in the weak operator topology (see (3.15) below). For this we need the following assumption on the operator A.

Let A be a closed densely defined operator on a Banach space X, and assume that the spectrum $\sigma(A)$ is contained in $\overline{\mathbb{C}}_+$ and

$$\sup_{\alpha > 0} \alpha \int_{\mathbb{R}} |\langle (\alpha + i\beta + A)^{-2} x, x^* \rangle| \, d\beta < \infty \tag{3.12}$$

for all $x \in X$ and $x^* \in X^*$. In other words, the functions

$$g_{x,x^*} : z \mapsto \langle (z + A)^{-1} x, x^* \rangle \tag{3.13}$$

belong to $\mathcal{E}$. These assumptions were first considered by Gomilko [17], and by Shi and Feng [34], who independently proved that they imply that $-A$ is the generator of a bounded C_0-semigroup on X. We shall call them the (GSF) *assumptions* on A.

By the Closed Graph Theorem, the supremum in (3.12) is bounded for $\|x\| \le 1$ and $\|x^*\| \le 1$. Hence, if the (GSF) assumptions are satisfied there is a minimal constant γ_A such that

$$\int_{\mathbb{R}} |\langle (\alpha + i\beta + A)^{-2} x, x^* \rangle| \, d\beta \le \frac{\gamma_A}{\alpha} \|x\| \, \|x^*\|, \qquad x \in X, \ x^* \in X^*. \tag{3.14}$$

Examples 3.12

1. Let $-A$ be the generator of a bounded C_0-semigroup $(T(t))_{t \ge 0}$ on a Hilbert space X. Then A satisfies the (GSF) assumptions with $\gamma_A \le 2K_A^2$ [8, Example 4.1(1)].

2. Let $A \in \mathrm{Sect}(\pi/2-)$. Then A satisfies the (GSF) assumptions, with $\gamma_A \le CM_A(\log M_A + 1)$, where C is an absolute constant. See Theorem 4.1 below.
3. Let A be the generator of the C_0-group of translations on $L^p(\mathbb{R})$, where $1 \le p < \infty$, $p \ne 2$. Then A does not satisfy the (GSF) assumptions [9, Corollary 6.7].

Let A satisfy the (GSF) assumptions, and let $f \in \mathcal{B}$. Define

$$\langle f(A)x, x^*\rangle := f(\infty)\langle x, x^*\rangle + \frac{2}{\pi}\langle g_{x,x^*}, f\rangle_{\mathcal{B}} \tag{3.15}$$
$$= f(\infty)\langle x, x^*\rangle - \frac{2}{\pi}\int_0^\infty \alpha \int_{\mathbb{R}} \langle(\alpha - i\beta + A)^{-2}x, x^*\rangle f'(\alpha + i\beta)\, d\beta\, d\alpha$$

for all $x \in X$ and $x^* \in X^*$. It is easily seen that this defines a bounded linear mapping $f(A) : X \to X^{**}$, and that the linear mapping

$$\Phi_A : \mathcal{B} \to L(X, X^{**}), \qquad f \mapsto f(A),$$

is bounded. Indeed, by (3.10) and (3.14),

$$\|f(A)\| \le |f(\infty)| + \gamma_A \|f\|_{\mathcal{B}_0} \le \gamma_A \|f\|_{\mathcal{B}}.$$

Theorem 3.13 *If A satisfies the* (GSF) *assumptions, then the map $\Phi_A : f \mapsto f(A)$ is a bounded algebra homomorphism from $\mathcal{B}$ into $\mathcal{L}(X)$, which extends the Hille-Phillips calculus. Moreover, $\|\Phi_A\| \le \gamma_A$.*

It is not immediately clear why (3.15) defines an operator $f(A)$ which maps X into X. The proof of this has three steps which we sketch here.

Firstly, consider $f \in \mathcal{LM}$. Then the calculation at the beginning of this section is justifiable in the weak operator topology, and it shows that $f(A)x$ as in (3.15) agrees with $f(A)x$ as defined in (2.3). In particular, $f(A)$ maps X into X.

Secondly, consider $f \in \mathcal{G}$. For $\delta > 0$, let

$$e_\delta(z) := \frac{1 - e^{-\delta z}}{\delta z}, \qquad z \in \mathbb{C}_+.$$

Then $e_\delta \in \mathcal{LM}$, $fe_\delta \in \mathcal{LM}$, and Lemma 3.11 can be used to show that $\lim_{\delta\to 0+}(fe_\delta)(A) = f(A)$ in the strong operator topology. Thus $f(A)$ maps X into X.

Finally, consider $f \in \mathcal{B}$. By Theorem 3.9, $f - f(\infty)$ is in the norm-closure of $\mathcal{G}$, so $f(A) - f(\infty)$ is the limit in operator norm of a sequence $(f_n(A))_{n\ge 1}$ where each $f_n \in \mathcal{G}$. Hence $f(A)$ maps X into X.

For more detailed proofs, including the fact that Φ_A is an algebra homomorphism, see [8, Lemma 4.2, Lemma 4.3 and Theorem 4.4].

There are converse results to Theorem 3.13 as follows.

Theorem 3.14 *Let A be a closed densely defined operator on a Banach space X, and assume that $\sigma(A) \subseteq \overline{\mathbb{C}}_+$ and there exists a $\mathcal{B}$-calculus Φ for A. Then A satisfies the* (GSF) *assumptions, and $\Phi = \Phi_A$ as defined in* (3.15).

The proof of necessity of the (GSF) assumptions uses the general method of constructing functions in $\mathcal{B}$ described in Proposition 3.4, and the proof that $\Phi = \Phi_A$ can be obtained by following the steps in the proof of Theorem 3.13. Details are given in [9, Theorems 6.1 and 6.2].

Theorems 3.13 and 3.14 establish that a densely defined operator A has a bounded functional calculus for $\mathcal{B}$ if and only if A satisfies the (GSF) assumptions. The unique functional calculus is Φ_A as above, and it extends the Hille-Phillips calculus. We will say that Φ_A is the *$\mathcal{B}$-calculus* for A.

The $\mathcal{B}$-calculus is compatible with the sectorial calculus in various ways. In particular if $A \in \mathrm{Sect}(\pi/2-)$, $f \in \mathcal{B}$ and $f(A)$ can be defined in the sectorial calculus (for example, if A is injective), then the definitions agree. There are also similar statements when A satisfies the (GSF) conditions, f is holomorphic on a sector larger than $\mathbb{C}_+$, and the restriction of f to $\mathbb{C}_+$ belongs to $\mathcal{B}$. The $\mathcal{B}$-calculus is compatible in similar ways with the half-plane calculus considered in [6]. We refer the reader to [8, Section 4.3] for details.

There is a Convergence Lemma for the $\mathcal{B}$-calculus, which can easily be deduced from Lemma 3.5 (see [9, Corollary 8.3]). It was first proved by a different argument in [8, Theorem 4.13].

Theorem 3.15 *Let A be an operator which satisfies the* (GSF) *assumptions, and let f_k and f_0 be as in Lemma 3.5. Then $f_k(A) \to f_0(A)$ in the strong operator topology as $k \to \infty$.*

There is a Spectral Inclusion Theorem for the $\mathcal{B}$-calculus, as follows. The proofs are simple variants of arguments used for the HP-calculus, involving Banach algebra techniques as in [24, Section 16]. The third statement uses the F-product as in [27, Section A.III.4]. See [8, Theorem 4.17] for the proofs of the other statements.

Theorem 3.16 *Let A be an operator which satisfies the* (GSF) *assumptions. Let $f \in \mathcal{B}$ and $\lambda \in \mathbb{C}$.*

1. *If $x \in D(A)$ and $Ax = \lambda x$, then $f(A)x = f(\lambda)x$.*
2. *If $x^* \in D(A^*)$ and $A^*x^* = \lambda x^*$, then $f(A)^*x^* = f(\lambda)x^*$.*
3. *If $(x_n)_{n\geq 1}$ are unit vectors in $D(A)$ and $\lim_{n\to\infty} \|Ax_n - \lambda x_n\| = 0$, then $\lim_{n\to\infty} \|f(A)x_n - f(\lambda)x_n\| = 0$.*
4. *If $\lambda \in \sigma(A)$ then $f(\lambda) \in \sigma(f(A))$.*
5. *If $A \in \mathrm{Sect}(\pi/2-)$, then $\sigma(f(A)) \cup \{f(\infty)\} = f(\sigma(A)) \cup \{f(\infty)\}$.*

Applications The $\mathcal{B}$-calculus has various applications, some of them involving fine estimates of norms of particular operators $f(A)$, and others providing new proofs and/or variants of known results in semigroup theory. We give some examples of each type. Recall that $\mathcal{L}\mu$ denotes the Laplace transform of a measure $\mu \in M(\mathbb{R}_+)$.

1. *Spectral Mapping Theorem*. The Spectral Mapping Theorem does not hold for C_0-semigroups even when X is a Hilbert space, so it does not hold for the $\mathcal{B}$-calculus. However the following result shows that it holds for a subalgebra of $\mathcal{B}$.

Theorem 3.17 *If $-A$ generates a bounded C_0-semigroup on a Hilbert space, and $f \in \mathcal{B} \cap C_0(\mathbb{C}_+)$, then*

$$\sigma(f(A)) \cup \{0\} = f(\sigma(A)) \cup \{0\}. \tag{3.16}$$

In particular, if the Fourier transform of $\mu \in M(\mathbb{R}_+)$ belongs to $C_0(\mathbb{R})$ then

$$\sigma(\mathcal{L}\mu(A)) \cup \{0\} = \mathcal{L}\mu(\sigma(A)) \cup \{0\}.$$

This result can be proved by using Proposition 3.7 and a result of Stafney [35, Lemma 6.1]. See [9, Theorem 7.3] for details. In fact, the result holds for all operators A satisfying the (GSF) assumptions.

2. *Bernstein functions and subordination*. Let g be a Bernstein function on $\mathbb{C}_+$. One of the many equivalent definitions of Bernstein functions says that there is a vaguely continuous convolution semigroup of subprobability measures $(\mu_t)_{t\ge 0}$ on $\mathbb{R}_+$ such that $\mathcal{L}\mu_t = e^{-tg}$ for all $t \ge 0$. See [31] for further information about Bernstein functions.

Phillips [30] showed that, if $-A$ is the generator of a bounded C_0-semigroup on X, then the operators $S(t)$ defined by

$$S(t)x := \int_0^\infty e^{-sA}x \, d\mu_t(s), \qquad x \in X, \ t \ge 0,$$

form a bounded C_0-semigroup on X, which is said to be *subordinate* to X. The negative generator of the subordinate semigroup is an operator $g_{\mathrm{Bern}}(A)$. This operator $g_{\mathrm{Bern}}(A)$ coincides with the operator $g(A)$ defined in other functional calculi, such as the extended sectorial calculus if $A \in \mathrm{Sect}(\pi/2-)$ and A is injective.

The following question was raised in [26]:

If $A \in \mathrm{Sect}(\theta)$ where $\theta \in [0, \pi/2)$, is $g_{\mathrm{Bern}}(A) \in \mathrm{Sect}(\theta)$?

In other words, if $-A$ generates a bounded holomorphic C_0-semigroup of angle θ, is the subordinate semigroup generated by $-g_{\mathrm{Bern}}(A)$ also holomorphic, of angle at least θ?

Following some partial answers, a positive answer to the full question was given in [20], followed by a second proof in [7]. The $\mathcal{B}$-calculus can be used to give a third proof [8, Proposition 3.8, Corollary 5.10]. A further proof, based on the $\mathcal{D}$- and $\mathcal{H}$-calculi, is briefly discussed at the end of Sect. 4.

3. *Short-time behaviour*. It is a starting point of semigroup theory that one defines the generator $-A$ of a C_0-semigroup as the right-hand derivative (at $t = 0$) of the

semigroup e^{-tA} on its natural domain. The result below, proved in [8, Theorem 8.9], shows that the negative exponential function can be replaced by many other functions. The proof uses the $\mathcal{B}$-calculus and other properties, so it is not clear that the result holds for all operators satisfying the (GSF) assumptions.

Theorem 3.18 *Let $-A$ be the generator of either a bounded C_0-semigroup on a Hilbert space X, or a bounded holomorphic C_0-semigroup on a Banach space X. Let $f \in D(A_{\mathcal{B}})$, so $f \in \mathcal{B}$ and $f' \in \mathcal{B}$. Then, for all $x \in D(A)$,*

$$\lim_{t \to 0+} t^{-1}(f(tA)x - f(0)x) = f'(0)Ax. \tag{3.17}$$

Conversely, if $f'(0) \neq 0$ and

$$\lim_{t \to 0+} t^{-1}(f(tA)x - f(0)x)$$

exists for some $x \in X$, then $x \in D(A)$ and (3.17) *holds. Thus, if $f'(0) = -1$, then $-A$ coincides with the right-hand derivative of $f(\cdot A)$ at 0, in the strong operator topology, on its natural domain.*

4. *Cayley transform problem on Hilbert spaces.* Let $-A$ be the generator of a bounded C_0-semigroup on a Hilbert space, and $V(A) = (A-1)(A+1)^{-1}$. If the semigroup is contractive, it is well-known that $V(A)$ is also a contraction. The discussion in Sect. 2 leaves open the specific question whether $V(A)$ must be power-bounded when the semigroup is bounded but not contractive. It was shown in [18] that $\|V(A)^n\|$ grows at most logarithmically in n. The proof of this involved intricate estimates of Laguerre polynomials.

   The $\mathcal{B}$-calculus shows that

$$\|V(A)^n\| \le C\|V^n\|_{\mathcal{B}}.$$

It is shown in [8, Lemma 3.7] (with relatively simple estimates) that $\|V^n\|_{\mathcal{B}} = O(\log n)$ and in [9, Lemma 5.1] that $\|V^n\|_{\mathcal{B}} \asymp \log n$. It remains unknown whether the estimate $\|V(A)^n\| = O(\log n)$ is optimal for bounded semigroups on Hilbert spaces (see Sect. 5 for further discussion).

5. *Inverse generator problem on Hilbert spaces.* This problem was discussed in Sect. 2. In the case of Hilbert spaces, the answer is unknown. Let A have dense range on a Hilbert space, and assume that $-A$ generates a bounded C_0-semigroup. If the inverse generator problem on Hilbert spaces has a positive answer, then the C_0-semigroup $(e^{-tA^{-1}})_{t\ge 0}$ must be bounded. Thus a subsidiary question is whether all semigroups of this type on Hilbert spaces are bounded. This question can be addressed by functional calculus.

Let $\varphi_t(z) = z(z+1)^{-1}e^{-t/z}$. Then $\varphi_t \in \mathcal{LM}$. The HP-norm and the $\mathcal{B}$-norm can be estimated as follows [9, Section 5.3].

Proposition 3.19 *Let φ_t, $t \geq 0$ be as above. Then*

$$\|\varphi_t\|_{\mathrm{HP}} \asymp t, \qquad \|\varphi_t\|_{\mathcal{B}} \asymp \log t, \qquad t \to \infty.$$

Applying the $\mathcal{B}$-calculus and the estimate above for $\|\varphi_t\|_{\mathcal{B}}$ to estimate the norm of $\varphi_t(A)$ leads easily to the following result, a version of which was obtained in [39, Theorem 2.2] under stronger assumptions on A. Although the logarithmic estimate in Proposition 3.19 is sharp, it does not follow that the estimate in Corollary 3.20 is sharp.

Corollary 3.20 *Let $-A$ be the generator of a bounded C_0-semigroup on a Hilbert space X, and assume that A has a bounded inverse. Then*

$$\|e^{-tA^{-1}}\| = O(\log t), \qquad t \to \infty.$$

6. *Exponentially stable semigroups.* Let $-A$ be the generator of an exponentially stable C_0-semigroup on a Hilbert space, and $f \in H^\infty(\mathbb{C}_+)$. In this situation, $f(A)$ can be defined as a closed operator via the half-plane calculus. Under a weak assumption on the boundary function of f, the following result shows that $f(A) \in L(X)$.

Corollary 3.21 *Let $-A$ be the generator of an exponentially stable C_0-semigroup on a Hilbert space X, so that*

$$\|e^{-tA}\| \leq Me^{-\omega t}, \qquad t \geq 0,$$

for some $M, \omega > 0$. Let $f \in H^\infty(\mathbb{C}_+)$, let

$$h(t) = \operatorname{ess\,sup}_{|s|\geq t} |f^b(s)|,$$

and assume that

$$\int_0^\infty \frac{h(t)}{1+t}\, dt < \infty.$$

Then $f(A) \in L(X)$ and

$$\|f(A)\| \leq 6M^2 \int_0^\infty \frac{h(t)}{\omega + t}\, dt.$$

The proof uses the $\mathcal{B}$-calculus, applied to the function $g(z) := f(z+\omega)$, which belongs to $\mathcal{B}_0$. For the details, see [8, Lemma 3.3 and Corollary 5.6]. The result had previously been proved by Schwenninger and Zwart [33] in the special case when

$|f^b(s)| \le (\log(|s|+e)^{-\alpha}$ for some $\alpha > 1$. It is valid on Banach spaces, provided that A satisfies the (GSF) assumptions.

4 $\mathcal{D}$- and $\mathcal{H}$-Calculi for Sectorial Operators

Let $A \in \mathrm{Sect}(\pi/2-)$, so that (2.2) holds and $-A$ is the generator of a bounded holomorphic C_0-semigroup on X. In [37], Vitse established that A has a bounded $\mathcal{B}$-calculus with an estimate $\|f(A)\| \le 31 M_A^3 \|f\|_{\mathcal{B}}$. An estimate of the form (4.2) was later obtained in [32, Theorem 5.2]. Both those papers used the sectorial functional calculus and Littlewood-Paley decompositions to obtain the estimates. The estimate (4.1) was shown in [8, Corollary 4.8] (with $C = 8(2+\log 2)$), using methods of complex analysis and functional analysis.

Theorem 4.1 *Let $A \in \mathrm{Sect}(\pi/2-)$. There is an absolute constant C such that*

$$\gamma_A \le C M_A(\log M_A + 1). \tag{4.1}$$

Hence A has a $\mathcal{B}$-calculus, and, for all $f \in \mathcal{B}$,

$$\|f(A)\| \le C M_A(\log M_A + 1)\|f\|_{\mathcal{B}}, \tag{4.2}$$

and

$$f(A) = f(\infty) - \frac{2}{\pi}\int_0^\infty \alpha \int_{\mathbb{R}} (\alpha - i\beta + A)^{-2} f'(\alpha + i\beta)\, d\beta\, d\alpha,$$

where the integral converges in operator norm.

For sectorial operators, it is possible to extend the $\mathcal{B}$-calculus to more functions in two ways. In the $\mathcal{B}$-calculus, the functions f are holomorphic on the half-plane $\mathbb{C}_+$, but $\sigma(A)$ is contained in a smaller sector (including the origin). Consequently it may be possible to define $f(A)$ as a bounded operator, for some functions f which are holomorphic, but not bounded, on $\mathbb{C}_+$. More significantly, one can define a bounded calculus for holomorphic functions which are defined only on subsectors of $\mathbb{C}_+$. In this section we describe two related calculi, the $\mathcal{D}$-calculus for functions defined on $\mathbb{C}_+$, and the $\mathcal{H}$-calculus for functions on sectors.

The Spaces $\mathcal{D}_s$ For $s > -1$, let $\mathcal{V}_s$ be the Banach space of (equivalence classes of) measurable functions $g : \mathbb{C}_+ \to \mathbb{C}$ such that the norm

$$\|g\|_{\mathcal{V}_s} := \int_0^\infty \alpha^s \int_{\mathbb{R}} \frac{|g(\alpha + i\beta)|}{(\alpha^2 + \beta^2)^{(s+1)/2}}\, d\beta\, d\alpha < \infty. \tag{4.3}$$

If $\sigma > s$, then

$$\mathcal{V}_s \subset \mathcal{V}_\sigma \qquad \text{and} \qquad \|g\|_{\mathcal{V}_\sigma} \le \|g\|_{\mathcal{V}_s}, \qquad g \in \mathcal{V}_s. \tag{4.4}$$

For $s > -1$, let $\mathcal{D}_s$ be the space of all holomorphic functions f on $\mathbb{C}_+$ such that $f' \in \mathcal{V}_s$, with the seminorm

$$\|f\|_{\mathcal{D}_{s,0}} := \|f'\|_{\mathcal{V}_s}.$$

Functions in this class appeared in [1, Corollary 4.2], but an extended theory first appeared in [10].

Examples 4.2

1. Let $\lambda \in \mathbb{C}_+$, and consider $r_\lambda(z) := (\lambda + z)^{-1}$. Then $r_\lambda \in \mathcal{D}_s$ for all $s > -1$, and $\|r_\lambda\|_{\mathcal{D}_{s,0}} \le C_s|\lambda|^{-1}$, where C_s depends only on s. See [10, Example 3.3].
2. Let $V(z) = (z-1)(z+1)^{-1}$. For $s > 0$,

$$\|V^n\|_{\mathcal{D}_{s,0}} \le 16\big(B(s/2, 1/2) + 2^{-s/2}\big), \qquad n \in \mathbb{N},$$

where B is the beta function. See [10, Lemma 12.1].

3. Let

$$f_\nu(z) := z^\nu e^{-z}, \quad z \in \mathbb{C}_+, \quad \nu \ge 0.$$

Then $f_\nu \in \mathcal{D}_s$ if and only if $s > \nu$. Moreover, if $s > \nu$, then

$$\|f_\nu\|_{\mathcal{D}_{s,0}} \le 2B\left(\frac{s-\nu}{2}, \frac{1}{2}\right)\Gamma(\nu+1). \tag{4.5}$$

In particular, the function $e^{-z} \in \mathcal{B}$ but it is not in $\mathcal{D}_0$. See [10, Example 3.4].

4. As noted in Example 3.3(2), the function arccot is not in $\mathcal{B}$. However it is in $\mathcal{D}_s$ for all $s > -1$. The function $g(z) := \exp(\operatorname{arccot} z)$ is bounded on $\mathbb{C}_+$ and it is in $\mathcal{D}_s$ for all $s > -1$. However the boundary function of g is not continuous, so $g \notin \mathcal{B}$. Moreover, $\lim_{\epsilon\to 0+} g(\epsilon + i)$ does not exist. See [10, Examples 3.5 and 3.19].

The following inclusions are easily seen from (4.4) and [10, Proposition 3.15].

Proposition 4.3

1. *Let $\sigma > s > -1$. Then $\mathcal{D}_s \subset \mathcal{D}_\sigma$.*
2. *Let $s > 0$. Then $\mathcal{B} \subset \mathcal{D}_s$.*

For $g \in \mathcal{V}_s$, let

$$(Q_s g)(z) := -\frac{2^s}{\pi}\int_0^\infty \alpha^s \int_{\mathbb{R}} \frac{g(\alpha + i\beta)}{(\alpha - i\beta + z)^{s+1}}\, d\beta\, d\alpha, \quad z \in \mathbb{C}_+ \cup \{0\}. \tag{4.6}$$

Note that the definition of Q_1 agrees with the operator Q defined in (3.5), but the domain of Q_1 is larger than the domain of Q.

The properties of $\mathcal{V}_s$, $\mathcal{D}_s$ and Q_s stated in Proposition 4.4 below generally correspond to the properties of $\mathcal{W}$, $\mathcal{B}$ and Q given in Proposition 3.4. One variation is that Q_s does not map all functions in $\mathcal{V}_s$ into $\mathcal{D}_s$ but Proposition 4.5 shows that Q_s does map all functions in $\mathcal{V}_s$ into $\mathcal{D}_\sigma$ for all $\sigma > s$. The representation formula (4.7) was shown in [1, Corollary 4.2]. A proof of the first statement in Proposition 4.4 is given in [10, Proposition 3.7], and the existence of the sectorial limits $f(0)$ and $f(\infty)$ and the representation formula (4.7) follow as corollaries.

A function f on $\mathbb{C}_+$ has *sectorial limits* at 0 and at ∞ if the following limits exist in $\mathbb{C}$ for all $\psi \in (0, \pi/2)$:

$$f(0) := \lim_{|z|\to 0, z\in\Sigma_\psi} f(z), \qquad f(\infty) := \lim_{|z|\to \infty, z\in\Sigma_\psi} f(z).$$

The function $Q_s g$ in (4.6) has sectorial limits $(Q_s g)(0)$ given by (4.6) for $z = 0$, and $(Q_s g)(\infty) = 0$.

Proposition 4.4 *Let $s > -1$.*

1. *Let g be a holomorphic function in $\mathcal{V}_s$. Then $Q_s g \in \mathcal{D}_s$, and $(Q_s g)' = g$.*
2. *Let $f \in \mathcal{D}_s$. Then $f' \in \mathcal{V}_s$, f has sectorial limits at 0 and ∞, and*

$$\begin{aligned} f(z) &= f(\infty) + (Q_s f')(z) \\ &= f(\infty) - \frac{2^s}{\pi} \int_0^\infty \alpha^s \int_{\mathbb{R}} \frac{f'(\alpha + i\beta)}{(\alpha - i\beta + z)^{s+1}} \, d\beta \, d\alpha. \end{aligned} \tag{4.7}$$

3. *There exists $g_s \in \mathcal{V}_s$ such that $Q_s g_s \notin \mathcal{D}_s$.*

There is a norm on $\mathcal{D}_s$ given by

$$\|f\|_{\mathcal{D}_s} := |f(\infty)| + \|f\|_{\mathcal{D}_{s,0}}.$$

With this norm, $\mathcal{D}_s$ is a Banach space [10, Corollary 3.11], and the inclusions in Proposition 4.3 are continuous. On the other hand, $\mathcal{D}_s$ is not an algebra, but $\mathcal{D}_\infty := \bigcup_{s>-1} \mathcal{D}_s$ is an algebra [10, Lemma 3.21].

Some functions in $\mathcal{D}_s$ are unbounded; arccot is an example. Let $\mathcal{D}_s^\infty := \mathcal{D}_s \cap H^\infty(\mathbb{C}_+)$. There is a norm on $\mathcal{D}_s^\infty$ defined by

$$\|f\|_{\mathcal{D}_s^\infty} = \|f\|_\infty + \|f\|_{\mathcal{D}_{s,0}}.$$

With this norm, $\mathcal{D}_s^\infty$ is a Banach algebra.

The following results are proved in [10, Proposition 3.6].

Proposition 4.5 *Let $\sigma > s > -1$.*

1. *The restriction of Q_σ to $\mathcal{V}_s$ is in $L(\mathcal{V}_s, \mathcal{D}_s)$.*
2. *$Q_s \in L(\mathcal{V}_s, \mathcal{D}_\sigma)$.*

The representation formula (4.7) is used to prove the second and third statements in the following proposition (see [10, Lemmas 3.17 and 3.22, Corollary 5.5]).

Proposition 4.6 *Let $f \in \mathcal{D}_s$, where $s > -1$.*

1. *The functions $\tilde{f}(z) := f(1/z)$ and $f_t(z) := f(tz)$, $t > 0$, are in $\mathcal{D}_s$, and $\|\tilde{f}\|_{\mathcal{D}_{s,0}} = \|f\|_{\mathcal{D}_{s,0}}$, $\|f_t\|_{\mathcal{D}_s} = \|f\|_{\mathcal{D}_s}$.*
2. *For $\tau \in \mathbb{C}_+$, let $T(\tau)f(z) = f(z+\tau)$. Then $T(\tau)f \in \mathcal{D}_s$. Moreover $(T(\tau))_{\tau\in\mathbb{C}_+}$ form a bounded holomorphic C_0-semigroup on $\mathcal{D}_s$.*
3. *For $n \in \mathbb{N}$, the function $z^n f^{(n)}(z) \in \mathcal{D}_{s+n}$.*

Note that $\mathcal{D}_s$ is not invariant under vertical shifts (see [10, Example 3.19] or Example 4.2(4) above).

The Spaces $\mathcal{H}_\psi$ Here we introduce spaces $\mathcal{H}_\psi$ of holomorphic functions on sectors Σ_ψ. Later in this section we will construct a bounded functional calculus for these spaces, for operators $A \in \mathrm{Sect}(\theta)$ where $0 \le \theta < \psi < \pi$.

Let $\psi \in (0, \pi)$. The Hardy space $H^1(\Sigma_\psi)$ on the sector Σ_ψ is defined to be the space of all holomorphic functions g on Σ_ψ such that

$$\|g\|_{H^1(\Sigma_\psi)} := \sup_{|\varphi|<\psi} \int_0^\infty \left(|g(te^{i\varphi})| + |g(te^{-i\varphi})|\right) dt < \infty. \tag{4.8}$$

For $\psi = \pi/2$, this definition may appear to differ from the classical definition of a Hardy space on a half-plane, but actually the two definitions coincide (a proof is given in [10, Corollary 4.3]).

We consider the space $\mathcal{H}_\psi$ of all holomorphic functions f on Σ_ψ such that $f' \in H^1(\Sigma_\psi)$. These spaces are sometimes called *Hardy-Sobolev spaces*.

It was shown in [8, Proposition 2.4] that $\mathcal{H}_{\pi/2} \subset \mathcal{B}$ and $\lim_{|z|\to\infty, z\in\mathbb{C}_+} f(z)$ exists in $\mathbb{C}$ for all $f \in \mathcal{H}_{\pi/2}$ (see also Theorem 4.11(1) below). The first two statements in Proposition 4.7 follow immediately for $\psi = \pi/2$. Since the map $f(z) \mapsto f(z^{2\psi/\pi})$ is a linear isomorphism of $\mathcal{H}_\psi$ onto $\mathcal{H}_{\pi/2}$, those statements also hold for all $\psi \in (0, \pi)$. The third statement is easily seen.

Proposition 4.7 *Let $f \in \mathcal{H}_\psi$, $\psi \in (0, \pi)$.*

1. *The function f extends to a continuous bounded function on $\overline{\Sigma}_\psi$.*
2. *The limit*

$$f(\infty) := \lim_{|z|\to\infty, z\in\Sigma_\psi} f(z)$$

exists in $\mathbb{C}$.

3. *For all $z \in \Sigma_\psi$,*

$$|f(z)| \le |f(\infty)| + \|f'\|_{H^1(\Sigma_\psi)}.$$

Consequently, there is a norm on $\mathcal{H}_\psi$ defined by

$$\|f\|_{\mathcal{H}_\psi} := \|f\|_{H^\infty(\Sigma_\psi)} + \|f'\|_{H^1(\Sigma_\psi)}.$$

Proposition 4.8 *For any $\psi \in (0, \pi)$, the space $\mathcal{H}_\psi$ is a Banach algebra. Moreover, for $\psi_1, \psi_2 \in (0, \pi)$, the map $f(z) \mapsto f(z^{\psi_1/\psi_2})$ is an isometric algebra homomorphism from $\mathcal{H}_{\psi_1}$ onto $\mathcal{H}_{\psi_2}$.*

Examples 4.9

1. Let $\psi \in (0, \pi)$, $\varphi \in (0, \pi - \psi)$ and $\lambda \in \Sigma_\varphi$. Then $r_\lambda \in \mathcal{H}_\psi$, and

$$\|r_\lambda\|_{\mathcal{H}_\psi} \le \frac{C_{\varphi,\psi}}{|\lambda|},$$

   where $C_{\varphi,\psi}$ depends only on φ and ψ. See [10, Example 4.10(1)].
2. Let $\gamma \in (0, \pi/(2\psi))$ and $e_{\gamma,\lambda}(z) := e^{-\lambda z^\gamma}$, $z \in \Sigma_\psi$. Then $e_{\gamma,\lambda} \in \mathcal{H}_\psi$ and

$$\begin{aligned}\|e'_{\gamma,\lambda}\|_{H^1(\Sigma_\psi)} = \|e'_{1,\lambda}\|_{H^1(\Sigma_{\gamma\psi})} &\le \int_{\partial\Sigma_{\gamma\psi}} |\lambda| e^{-\operatorname{Re}\lambda z}\,|dz| \\ &\le \frac{1}{\cos(\varphi + \gamma\psi)} + \frac{1}{\cos(\varphi - \gamma\psi)}.\end{aligned}$$

The estimates for resolvents of Bernstein functions in the next proposition illustrate the value of the spaces $\mathcal{D}_s$ and $\mathcal{H}_\psi$, and they are applied at the end of this section. They are proved in [10, Lemma 3.20 and Corollary 4.16].

Proposition 4.10 *Let g be a Bernstein function.*

1. *Let $\lambda \in \mathbb{C}_+$ and $f(z) = (\lambda + g(z))^{-1}$ for $z \in \mathbb{C}_+$. Then $f \in \mathcal{D}_s$ for $s > 2$ and*

$$\|f\|_{\mathcal{D}_s} \le \frac{C_s}{|\lambda|},$$

   where C_s is a constant depending only on s.
2. *Let $\psi \in (0, \pi/2)$, $\varphi \in (\psi, \pi)$ and $\lambda \in \Sigma_{\pi-\varphi}$. Let $f(z) = (\lambda + g(z))^{-1}$ for $z \in \Sigma_\psi$. Then $f \in \mathcal{H}_\psi$ and*

$$\|f\|_{\mathcal{H}_\psi} \le 2\left(\frac{1}{\sin(\min(\varphi, \pi/2))} + \frac{2}{\cos\psi \sin^2((\varphi - \psi)/2)}\right)\frac{1}{|\lambda|}. \tag{4.9}$$

We now lay out relations between $\mathcal{B}$, $\mathcal{D}_s$ and $\mathcal{H}_\psi$ which are proved in [10, Theorem 4.12 and Lemma 4.13].

Theorem 4.11

1. *If* $f \in \mathcal{H}_{\pi/2}$, *then there exists* $g \in L^1(\mathbb{R}_+)$ *such that* $f = f(\infty) + \mathcal{L}g$. *Hence* $\mathcal{H}_{\pi/2} \subset \mathcal{LM} \subset \mathcal{B}$, *and the inclusions are continuous.*
2. *If* $f \in \mathcal{H}_{\pi/2}$, *then* $f \in \mathcal{D}_s$ *for all* $s > -1$, *and the inclusions are continuous.*
3. *If* $f \in \mathcal{D}_s$ *for some* $s > -1$, *then (the restriction of)* f *belongs to* $\mathcal{H}_\psi$ *for every* $\psi \in (0, \pi/2)$, *and the restriction maps are continuous.*

The representation formula (4.7) for $\mathcal{D}_s$ in Proposition 4.4(2) can be transformed into the formula (4.10) below for functions in $\mathcal{H}_\psi$, by means of Proposition 4.8, Theorem 4.11 and Proposition 4.7.

Proposition 4.12 *Let* $f \in \mathcal{H}_\psi$, *where* $\psi \in (0, \pi)$. *Let* $\nu := 2\psi/\pi$ *and*

$$f_\nu(z) := f(z^\nu), \qquad z \in \mathbb{C}_+.$$

Then

$$f(z) = f(\infty) - \frac{1}{\pi}\int_0^\infty \int_{\mathbb{R}} \frac{f_\nu'(\alpha + i\beta)}{\alpha - i\beta + z^{1/\nu}}\, d\beta\, d\alpha, \qquad z \in \Sigma_\psi \cup \{0\}. \tag{4.10}$$

The formula (4.10) can be recast by replacing the double (area) integral with a line integral involving boundary values of f' instead of scalings of f', and offering a kernel of a different type which may sometimes be more useful. The reproducing formula (4.12) for functions in $\mathcal{H}_\psi$ was formulated and proved in [10, Proposition 4.19], but it resembles, and was inspired by [11, Lemma 7.4]. The arccot function is defined in (3.2).

Proposition 4.13 *Let* $f \in \mathcal{H}_\psi$, $\psi \in (0, \pi)$. *Let* $\gamma := \pi/(2\psi)$, *and*

$$f_\psi(t) := \frac{f(e^{i\psi}t) + f(e^{-i\psi}t)}{2}, \qquad t > 0. \tag{4.11}$$

Then

$$f(z) = f(\infty) - \frac{2}{\pi}\int_0^\infty f_\psi'(t)\,\operatorname{arccot}(z^\gamma/t^\gamma)\, dt, \qquad z \in \Sigma_\psi \cup \{0\}. \tag{4.12}$$

Density of Rational Functions Let $\mathcal{R}(\mathbb{C}_+)$ be the linear span of $\{r_\lambda : \lambda \in \mathbb{C}_+\}$ and the constant functions. The first statement in the following theorem plays an important role in the $\mathcal{D}$- and $\mathcal{H}$-calculi.

Theorem 4.14

1. *The space* $\mathcal{R}(\mathbb{C}_+)$ *is dense in* $\mathcal{D}_s$ *for* $s > -1$.
2. *The space* $\mathcal{B}$ *is not dense in* $\mathcal{D}_s^\infty$ *for* $s > 0$.

We outline the proof of the first statement in Theorem 4.14 in the case when $s \in (-1, 0)$. For $f \in \mathcal{D}_s$, let

$$R_f(\lambda) := -\frac{1}{\pi} f'(\lambda) r_{\overline{\lambda}}, \qquad \lambda \in \mathbb{C}_+.$$

Then, R_f is a continuous function from $\mathbb{C}_+$ to $\mathcal{D}_s$, and by Example 4.2(1) it is Bochner integrable with respect to area measure S on $\mathbb{C}_+$. Since the point evaluations are continuous on $\mathcal{D}_s$, it follows that

$$Q_0 f' = \int_{\mathbb{C}_+} R_f(\lambda)\, dS(\lambda).$$

This is the Bochner integral of a continuous function, so $Q_0 f'$ belongs to the closure in $\mathcal{D}_s$ of the span of the integrand which is contained in the closure of $\mathcal{R}_0(\mathbb{C}_+)$. Since $f = f(\infty) + Q_0 f'$, it follows that f is in the closure of $\mathcal{R}(\mathbb{C}_+)$ in $\mathcal{D}_s$.

A similar argument may be used to prove the first statement for $s \geq 0$.

The second statement in Theorem 4.14 follows from the fact that any function in the closure of $\mathcal{B}$ has a continuous extension to $i\mathbb{R}$. See [10, Theorem 5.1, Corollary 5.2] for details.

There is a similar result for the spaces $\mathcal{H}_\psi$, where $\psi \in (0, \pi)$. Let $\mathcal{R}_0(\Sigma_\psi)$ be the linear span of $\{r_\lambda : \lambda \in \Sigma_{\pi-\psi}\}$, and $\mathcal{R}_{\mathcal{H}}(\Sigma_\psi)$ be the closure of $\mathcal{R}_0(\Sigma_\psi)$ in $\mathcal{H}_\psi$. See [10, Theorem 5.10] for a proof of the following density result.

Theorem 4.15 *For $\psi \in (0, \pi)$,*

$$\mathcal{R}_{\mathcal{H}}(\Sigma_\psi) = \left\{ f \in \mathcal{H}_\psi : f(\infty) = 0 \right\}.$$

Convergence Lemmas To obtain Convergence Lemmas for functions in $\mathcal{D}_s$ and $\mathcal{H}_\psi$, we need to consider functions of the form $f(z^\gamma)$, where $\gamma \in (0, 1)$. See [10, Corollary 4.14 and Lemmas 6.1, 6.2] for proofs.

Lemma 4.16 *Let $s > -1$, $f \in \mathcal{D}_s$ and $\gamma \in (0, 1)$. Let $f_\gamma(z) = f(z^\gamma)$, $z \in \mathbb{C}_+$. Then $f_\gamma \in \mathcal{D}_\sigma^\infty \cap \mathcal{H}_{\pi/2}$ for all $\sigma > -1$. Moreover, the maps $f \mapsto f_\gamma$ are continuous from $\mathcal{D}_s$ into $\mathcal{D}_\sigma^\infty$ and from $\mathcal{D}_s$ into $\mathcal{H}_{\pi/2}$.*

Now we state simultaneously Convergence Lemmas for $\mathcal{D}_s$ and $\mathcal{H}_\psi$.

Lemma 4.17 (Convergence Lemmas) *Let $\mathfrak{A}$ denote either $\mathcal{D}_s$ for some $s > -1$ or $\mathcal{H}_\psi$ for some $\psi \in (0, \pi)$. Let $(f_k)_{k=1}^\infty \subset \mathfrak{A}$ be such that*

$$\sup_{k \geq 1} \|f_k\|_{\mathfrak{A}} < \infty,$$

and for every $z \in \mathbb{C}_+$ there exists

$$f(z) := \lim_{k \to \infty} f_k(z).$$

Let $g \in \mathfrak{A}$ satisfy

$$g(0) = g(\infty) = 0.$$

Let $\gamma \in (0, 1)$ and

$$f_{k,\gamma}(z) := f_k(z^\gamma), \quad g_\gamma(z) := g(z^\gamma), \qquad z \in \mathbb{C}_+.$$

Then

$$\lim_{k\to\infty} \|(f_{k,\gamma} - f_\gamma)g_\gamma\|_{\mathfrak{A}} = 0.$$

The $\mathcal{D}$- and $\mathcal{H}$-Calculi Let $A \in \mathrm{Sect}(\pi/2-)$. To define the $\mathcal{D}$-calculus, we adapt the formula in (4.7), replacing z by A. If $f \in \mathcal{D}_s$ for some $s > -1$, define

$$f_{\mathcal{D}}(A) := f(\infty) - \frac{2^s}{\pi}\int_0^\infty \alpha^s \int_{\mathbb{R}} f'(\alpha + i\beta)(\alpha - i\beta + A)^{-(s+1)}\, d\beta\, d\alpha,$$

where the fractional power $(\alpha + i\beta + A)^{-(s+1)}$ can be defined by the holomorphic functional calculus or other methods. The integral is absolutely convergent in operator norm. If $f \in \mathcal{B}$ and $s = 1$, this definition agrees with the formula in (3.15).

Theorem 4.18 *Let $A \in \mathrm{Sect}(\pi/2-)$.*

1. *If A is injective, then $f_{\mathcal{D}}(A)$ as defined above coincides with $f(A)$ as defined in the sectorial functional calculus.*
2. *If $f \in \mathcal{D}_\infty$, then the definition of $f_{\mathcal{D}}(A)$ does not depend on s.*
3. *If $f \in \mathcal{B} \cap \mathcal{D}_\infty$, then $f_{\mathcal{D}}(A) = \Phi_A(f)$.*
4. *Let $\lambda \in \mathbb{C}_+$. Then $(r_\lambda)_{\mathcal{D}}(A) = (\lambda + A)^{-1}$.*
5. *The map $\Psi_A : f \mapsto f_{\mathcal{D}}(A)$ is an algebra homomorphism from $\mathcal{D}_\infty$ to $L(X)$.*
6. *For each $s > -1$, the map $f \mapsto f_{\mathcal{D}}(A)$ is bounded from $\mathcal{D}_s$ to $L(X)$.*

Moreover, Ψ_A is the unique linear map from $\mathcal{D}_\infty$ to $L(X)$ satisfying (4) and (6).

Proofs of the statements in Theorem 4.18 may be found in [10, Theorems 7.4 and 7.6]. The map Ψ_A is called the *$\mathcal{D}$-calculus* for A. An estimate for the boundedness of Ψ_A on $\mathcal{D}_s$ is:

$$\|f(A)\| \le |f(\infty)| + \frac{2M_A^{\lceil s\rceil+1}}{\pi}\|f\|_{\mathcal{D}_{s,0}}. \tag{4.13}$$

Now suppose that $A \in \mathrm{Sect}(\theta)$, where $\theta \in [0, \pi)$, $\psi \in (\theta, \pi)$, $\gamma := \pi/(2\psi)$ and $f_{1/\gamma}(z) := f(z^{1/\gamma})$. Then $f_{1/\gamma} \in \mathcal{H}_{\pi/2} \subset \mathcal{D}_0$, by Proposition 4.8 and Theorem 4.11(2). Let

$$f_{\mathcal{H}}(A) := f(\infty) - \frac{1}{\pi} \int_0^{\infty} \int_{\mathbb{R}} f'_{1/\gamma}(\alpha + i\beta)(\alpha - i\beta + A^{\gamma})^{-1} \, d\beta \, d\alpha, \tag{4.14}$$

where A^{γ} is the fractional power of A. This definition does not depend on the particular choice of ψ and the consequent value of γ. Moreover it agrees with $f(A)$ as defined by the sectorial functional calculus (assuming that A is injective), and it agrees with $f_{\mathcal{D}}(A)$ (assuming that $\theta < \pi/2$ and $f \in \mathcal{D}_{\infty}$). See [10, Proposition 8.1].

Now we define the $\mathcal{H}$-calculus for A, as in [10, Theorem 8.2].

Theorem 4.19 *Let $A \in \mathrm{Sect}(\theta)$, where $\theta \in (0, \pi)$. For any $\psi \in (\theta, \pi)$ the formula* (4.14) *defines a bounded algebra homomorphism:*

$$\Upsilon_A : \mathcal{H}_{\psi} \mapsto L(X), \qquad \Upsilon_A(f) = f_{\mathcal{H}}(A).$$

The homomorphism Υ_A satisfies $\Upsilon_A(r_\lambda) = (\lambda + A)^{-1}$ for all $\lambda \in \Sigma_{\pi-\psi}$, and it is the unique homomorphism with these properties.

Remark 4.20 If A is any operator which admits a $\mathcal{D}$-calculus satisfying the properties (4) and (6) of Theorem 4.18, then it follows from Example 4.2(1) that $A \in \mathrm{Sect}(\pi/2-)$. Similarly, it follows from Example 4.9 that if A admits an $\mathcal{H}_{\psi}$-calculus as in Theorem 4.19, then $A \in \mathrm{Sect}(\theta)$ for some $\theta \in [0, \psi)$.

Theorem 4.21 provides an alternative to the formula (4.14), based on (4.12) and an estimate

$$\| \operatorname{arccot}(A^{\gamma}) \| \le \frac{\pi}{2} M_{\psi}(A) \| f \|_{\mathcal{H}_{\psi}}.$$

There is a proof of the theorem in [10, Lemma 8.4, Theorem 8.6], followed by a discussion of relations between the theorem and [11].

Theorem 4.21 *Let $A \in \mathrm{Sect}(\theta)$ and $\gamma = \pi/(2\psi)$, where $0 \le \theta < \psi < \pi$. If $f \in \mathcal{H}_{\psi}$, and f_{ψ} is given by* (4.11), *then*

$$f_{\mathcal{H}}(A) = f(\infty) - \frac{2}{\pi} \int_0^{\infty} f'_{\psi}(t) \operatorname{arccot}(A^{\gamma}/t^{\gamma}) \, dt,$$

where the integral converges in the operator norm topology, and

$$\| f_{\mathcal{H}}(A) \| \le |f(\infty)| + \frac{M_{\psi}(A)}{2} \| f' \|_{H^1(\Sigma_{\psi})} \le M_{\psi}(A) \| f \|_{\mathcal{H}_{\psi}}.$$

Remark 4.22 If $\psi < \pi/2$ and $M_\psi(A) = 1$ in Theorem 4.21, that is, if $-A$ generates a holomorphic C_0-semigroup which is contractive on $\Sigma_{(\pi/2)-\psi}$, then the $\mathcal{H}_\psi$-calculus is contractive. We are not aware of such estimates in constructions of other calculi in the literature.

Convergence Lemmas and Spectral Mapping Theorems There are Convergence Lemmas and Spectral Mapping Theorems for both the $\mathcal{D}$-calculus and the $\mathcal{H}$-calculus. In the statements below, $f(A)$ may be interpreted as being either $f_{\mathcal{D}}(A)$ or $f_{\mathcal{H}}(A)$, as appropriate. The Convergence Lemma in Theorem 4.23 follows easily from Lemma 4.17 applied with a suitable value of γ and $g(z) = z(1+z)^{-2}$, for example. See [10, Section 6.1] for details.

Theorem 4.23 *Let $A \in \mathrm{Sect}(\pi/2-)$, with dense range. Let f_k and f be as in Lemma 4.17. Then $f_k(A) \to f(A)$ in the strong operator topology as $k \to \infty$.*

The Spectral Mapping Theorem is proved in a similar way to Theorem 3.16.

Theorem 4.24 *Let $A \in \mathrm{Sect}(\theta)$ where either (a) $\theta < \pi/2$ and $f \in \mathcal{D}_\infty$, or (b) $\theta < \psi < \pi$ and $f \in \mathcal{H}_\psi$. Let $\lambda \in \mathbb{C}$.*

1. *If $x \in D(A)$ and $Ax = \lambda x$, then $f(A)x = f(\lambda)x$.*
2. *If $x^* \in D(A^*)$ and $A^*x^* = \lambda x^*$, then $f(A)^*x^* = f(\lambda)x^*$.*
3. *If $(x_n)_{n\ge 1}$ are unit vectors in $D(A)$ and $\lim_{n\to\infty} \|Ax_n - \lambda x_n\| = 0$, then $\lim_{n\to\infty} \|f(A)x_n - f(\lambda)x_n\| = 0$.*
4. *One has $\sigma(f(A)) \cup \{f(\infty)\} = f(\sigma(A)) \cup \{f(\infty)\}$.*

Applications The $\mathcal{D}$-calculus provides explicit norm-estimates for various known facts about an operator $A \in \mathrm{Sect}(\pi/2-)$, as the bounded holomorphic semigroup generated by $-A$ is $(e^{-tA})_{t\ge 0}$ where e^{-tA} is defined by any of the functional calculi. The following corollary exhibits some estimates. The power-boundedness of $V(A)$ was first proved in [13] and [28] using different methods. The general form of the second and third statements are now classical. The results follow from estimates in Examples 4.2 and (4.13), together with some compatibility arguments and Proposition 4.6(1). See [10, Corollary 10.1] for details.

Corollary 4.25 *Let $A \in \mathrm{Sect}(\pi/2-)$.*

1. *For all $n \in \mathbb{N}$,*

$$\|V(A)^n\| \le 1 + 32(1 + (\sqrt{2}\pi)^{-1})M_A^2.$$

2. *For all $\nu \ge 0$ and $t > 0$,*

$$\|A^\nu e^{-tA}\| \le 2^{\nu+2} t^{-\nu} \Gamma(\nu+1) M_A^{\lceil \nu \rceil + 2}.$$

3. *If A has dense range, then A^{-1} generates a bounded holomorphic C_0-semigroup $(e^{-tA^{-1}})_{t\geq 0}$ satisfying*

$$\|A^{-\nu}e^{-tA^{-1}}\| \leq 2^{\nu+2}t^{-\nu}\Gamma(\nu+1)M_A^{\lceil\nu\rceil+2}$$

for all $\nu \geq 0$ and $t > 0$.

The $\mathcal{H}$-calculus also provides norm-estimates, such as the following which can be deduced from Example 4.9(2) and (4.21).

Corollary 4.26 *Let $A \in \mathrm{Sect}(\theta)$, $\theta \in (0,\pi)$, and $\gamma \in (0,\pi/(2\theta))$. Then $(e^{-tA^\gamma})_{t\geq 0}$ is a bounded holomorphic C_0-semigroup of angle $(\pi/2) - \gamma\theta$. More precisely, if $\psi \in (\theta, \pi/(2\gamma))$ and $\lambda = |\lambda|e^{i\varphi} \in \Sigma_{(\pi/2)-\gamma\psi}$, then*

$$\|e^{-\lambda A^\gamma}\| \leq \frac{1}{2}\left(\frac{1}{\cos(\gamma\psi+\varphi)} + \frac{1}{\cos(\gamma\psi-\varphi)}\right) M_{\gamma\psi}(A). \tag{4.15}$$

Another application of the $\mathcal{D}$- and $\mathcal{H}$-calculi provides a new proof that the question raised in Application 2 of Sect. 3 has a positive answer. Let g be a Bernstein function, and $A \in \mathrm{Sect}(\theta)$ where $\theta \in [0,\pi/2)$. Applying the $\mathcal{D}$-calculus and Proposition 4.10(1) to functions of the form $(\lambda+g)^{-1}$ establishes that the operator $g_{\mathrm{Bern}}(A) \in \mathrm{Sect}(\pi/2-)$. Applying the $\mathcal{H}$-calculus and the estimate in (4.9) shows that $g_{\mathrm{Bern}}(A) \in \mathrm{Sect}(\theta)$. See [10, Theorem 10.5].

5 Further Extensions on Hilbert Spaces

In [29], Peller indicated two extensions of the Besov calculus for power-bounded operators on Hilbert spaces to larger classes of functions. Here we briefly describe corresponding extensions in the case of bounded C_0-semigroups on Hilbert spaces.

To put this in context, we need to explain the relationship between the space $\mathcal{B}$ of so-called analytic Besov functions and a conventional Besov space. Briefly, the reason is that the space $\mathcal{B}_0$ is isometrically isomorphic to a space $\mathcal{B}_{\mathrm{dyad}}$ which is a homogeneous Besov space of functions on $\mathbb{R}$ defined by means of Littlewood-Paley decompositions. This is described in the first subsection, and it is followed by outlines of results from [3] and [38], which provide sharper extimates for the norm of $f(A)$ than the $\mathcal{B}$-norm does. Although [38] predated [3] by three decades, it was accessible to other researchers only very recently, and we discuss [3] before [38] for presentational reasons. Finally we discuss the effect of these results on the Cayley transform question for Hilbert spaces.

$\mathcal{B}$ as a Besov Space It is a standard fact that $H^\infty(\mathbb{C}_+)$ is isometrically isomorphic to the space

$$H^\infty(\mathbb{R}) := \left\{ g \in L^\infty(\mathbb{R}) : \operatorname{supp} \mathcal{F}^{-1}g \subset \mathbb{R}_+ \right\},$$

with the sup-norm, where $\mathcal{F}^{-1}$ denotes the inverse (distributional) Fourier transform. The isomorphism is obtained by passage to the boundary function, and its inverse is the standard extension via Poisson integrals::

$$\begin{aligned} f \in H^\infty(\mathbb{C}_+) &\mapsto f^b \in H^\infty(\mathbb{R}), \\ g \in H^\infty(\mathbb{R}) &\mapsto \mathcal{P}g \in H^\infty(\mathbb{C}_+) \\ (\mathcal{P}g)(t+is) &:= \frac{1}{\pi}\int_{\mathbb{R}} \frac{tg(y)}{t^2+(y-s)^2}\,dy, \qquad t>0, s \in \mathbb{R}. \end{aligned}$$

The isomorphism $f \mapsto f^b$ maps $\mathcal{B}_0$ onto the space $\mathcal{B}_{\mathrm{dyad}} := H^\infty(\mathbb{R}) \cap B^0_{\infty,1}(\mathbb{R})$, where $B^0_{\infty,1}(\mathbb{R})$ is a conventional homogeneous Besov space, defined in terms of Littlewood-Paley decomposition with respect to a suitable partition of unity $(\varphi_k)_{k\in\mathbb{Z}}$, and equipped with the norm

$$\|g\|_{\mathcal{B}_{\mathrm{dyad}}} := \sum_{k\in\mathbb{Z}} \|g * \varphi_k\|_\infty < \infty.$$

The space $\mathcal{B}_{\mathrm{dyad}}$ can be described as follows:

$$\mathcal{B}_{\mathrm{dyad}} = \left\{ g \in \mathrm{BUC}(\mathbb{R}) : g = \sum_{k\in\mathbb{Z}} g * \varphi_k,\ \|g\|_{\mathcal{B}_{\mathrm{dyad}}} := \sum_{k\in\mathbb{Z}} \|g*\varphi_k\|_\infty < \infty \right\}.$$

The restriction map and its inverse are both bounded for the norms on the two spaces. Thus $\mathcal{B}_0$ is isomorphic to $\mathcal{B}_{\mathrm{dyad}}$.

This relationship between $\mathcal{B}$ and $\mathcal{B}_{\mathrm{dyad}}$ was known and used in [37], but the proof there is not very clear. A precise statement and proof are given in [8, Proposition 6.2], using the density of $\mathcal{G}$ in $\mathcal{B}_0$ as established in Theorem 3.9 by means of Arveson's spectral theory for C_0-groups.

The $\mathcal{A}$-Calculus In [3] Arnold and Le Merdy define an $\mathcal{A}$-calculus for negative generators of bounded C_0-semigroups on Hilbert spaces, by adapting ideas from [29]. They introduce the space

$$\mathcal{A}(\mathbb{R}) := \left\{ \sum_{k=1}^\infty g_k * h_k : g_k \in \mathrm{BUC}(\mathbb{R}),\ h_k \in H^1(\mathbb{R}),\ \sum_{k=1}^\infty \|g_k\|_\infty \|h_k\|_{H^1} < \infty \right\},$$

with the norm of $f \in \mathcal{A}(\mathbb{R})$ given by

$$\|f\|_{\mathcal{A}} := \inf \left\{ \sum_{k=1}^\infty \|g_k\|_\infty \|h_k\|_{H^1} : f = \sum_{k=1}^\infty g_k * h_k,\ g_k \in \mathrm{BUC}(\mathbb{R}),\ h_k \in H^1(\mathbb{R}) \right\},$$

where $H^1(\mathbb{R}) = \{f^b : f \in H^1(\mathbb{C}_+)\}$. Using a quotient of the projective tensor product of the Banach algebras BUC($\mathbb{R}$) (with pointwise multiplication) and $H^1(\mathbb{R})$ (under convolution), they show that $\mathcal{A}(\mathbb{R})$ is complete. Using an idea from [29] they prove that $\mathcal{A}(\mathbb{R})$ is an algebra under pointwise multiplication, so it is a Banach algebra which is continuously included in BUC($\mathbb{R}$) $\cap$ $H^\infty(\mathbb{R})$. A corresponding Banach algebra $\mathcal{A}(\mathbb{C}_+)$ contained in $H^\infty(\mathbb{C}_+)$ is obtained from $\mathcal{A}(\mathbb{R})$ by the Poisson extension process above. There is a similar construction of spaces $\mathcal{A}_0(\mathbb{R})$ and $\mathcal{A}_0(\mathbb{C}_+)$, with BUC($\mathbb{R}$) replaced by $C_0(\mathbb{R})$. The algebras $\mathcal{B}$ and $\mathcal{B}_0$ are densely and continuously included in $\mathcal{A}(\mathbb{C}_+)$ and $\mathcal{A}_0(\mathbb{C}_+)$ respectively.

Theorem 5.1 *Let $-A$ be the generator of a bounded C_0-semigroup on a Hilbert space X. There is a unique bounded homomorphism $\Xi_{\mathcal{A}} : \mathcal{A}(\mathbb{C}_+) \to L(X)$ which extends the Hille-Phillips calculus and the $\mathcal{B}$-calculus for A.*

The authors use the half-plane calculus to define $f(A)$ for $f \in \mathcal{A}(\mathbb{C}_+)$. They first obtain a fundamental estimate showing that if $f = \mathcal{L}(\mathcal{F}^{-1}g\mathcal{F}^{-1}h)$ with $g \in H^1(\mathbb{R})$ and $h, \mathcal{F}^{-1}h \in L^1(\mathbb{R})$, then $\|f(A)\| \le K_A^2\|g\|_{L^1}\|h\|_\infty$. This depends on approximation of $H^2(\mathbb{R})$-functions by Schwartz functions, factorisation of H^1-functions as products of two H^2-functions, and the vector-valued Plancherel theorem (the only place where the Hilbert space structure is used). By embedding $\mathcal{A}_0(\mathbb{R})^*$ into the space of Fourier multipliers on $H^1(\mathbb{R})$, the authors prove that $\mathcal{L}L^1 := \{\mathcal{L}h : h \in L^1(\mathbb{R}_+)\}$ is dense in $\mathcal{A}_0(\mathbb{C}_+)$. It follows from this and the fundamental estimate that there is a homomorphism from $\mathcal{A}_0(\mathbb{C}_+)$ into $L(X)$, which coincides with the Hille-Phillips calculus on $\mathcal{L}L^1$. Finally the homomorphism extends to $\mathcal{A}(\mathbb{C}_+)$ by an approximation procedure similar in spirit to the extension of the $\mathcal{B}$-calculus from $\mathcal{L}\mathcal{M}$ to $\mathcal{B}$ in the second step of the proof of Theorem 3.13.

White's Approach White considers negative generators of bounded C_0-semigroups on Hilbert spaces in Section 5 of his thesis [38]. He follows Peller's strategy in [29], but the arguments are more complicated in the continuous case, and they rely on a duality method and several quite subtle identifications. Using the injective and projective products of $L^1(\mathbb{R}_+)$ along with the duals of the products, he derives a continuous-time version of Grothendieck's inequality and obtains several estimates for $f(A)$ involving tensor product norms. Passing to Schur multipliers he obtains an estimate

$$\|f(A)\| \le K_G K_A^2 \sup\left\{\left|\int_0^\infty (\mathcal{F}^{-1}f^b)(t)h(t)\,d\mu(t)\right| : \|h\|_{M(\mathrm{BHK})} \le 1\right\}. \tag{5.1}$$

Here $f \in H^1(\mathbb{C}_+)$, $f^b \in L^1(\mathbb{R})$ and $\operatorname{supp}(\mathcal{F}^{-1}f^b)$ is a compact subset of $\mathbb{R}_+$, and $h \in L^\infty(\mathbb{R}_+)$ is a pointwise (Schur) multiplier on a space BHK of Hankel type kernels of bounded integral operators on $H^2(\mathbb{R})$, $\|\cdot\|_{M(\mathrm{BHK})}$ stands for the multiplier norm, and K_G denotes Grothendieck's constant. He constructs an isometric embedding q of BHK into $H^1(\mathbb{R})^*$, and then he shows that f^b belongs to a "projective convolution space", $C := H^1(\mathbb{R}) * q(\mathrm{BHK})$, which is defined similarly to the space $\mathcal{A}(\mathbb{R})$ above. Moreover, $\|f^b\|_C \le C\|f^b\|_{\mathcal{B}_{\mathrm{dyad}}}$, where C

is an absolute constant. From these identifications, the estimate (5.1) and the isomorphism between $\mathcal{B}_{\mathrm{dyad}}$ and $\mathcal{B}_0$, it follows that there is an absolute constant C_1 such that

$$\|f(A)\| \leq C_1 K_A^2 \|f\|_{\mathcal{B}_0}$$

if $f \in \mathcal{G}$ (as in (3.7)) and f^b is a Schwartz function on $\mathbb{R}$. This estimate does not appear explicitly in the thesis, but it can be seen from the arguments leading to [38, Theorem 5.5.12].

Cayley Transform Question The various estimates in [3] and [38, Section 5] involve new norms on $\mathcal{B}$ which are dominated by the $\mathcal{B}$-norm, and which provide upper bounds for rational functions of A. This is relevant to the Cayley transform question on Hilbert spaces in the following way. Here we will assume that the norm is the $\mathcal{A}$-norm, but the same comments apply to other norms of this type.

If $-A$ generates a bounded C_0-semigroup, then it follows from the boundedness of the Hille-Phillips calculus with respect to the $\mathcal{A}$-norm that

$$\|V(A)^n\| \leq C\|V^n\|_{\mathcal{A}}, \qquad n \in \mathbb{N},$$

where C is a constant. Thus it would be interesting to determine the asymptotic behaviour of the norms $\|V^n\|_{\mathcal{A}}$ for large n. As seen in Application 4 of Sect. 3, they cannot grow faster than logarithmically, and they are bounded below by $1/C$. If the norms are bounded above by a constant, then the Cayley transform question for Hilbert spaces has a positive answer. On the other hand, no new information is obtained if the norms grow logarithmically. There is also the possibility that the norms grow at a rate which is slower than logarithmic.

Unfortunately, the $\mathcal{A}$-norms of simple functions such as V^n are difficult to estimate, because of the indirect way that the $\mathcal{A}$-norm is defined. Moreover we do not know of a simple reproducing formula for the algebra $\mathcal{A}$ of the style that we have for $\mathcal{B}$, $\mathcal{D}_s$ and $\mathcal{H}_\psi$. Thus the asymptotic behaviour of $\|V^n\|_{\mathcal{A}}$ for large n remains unknown to us. This also applies to the other norms of this type, but not to the $\mathcal{B}$-norm for which the formula (3.1) led to a sharp estimate.

References

1. A.B. Aleksandrov, Operator Lipschitz functions and model spaces. J. Math. Sci. (N.Y.) **202**, 485–518 (2014)
2. W. Arendt, C.J.K. Batty, M. Hieber, F. Neubrander, *Vector-Valued Laplace Transforms and Cauchy Problems*, 2nd edn. Monogr. Math., vol. 96 (Birkhäuser, Basel, 2011)
3. L. Arnold, C. Le Merdy, Functional calculus for a bounded C_0-semigroup on Hilbert space. J. Funct. Anal. **284**, 49 (2023)
4. W. Arveson, The harmonic analysis of automorphism groups, in *Operator Algebras and Applications*, vol. I. Proc. Symp. Pure Math., vol. 38 (American Mathematical Society, Providence, 1982), pp. 199–269

5. W.G. Bade, An operational calculus for operators with spectrum in a strip. Pacific J. Math. **3**, 257–290 (1953)
6. C. Batty, M. Haase, J. Mubeen, The holomorphic functional calculus approach to operator semigroups. Acta Sci. Math. (Szeged) **79**, 289–323 (2013)
7. C. Batty, A. Gomilko, Yu. Tomilov, Resolvent representations for functions of sectorial operators. Adv. Math. **308**, 896–940 (2017)
8. C. Batty, A. Gomilko, Yu. Tomilov, A Besov algebra calculus for generators of operator semigroups and related norm-estimates. Math. Ann. **379**, 23–93 (2021)
9. C. Batty, A. Gomilko, Yu. Tomilov, The theory of Besov functional calculus: developments and applications to semigroups. J. Funct. Anal. **281**, 109089, 60 pp. (2021)
10. C. Batty, A. Gomilko, Yu. Tomilov, Functional calculi for sectorial operators and related function theory. J. Inst. Math. Jussieu, 1–81 (2021)
11. K.N. Boyadzhiev, Integral representations of functions on sectors, functional calculus and norm estimates. Collect. Math. **53**, 287–302 (2002)
12. P. Brenner, V. Thomée, On rational approximations of semigroups. SIAM J. Numer. Anal. **16**, 683–694 (1979)
13. N. Crouzeix, S. Larsson, S. Piskarev, V. Thomée, The stability of rational approximations of analytic semigroups. BIT **33**, 74–84 (1993)
14. E.B. Davies, *One-Parameter Semigroups* (Academic, London, 1980)
15. R. deLaubenfels, Inverses of generators. Proc. Am. Math. Soc. **104**, 443–448 (1988)
16. R. deLaubenfels, Inverses of generators of nonanalytic semigroups. Studia Math. **191**, 11–38 (2009)
17. A. Gomilko, On conditions for the generating operator of a uniformly bounded C_0-semigroup of operators. Funct. Anal. Appl. **33**, 294–296 (1999)
18. A. Gomilko, The Cayley transform of the generator of a uniformly bounded C_0-semigroup of operators. Ukrain. Mat. Zh. **56**, 1018–1029 (2004). Transl. in Ukrainian Math. J. **56**, 1212–1226 (2004)
19. A. Gomilko, Inverses of semigroup generators: a survey and remarks, in *Études opératorielles*. Banach Center Publ., vol. 112 (Polish Academy of Sciences, Institute of Mathematics, Warszawa, 2017), pp. 107–142
20. A. Gomilko, Yu. Tomilov, On subordination of holomorphic semigroups. Adv. Math. **283**, 155–194 (2015)
21. M. Haase, *The Functional Calculus for Sectorial Operators*. Operator Theory: Advances and Applications, vol. 169 (Birkhäuser, Basel, 2006)
22. M. Haase, Transference principles for semigroups and a theorem of Peller. J. Funct. Anal. **261**, 2959–2998 (2011)
23. V. Havin, B. Jöricke, *The Uncertainty Principle in Harmonic Analysis*. Ergebnisse der Math. und ihrer Grenzgeb. vol. 28 (Springer, Berlin, 1994)
24. E. Hille, R.S. Phillips, *Functional Analysis and Semi-groups*. Colloquium Publications, vol. 31 (American Mathematical Society, Providence, 1957)
25. Y. Katznelson, *An Introduction to Harmonic Analysis* (Dover, New York, 1976)
26. A. Kishimoto, D.W. Robinson, Subordinate semigroups and order properties. J. Aust. Math. Soc. A **31**, 59–76 (1981)
27. R. Nagel (ed.), *One-Parameter Semigroups of Positive Operators*. Lecture Notes in Mathematics, vol. 1184 (Springer, Berlin, 1986)
28. C. Palencia, A stability result for sectorial operators in Banach spaces. SIAM J. Numer. Anal. **30**, 1373–1384 (1993)
29. V. Peller, Estimates of functions of power bounded operators on Hilbert spaces. J. Operator Theory **7**, 341–372 (1982)
30. R.S. Phillips, On the generation of semigroups of linear operators. Pacific J. Math. **2**, 343–369 (1952)
31. R.L. Schilling, R. Song, Z. Vondraček, *Bernstein Functions*, 2nd edn. De Gruyter Studies in Mathematics, vol. 37 (de Gruyter, Berlin, 2012)

32. F. Schwenninger, On measuring unboundedness of the H^∞-calculus for generators of analytic semigroups. J. Funct. Anal. **271**, 49–84 (2016)
33. F. Schwenninger, H. Zwart, Bounded ideals of the H^∞-calculus for semigroup generators, private communication
34. D.-H. Shi, D.-X. Feng, Characteristic conditions of the generation of C_0-semigroups in a Hilbert space. J. Math. Anal. Appl. **247**, 356–376 (2000)
35. J.D. Stafney, A spectral mapping theorem for generating operators in second-order linear differential equations. J. Math. Anal. Appl. **77**, 433–450 (1980)
36. M.H. Taibleson, On the theory of Lipschitz spaces of distributions on Euclidean n-space: I. Principal properties. J. Math. Mech. **13**, 407–479 (1964)
37. P. Vitse, A Besov class functional calculus for bounded holomorphic semigroups. J. Funct. Anal. **228**, 245–269 (2005)
38. S.J. White, Norm-estimates for functions of semigroups of operators. PhD thesis, Univ. of Edinburgh (1989). https://www.era.lib.ed.ac.uk/handle/1842/11552
39. H. Zwart, Growth estimates for $\exp(A^{-1}t)$ on a Hilbert space. Semigroup Forum **74**, 487–494 (2007)

A Noncommutative Bishop Peak Interpolation-Set Theorem

David P. Blecher

Abstract We prove a noncommutative version of Bishop's peak interpolation-set theorem.

1 Introduction

For us, an operator algebra is a norm closed algebra of operators on a Hilbert space, or equivalently a closed subalgebra A of a C^*-algebra B. In this paper we shall assume for simplicity that A and B share a common identity element. In 'noncommutative peak interpolation' as surveyed briefly in [3] say, one generalizes the classical peak interpolation theory to the setting of operator algebras, using Akemann's noncommutative topology (see [1] and e.g. other references in [3]).

In classical peak interpolation[1] the setting is a subalgebra A of $B = C(K)$, the continuous scalar functions on a compact Hausdorff space K, and one tries to build functions in A which have prescribed values or behaviour on a fixed closed subset $E \subset K$ (or on several disjoint subsets). The sets that 'work' for this are the p-sets, namely the intersections of *peak sets*, where the latter term means a set of form $f^{-1}(\{1\})$ for $f \in A$, $\|f\| = 1$ (in the separable case, they are just the p-sets).

A typical 'peak interpolation result' says that if $f \in C(K)$ is strictly positive, and E is a p-set, then the continuous functions g on E which are restrictions of functions in A, and which are dominated in modulus by the 'control function' f on E, have extensions h in A satisfying $|h| \leq f$ on all of K (see e.g. II.12.5 in [16]; there are nice pictures illustrating this result in [3]). We shall call this the 'Gamelin-

[1] When we use the term 'peak interpolation' we mean in this sense. Others sometimes use this term to refer to what we call peak interpolation-sets below.

D. P. Blecher (✉)
Department of Mathematics, University of Houston, Houston, TX, USA
e-mail: dpbleche@central.uh.edu

Y. Choi et al. (eds.), *Operators, Semigroups, Algebras and Function Theory*, Operator Theory: Advances and Applications 292,
https://doi.org/10.1007/978-3-031-38020-4_3

Bishop theorem' below. The special case of this where $g = 1$ in fact characterizes p-sets among the closed subsets of K (e.g. see [15]). Combining this with a result of Glicksberg (see [16, Theorem II.12.7] and [17]), one obtains that the p-sets are the closed sets E in K with $\mu_E \in A^\perp$ for all measures $\mu \in A^\perp$. Equivalently, if and only if the characteristic function χ_E is in $A^{\perp\perp}$.

An *interpolation-set* for A is a closed set $E \subset K$ such that $A_{|E} = C(E)$. A *peak interpolation-set* (resp. *p-interpolation set*) is a peak (resp. p-) set which is also an interpolation-set. In the light of the above it is clear and obvious that the p-interpolation sets may be characterized by the appropriate variant of the Gamelin-Bishop theorem above: if $f \in C(K)$ is strictly positive, then the continuous functions g on E which are dominated in modulus by the 'control function' f on E, have extensions h in A satisfying $|h| \leq f$ on all of K. There are other characterizations of p-interpolation sets, e.g. as the closed sets in K with $\mu_E = 0$ for all measures $\mu \in A^\perp$. By basic measure theory, the latter is equivalent to: $|\mu|(E) = 0$ for all such μ. By the *Bishop peak interpolation-set theorem* we shall mean the result that Bishop proved in [2]: namely that if $\mu_E = 0$ for all measures $\mu \in A^\perp$ then the extension theorem stated a few lines back involving f, g, h, holds. See Section 20 in [22] for more information on peak interpolation-sets.

A special case of interest is when f above is 1, and when this case is applied to the disk algebra, together with the F. & M. Riesz theorem, one obtains the famous Rudin-Carleson theorem (see e.g. II.12.6 in [16]).

The noncommutative analogue of p-sets are the p-projections. Analogously to the classical case above they have been characterized as the infima ('meet') of peak projections, or as the closed projections in B^{**} that lie in $A^{\perp\perp}$, if A is a unital subalgebra of a C^*-algebra B (see the start of Section 2 for references). See [3] for a survey of part of the 'noncommutative peak set' theory, and see also references therein for dozens of our other results. See also e.g. [8] and Sections 8 and 9 in the more recent work of the author and Neal on noncommutative topology and Jordan operator algebras. More recently Davidson and Clouâtre and Hartz and others, have studied forms of noncommutative peak interpolation-sets in specific classes of operator algebras (see e.g. [10–14] and references therein). Their aim is often to generalize to such classes the classical theory of interpolation sets for the ball algebra (see [21, Chapter 10]), the Rudin-Carleson theorem, etc. This work has strong connections to our general 'noncommutative peak interpolation' theory (from [3] and references therein, or in the present paper); indeed some of the main theorems in [14] follow quickly from our more general theory (such as Theorem 1.1 below), as we mention at the end of this Introduction. Exploring these connections raises many interesting questions and should lead to further important progress.

The following is the unital case of a very general noncommutative variant of the Gamelin-Bishop theorem [3, Theorem 3.4]:

Theorem 1.1 ([3]) *Suppose that A is a subalgebra of a unital C^*-algebra B, with $1_B \in A$. Suppose that q is a closed projection in B^{**} that lies in $A^{\perp\perp}$. If $b \in A$ with $bq = qb$, and $qb^*bq \leq qd$ for an invertible positive $d \in B$ which commutes with q, then there exists an element $a \in A$ with $aq = qa = bq$, and $a^*a \leq d$.*

This result may fail without the 'commuting hypothesis' $bq = qb$, and even in the case $d = 1$. See [7, Corollary 2.4] for an example of a peak projection q and

$b \in A$ with $\|bq\| \leq 1$, but there is no $a \in \text{Ball}(A)$ with $aq = bq$. Thus as stated after that result,

> Clearly the way to proceed from this point, in noncommutative peak interpolation, is to insist on a commutativity assumption of [this] type.

Let us apply this principle in an attempt to find a noncommutative version of the classic Bishop peak interpolation-set theorem above. We first note that [19, Propositions 3.2 and 3.4] may be regarded as such, however there is an ϵ parameter there that one does not see in Bishop's result. Our next observation is that simple noncommutative versions of the Bishop peak interpolation-set theorem follow immediately from Theorem 1.1 and [19, Proposition 3.4]. Indeed we alluded to this right at the end of [3]:

> Finally, we remark that simple Tietze theorems of the flavour of the Rudin-Carleson theorem mentioned on the first page, follow from our interpolation theorems by adding a hypothesis of the kind in Proposition 3.4 of [19].

For example, suppose that q is in the center of B^{**}. If $A^{\perp} \subset (qB)^{\perp}$, and $b \in B$ then [19, Proposition 3.4] provides $a \in A$ with $aq = bq = qa$ and $qb^*bq = qa^*aq$. Thus we may apply Theorem 1.1 to obtain $g \in A$ with $gq = qg = bq = aq$, and $g^*g \leq d$ as desired. However the assumption that q is central in B^{**} is very strong. If q is not central but $bq = qb$ then one may modify the argument above, but we were unable to see at the time of [3] how to get the desired conclusion without adding a strong or unappealing hypothesis. For example if A is commutative (which is the case in many results discussed in e.g. [10, 11, 14], and of course it does not imply that B is also commutative) then this modified argument works (see the proof of Corollary 2.4 (5) below).

In any case, the desired noncommutative Bishop peak interpolation-set theorem suggested by the above would be a characterization of the closed projections q in $A^{\perp\perp}$ with the following property: If $b \in B$ with $bq = qb$, and $qb^*bq \leq qd$ for an invertible positive $d \in B$ which commutes with q, then there exists an element $a \in A$ with $aq = qa = bq$, and $a^*a \leq d$. In the present paper we supply such a theorem.

Turning to notation, the reader is referred for example to [3, 4, 6, 9] for more details on some of the topics below if needed. We will use silently the fact from basic analysis that $X^{\perp\perp}$ is the weak* closure in Y^{**} of a subspace $X \subset Y$, and is isometric to X^{**}. For us a *projection* is always an orthogonal projection. If A is a unital operator algebra then its second dual A^{**} is a unital operator algebra with its (unique) Arens product, this is also the product inherited from the von Neumann algebra B^{**} if A is a subalgebra of a C^*-algebra B. Via semicontinuity, it is natural to declare a projection $q \in B^{**}$ to be *open* if it is an increasing (weak*) limit of positive elements in B, and closed if its 'perp' $1 - q$ is *open*. We write χ_E for the characteristic function of E. In the case that $B = C(K)$, and E is an open or closed set in K, the projection $q = \chi_E$ may be viewed as an element of $C(K)^{**}$ in a natural way since $C(K)^*$ is a certain space of measures on K. Thus if $B = C(K)$ the open or closed projections are precisely the characteristic functions of open or closed sets.

We thank Ken Davidson and Michael Hartz for several stimulating conversations, some of which are explicitly referenced in the body of this paper. This project was initiated during the IWOTA conference at Lancaster in 2021 when Davidson asked us if we had a noncommutative Bishop peak interpolation-set theorem. We replied by mentioning the quote from the end of [3] stated above, and the observation preceding it, and this led to a correspondence in 2021 involving a variant of the present manuscript. As we said above, the connections between our 'noncommutative peaking theory' and the work of Clouâtre, Davidson, Hartz, Timko and others, raise many interesting questions to explore further. For example using our noncommutative peak theory (and results of Hay, etc.) and some additional work to prove some operator algebraic peak interpolation results in [14]. Indeed in the course of our discussions Michael Hartz developed some ideas for the latter (and Michael and Ken had many other helpful perspectives and ideas). Shortly thereafter Raphael Clouâtre was able to find a very short argument showing how Theorems 1.4 and 8.1 in [14] follow from our earlier peak interpolation theorem (surveyed above or in [3]). We thank him for his careful explanation.

2 On the 'Bishop Peak Interpolation-Set' Result

In this paper A is a subalgebra of a unital C^*-algebra B with $1_B \in A$. Noncommutative peak sets for A were introduced in the thesis of Damon Hay [19]. A projection $p \in A^{**}$ is called open with respect to A or A-open if there exists a net $x_t \in A \cap pA^{**}p$ with $x_t \to p$ weak* (see [9, Section 2]). It is A-closed if $1 - p$ is A-open. These coincide with the projections in B^{**} that are open or closed in the C^*-algebra sense, that also lie in $A^{\perp\perp}$ [9, Theorem 2.4]. They also coincide with the p-projections for A (see e.g. [6, Theorem 1.2] and [5, Theorem 3.4]; these results have also been generalized to Jordan operator algebras by the author and Neal).

Lemma 2.1 *Suppose that A is a unital operator algebra, a subalgebra of a unital C^*-algebra B. Let q be a closed projection in A^{**}. If A_0 is the subalgebra $\{a \in A : aq = qa\}$ then q is a closed projection in A_0^{**} (and in B_0^{**}). Moreover, qB_0 and qA_0 are norm closed.*

Proof Indeed suppose that $x_t \in A \cap (1-q)A^{**}(1-q)$ with $x_t \to 1-q$ weak*. Then $qx_t = x_t q = 0$ so that $1 - x_t \in A_0$ and $1 - x_t \to q$ weak*. So q and $1-q$ are in $A_0^{\perp\perp}$ and $1-q$ is an open projection for A_0 by the latter mentioned definition.

It follows from [19, Proposition 3.1] with $X = A_0$ that qA_0 is closed. Indeed if $\varphi \in A_0^{\perp}$ then since $qA_0 \subset A_0^{\perp\perp}$ it follows that $\varphi \in (qA_0)_{\perp}$. So qA_0 is closed by [19, Proposition 3.1]. Hence, or similarly, $q \in B_0^{\perp\perp}$ and is a closed projection there, and qB_0 is closed. □

Remark 2.2 Note that $B_0^{\perp\perp} \subset B^{**} \cap \{q\}'$ clearly, however one can show that these sets differ in general. Similarly, $qB_0^{\perp\perp} \neq q(B^{**} \cap \{q\}') = qB^{**}q$.

The following contains the desired characterization of 'peak interpolation-sets', as discussed in the second paragraph after Theorem 1.1. Indeed (iii) is precisely the class of projections discussed there, namely the projections corresponding to the (noncommutative version of the) extension property in the Bishop peak interpolation-set theorem. Item (vi) is a weaker version of this peak interpolation-set extension property, while (v) is saying that q is a noncommutative 'interpolation-set'. Items (i), (ii), and (iv) are noncommutative reformulations of the classical condition $\mu_E = 0$ (or equivalently $\chi_E \, d\mu = 0$) for all measures $\mu \in A^\perp$ in Bishop's original theorem. We remark that the 'noncommutative null sets' in (iv) were first considered by Clouâtre and Timko in [12].

Theorem 2.3 (Noncommutative 'Bishop Peak Interpolation-Set' Result) *Suppose that A is a unital operator algebra, a subalgebra of a unital C^*-algebra B. Suppose that q is a closed projection in B^{**}, and that $B_0 = \{b \in B : bq = qb\}$, and $A_0 = A \cap B_0$. The following are equivalent:*

(i) $qB_0 \subset A_0^{\perp\perp}$.
(ii) $A_0^\perp \subset (qB_0)_\perp$ (or equivalently $A_0^\perp \subset (qB^{**})_\perp$).
(iii) If $b \in B_0$ and $qb^*bq \leq qd$ for an invertible positive $d \in B$ which commutes with q, then there exists an element $a \in A$ with $aq = qa = bq$, and $a^*a \leq d$.
(iv) q is A_0-null (that is, if $\varphi \in A_0^\perp$ then $|\varphi|(q) = 0$).
If the above all holds, then q is a p-projection for A (or equivalently $q \in A^{\perp\perp}$, or equivalently q is A-closed), qA_0 is a C^*-algebra, and we also have
(v) $qB_0 = qA_0$.
(vi) If $b \in B_0$ and $\|bq\| \leq 1$ then there exists an element $a \in$ Ball(A) with $aq = qa = bq$.

Item (vi) implies (v). If q is a p-projection for A then items (i)–(vi) are all equivalent.

Proof Note that $qB_0 \subset A_0^{\perp\perp}$ if and only if $\overline{qB_0}^{w*} = qB_0^{\perp\perp} \subset A_0^{\perp\perp}$; and if and only if $A_0^\perp \subset (qB_0^{\perp\perp})_\perp$. The equivalence of (i) and (ii) follow from the bipolar theorem, or by taking upper and lower $\perp$'s. Note that since 1 is in B_0 (or $B_0^{\perp\perp}$), (i) and (ii) force $q \in A^{\perp\perp}$. At the start of this section we discussed the equivalences between B-closed projections in $A^{\perp\perp}$, A-closed projections, and p-projections for A.

By [19, Proposition 3.4], (ii) implies (v). Conversely, suppose that $q \in A^{\perp\perp}$. Then q is A-closed as we just said, and $q \in A_0^{\perp\perp}$ by Lemma 2.1. Then (v) implies that

$$qB_0 = qA_0 \subset A_0^{\perp\perp} A_0 \subset A_0^{\perp\perp}.$$

Thus (i) holds. Suppose in addition that $b \in B_0$ and $qb^*bq \leq qd$ for an invertible positive $d \in B$ which commutes with q. Then there exists an element $a_0 \in A_0$ with $a_0 q = bq$, so that $qa_0^* a_0 q \leq qd$. By Theorem 1.1 there exists an element $a \in A_0$ with $aq = a_0 q = bq$, and $a^*a \leq d$. We have shown that (v) implies (iii) if $q \in A^{\perp\perp}$, and hence also that (ii) implies (iii) (since (ii) implies (v) and $q \in A^{\perp\perp}$).

That (iii) implies that $q \in A^{\perp\perp}$ follows essentially from the principle that the Gamelin-Bishop theorem characterizes p-sets, or that in the noncommutative case the condition in the last sentence of Theorem 1.1 with $b = 1$ characterizes p-projections (see e.g. [19, Theorem 5.10] for an even better result). So by the above, (iii) implies (i). Clearly (iii) with $d = 1$ implies (vi). Conversely, (vi) implies (v) by scaling. Note that qB_0 is a C^*-algebra since it is the range of the map of multiplication by q, which is a $*$-homomorphism from B_0 into its bidual.

The equivalences involving (iv) may be deduced from Clouâtre and Timko's proof of [12, Theorem 6.2], and follow from properties of the polar decomposition of a linear functional as may be found in basic C^*-algebra texts (e.g. 3.6.7 in [20] applied to the bidual). We include the proofs for completeness. Suppose that q is A_0-null and $\varphi \in A_0^{\perp}$ with polar decomposition $\varphi = u|\varphi|$ with $u \in B_0^{**}$. By Cauchy-Schwarz we have

$$|\varphi(qb)|^2 = |\varphi|(qbu)^2 \leq C|\varphi|(q) = 0, \qquad b \in B_0.$$

So (ii) holds. Conversely if (ii) holds then, as we said above, we have (i) and (v) and also $q \in A_0^{\perp\perp}$. Suppose that $\varphi \in A_0^{\perp}$ with polar decomposition $\varphi = u|\varphi|$ as above. We have $qu^* \in qB_0^{**} \subset A_0^{\perp\perp}$. Thus $|\varphi|(q) = \varphi(qu^*) = 0$. So (iv) holds. □

Note that A_0q is actually a C^*-algebra in the setting above, just as in the commutative case where $q = \chi_E$ for a closed set $E \subset K$, and $A_0q = C(E)$. Unfortunately Aq is not generally a C^*-algebra, which seems to be further evidence for the consideration of A_0 in place of A in certain such results.

Corollary 2.4 (Alternative Noncommutative 'Bishop Peak Interpolation-Set' Result in Special Cases) *Suppose that A is a unital operator algebra, a subalgebra of a unital C^*-algebra B. Suppose that q is a closed projection in B^{**}. As in the previous result (see also* [12, Theorem 6.2]*), $qB \subset A^{\perp\perp}$ if and only if $A^{\perp} \subset (qB)_{\perp} = (qB^{**})_{\perp}$, and if and only if q is A-null (that is, if $\varphi \in A^{\perp}$ then $|\varphi|(q) = 0$). Moreover these conditions are equivalent to: $qB = qA$ with $q \in A^{\perp\perp}$. If these conditions all hold then so does the 'Bishop peak interpolation-set' result in* (iii) *of the previous theorem, under any one of the following extra hypotheses:*

(1) $q \in B$.
(2) $A + B_0$ is norm closed (or equivalently: every functional in $A_0^{\perp}$ extends to a functional in $A^{\perp}$).
(3) q is central in B^{**}.
(4) q is a minimal projection in B^{**}.
(5) A is commutative.

Proof The first two 'if and only ifs' follow almost exactly as in the previous proof. Note that $qB \subset A^{\perp\perp}$ if and only if $\overline{qB}^{w*} = qB^{**} \subset A^{\perp\perp}$, and if and only if $A^{\perp} \subset (qB^{**})_{\perp}$. The equivalence with $q \in A^{\perp\perp}$ and $qB = qA$ is noted in [12, Theorem 6.2]. Indeed one direction of this is obvious and as in the last proof: these conditions imply $qB = qA \subset A^{\perp\perp}A \subset A^{\perp\perp}$. The other direction is immediate from [19,

Proposition 3.4]. (There was a mistake in the statement and proof of this equivalence in the arXiv version of [12], which we pointed out to Clouâtre and has been corrected in the published version.) Finally we show that each of conditions (1)–(5) imply that the conditions in the last theorem are satisfied.

(2) If $A + B_0$ is norm closed, then by a principle in functional analysis, $A^{\perp\perp} \cap B_0^{\perp\perp} = (A \cap B_0)^{\perp\perp} = A_0^{\perp\perp}$ (this follows easily for example from [18, Lemma I.1.14(a)]). Hence $qB_0 \subset A^{\perp\perp}$ or $qB \subset A^{\perp\perp}$ implies

$$qB_0 \subset A^{\perp\perp} \cap B_0^{\perp\perp} = A_0^{\perp\perp}.$$

So the conditions in the last theorem are satisfied. The other equivalence here is no doubt also a principle in functional analysis known in some quarters, which we now explain. Indeed if the restriction map from B^* to B_0^* maps $A^\perp$ onto $A_0^\perp$, then it is the dual of a bicontinuous injection $(A_0^\perp)_* = B_0/A_0$ into $(A^\perp)_* = B/A$. This injection is the canonical map $b_0 + A_0 \mapsto b_0 + A$, since the dual of the latter is the restriction map. This argument is reversible: if the canonical map $b_0 + A_0 \mapsto b_0 + A$ is bicontinuous then the restriction map from B^* to B_0^* maps $A^\perp$ onto $A_0^\perp$. By the closed range theorem applied to the canonical map $A + B_0 \to (A + B_0)/A \subset B/A$, if $A + B_0$ is closed then $(B_0 + A)/A$ is isomorphic to B_0/A_0 (see [18, Lemma I.1.14] if needed). Conversely, if the canonical map $b_0 + A_0 \mapsto b_0 + A$ is bicontinuous then $(B_0 + A)/A$ is closed in B/A, so that $A + B_0$ is closed.

(1) If $q \in B$ and $qB \subset A^{\perp\perp}$ then $q \in B \cap A^{\perp\perp} = A$. Then

$$qB_0 \subset qA^{\perp\perp}q = (qAq)^{\perp\perp} = \{a \in A : a = qa = aq\}^{\perp\perp} \subset A_0^{\perp\perp}.$$

So again the conditions in the theorem are satisfied. Alternatively, we can apply [19, Proposition 3.4] and [3, Theorem 3.4] to B_0 and qAq.

(3) This is obvious e.g. from Theorem 2.3.

(4) We have $q \in A^{\perp\perp}$ and so $qB_0 = \mathbb{C}q \subset A_0^{\perp\perp}$ by the lemma.

(5) Suppose that A is commutative and $A^\perp \subset (qB)_\perp$. Then $A = A_0$ and $A_0^\perp \subset (qB)_\perp \subset (qB_0)_\perp$. This does it. Alternatively, by the argument towards the end of the paragraph after Theorem 2.3 there exists $a \in A$ with $aq = qa = bq$ and we may conclude as in that argument. □

Remarks 2.5

1. In Bishop's result A need not be an algebra. One may therefore hope to extend some of our results above to the unital operator space setting. Note that [19, Proposition 3.4], an ingredient above, does not need A to be an algebra.
2. One approach to replacing A_0 by A, is to try to find conditions under which $\varphi \in A_0^\perp$ implies that there is an extension $\tilde{\varphi} \in A^\perp$. If this holds and if $A^\perp \subset (qB)_\perp$, then it is easy to check condition (ii) in Theorem 2.3 directly. Indeed suppose that $\varphi \in A_0^\perp$. If it has an extension $\tilde{\varphi} \in A^\perp$ then by hypothesis $\tilde{\varphi} \in (qB)_\perp$, so that $\varphi \in (qB_0)_\perp$ (using also the lemma above).

We thank Ken Davidson for suggesting to us in the discussions mentioned earlier that perhaps there always exists an extension of functionals in $A_0^\perp$ to $A^\perp$, if and only if $A + B_0$ is closed.

3. We do not know if $qB \subset A^{\perp\perp}$ implies the conditions in Theorem 2.3 in all cases, nor do we have a counterexample at this time. By symmetry if q is A-null then we have both $qB \subset A^{\perp\perp}$ and $Bq \subset A^{\perp\perp}$. A main difficulty that arises is that by the argument towards the end of the paragraph after Theorem 2.3 one may obtain 'left interpolating' elements a_1 and 'right interpolating' elements a_2. We have $a_1 q = q a_2$, but one really needs $a_1 = a_2$ and we have not been able to spot the trick to ensure this (except under strong hypotheses).
4. It seems possible that the ideas in [7, Corollary 5.4] (taking b there in B) might give another noncommutative variation of Bishop's peak interpolation-set theorem.
5. In many cases discussed in [10, 11, 14] the algebra A is commutative so that (5) in the last theorem applies.

Acknowledgments The author is supported by a Simons Foundation Collaboration Grant.

References

1. C.A. Akemann, Left ideal structure of C^*-algebras. J. Funct. Anal. **6**, 305–317 (1970)
2. E. Bishop, A general Rudin-Carleson theorem. Proc. Am. Math. Soc. **13**, 140–143 (1962)
3. D.P. Blecher, Noncommutative peak interpolation revisited. Bull. Lond. Math. Soc. **45**, 1100–1106 (2013)
4. D.P. Blecher, C. Le Merdy, *Operator Algebras and Their Modules—An Operator Space Approach* (Oxford University Press, Oxford, 2004)
5. D.P. Blecher, M. Neal, Open projections in operator algebras II: Compact projections. Studia Math. **209**, 203–224 (2012)
6. D.P. Blecher, C.J. Read, Operator algebras with contractive approximate identities. J. Funct. Anal. **261**, 188–217 (2011)
7. D.P. Blecher, C.J. Read, Operator algebras with contractive approximate identities II. J. Funct. Anal. **264**, 1049–1067 (2013)
8. D.P. Blecher, C.J. Read, Order theory and interpolation in operator algebras. Studia Math. **225**, 61–95 (2014)
9. D.P. Blecher, D.M. Hay, M. Neal, Hereditary subalgebras of operator algebras. J. Operator Theory **59**, 333–357 (2008)
10. R. Clouâtre, K.R. Davidson, Duality, convexity and peak interpolation in the Drury-Arveson space. Adv. Math. **295**, 90–149 (2016)
11. R. Clouâtre, K.R. Davidson, Ideals in a multiplier algebra on the ball. Trans. Am. Math. Soc. **370**, 1509–1527 (2018)
12. R. Clouâtre, E.J. Timko, Non-commutative measure theory: Henkin and analytic functionals on C^*-algebras. Math. Ann. **386**, 415–453 (2023)
13. R. Clouâtre, I. Thompson, Minimal boundaries for operator algebras. Trans. Am. Math. Soc. **10**, 807–832 (2023). https://doi.org/10.1090/btran/154
14. K. Davidson, M. Hartz, Interpolation and duality in algebras of multipliers on the ball. J. Eur. Math. Soc. (2022; published online). arXiv:2003.14341. https://doi.org/10.4171/JEMS/1245
15. T.W. Gamelin, Restrictions of subspaces of C(X). Trans. Am. Math. Soc. **112**, 278–286 (1964)

16. T.W. Gamelin, *Uniform Algebras* (Prentice-Hall, Englewood Cliffs, 1969)
17. I. Glicksberg, Measures orthogonal to algebras and sets of antisymmetry. Trans. Am. Math. Soc. **105**, 415–435 (1962)
18. P. Harmand, D. Werner, W. Werner, *M-ideals in Banach Spaces and Banach Algebras*. Lecture Notes in Mathematics, vol. 1547 (Springer, Berlin, 1993)
19. D.M. Hay, Closed projections and peak interpolation for operator algebras. Integr. Equ. Oper. Theory **57**, 491–512 (2007)
20. G.K. Pedersen, *C*-algebras and Their Automorphism Groups* (Academic, London, 1979)
21. W. Rudin, *Function Theory in the Unit Ball of* $\mathbb{C}^n$. Reprint of the 1980 edition. Classics in Mathematics (Springer, Berlin, 2008)
22. E.L. Stout, *The Theory of Uniform Algebras* (Bogden & Quigley, Inc., Publishers, Tarrytown-on-Hudson, 1971)

Non-autonomous Desch–Schappacher Perturbations

Christian Budde and Christian Seifert

Abstract We consider time-dependent Desch–Schappacher perturbations of non-autonomous abstract Cauchy problems and apply our result to non-autonomous uniformly strongly elliptic differential operators on L^p-spaces.

Keywords Non-autonomous Cauchy problems · Desch–Schappacher perturbations · Evolution families · Uniformly strongly elliptic differential operators · 37B55 · 34G10 · 47D06 · 35B20

1 Introduction

For many processes in sciences, the coefficients of the partial differential equation describing a dynamical system as well as the boundary conditions of it may vary with time. In such cases one speaks of *non-autonomous* (or time-varying) evolution equations. For applications of non-autonomous dynamical systems we refer for example to [29]. From an operator theoretical point of view one considers families of Banach space operators which depend on the time parameter and studies the associated non-autonomous abstract Cauchy problem. In particular, for fixed $T > 0$ and a family of linear (and typically unbounded) operators $(A(t), \mathrm{D}(A(t)))_{t\in[0,T]}$

C. Budde (✉)
Department of Mathematics and Applied Mathematics, University of the Free State, Bloemfontein, South Africa
e-mail: BuddeCJ@ufs.ac.za

C. Seifert
Mathematisches Seminar, Christian-Albrechts-Universität zu Kiel, Kiel, Germany

Institut für Mathematik, Technische Universität Hamburg, Hamburg, Germany
e-mail: christian.seifert@tuhh.de

Y. Choi et al. (eds.), *Operators, Semigroups, Algebras and Function Theory*, Operator Theory: Advances and Applications 292,
https://doi.org/10.1007/978-3-031-38020-4_4

on a Banach space X one considers the non-autonomous abstract Cauchy problem given by

$$\begin{aligned} \dot{u}(t) &= A(t)u(t), \quad T \geq t \geq s \geq 0, \\ u(s) &= x \in X. \end{aligned} \tag{nACP}$$

While in the autonomous case operator semigroups yield fundamental solutions and thus provide an appropriate solution concept, for the non-autonomous case one needs to make use of so-called evolution families $(U(t, s))_{t \geq s}$, which give rise to a notion of well-posedness [37, Sect. 3.2]. Unfortunately, the existence of solutions of (nACP) is not as simple as in the autonomous case which, among others, comes from the lack of a differentiable structure of an evolution family as opposed to C_0-semigroups [37, Ex. 3.5]. Nevertheless, there are several independent and technical attempts to find sufficient conditions for (unique) solutions of (nACP), for example by Acquistapace and Terreni [3] or Kato and Tanabe [23, 47].

Perturbation theory is a powerful tool in order to study existence, uniqueness and other qualitative properties and allows for a more general and abstract view on non-autonomous abstract Cauchy problems. More precisely, additionally given a family $(B(t), \mathrm{D}(B(t)))_{t \in [0,T]}$ of linear operators in X one studies the perturbed Cauchy problem

$$\begin{aligned} \dot{u}(t) &= (A(t) + B(t))u(t), \quad T \geq t \geq s \geq 0, \\ u(s) &= x \in X, \end{aligned}$$

trying to make use of information of the unperturbed Cauchy problem (nACP). Time-dependent perturbations of evolution equations have attracted a lot of interest in the past, see for example [14, 41].

In this article, we consider non-autonomous Desch–Schappacher perturbations, i.e., we demand that the perturbation is merely bounded with respect to suitable extrapolation spaces. In our context, there are several technicalities to deal with. First of all, one has to consider extrapolation of evolution families. For this purpose, we will assume that the extrapolation spaces regarding the given family $(A(t))_{t \in [0,T]}$ of operators are uniformly equivalent. This assumption is reasonable and has been made by several authors, see for example [5, 22]. We notice that the uniformity of the extrapolation spaces does not imply that the domains $\mathrm{D}(A(t))$ are independent of time as well. The usage of extrapolation spaces first yields evolution families in this extrapolation space. In order to obtain an evolution family for the perturbed system in the original space X we ask for a bit more regularity of the perturbation, namely by suitable mapping properties into the so-called Favard class. Our perturbation result is motivated by the work of Rhandi [41] for autonomous A and non-autonomous perturbation. In order to treat the non-autonomous case, we make use of the work of Bertoni [7–9], who studied perturbations of so-called CD-systems.

It is worthwhile to review the already existing literature regarding non-autonomous Desch–Schappacher perturbations. In [21], Hinrichsen and Pritchard consider stability for time-infinite systems of closed-loop output-feedback type by means of so-called weak evolution operators and piecewise continuous operator functions. They are particularly interested in perturbed system equations of the form

$$\dot{x}(t) = [A(t) + D(t)\Delta(t)E(t)]x(t), \quad t \geq 0.$$

They obtain non-autonomous Desch–Schappacher type perturbations by specifying $\Delta(t) = \mathrm{I}$ and $E(t) = \mathrm{I}$. A few years later, Schnaubelt also studied time-infinite systems of closed-loop output-feedback type in [44] where he makes use of a Duhamel type formula in an approximative sense due to the lack of extrapolation theory for non-autonomous evolution equations. Within this framework he gets joint nonautonomous extension of the Desch–Schappacher and Miyadera perturbation theorem, cf. [44, Rem. 4.6(a)]. Last but not least, we want to mention the joint work of Maniar and Schnaubelt [32] where they follow the most general approach due to Acquistapace and Terreni [3] to develop a non-autonomous Desch–Schappacher perturbation theorem. In particular, the domains of the operators $(A(t), \mathrm{D}(A(t)))$ are allowed to vary in time and to be non-dense and they allow Hölder continuity of the resolvent.

In this paper, we consider a "light" version of a non-autonomous Desch–Schappacher perturbation theorem. In particular, we ask for reasonable conditions which are easy to check in applications. Moreover, our approach, which relies on the works of Bertoni [7–9], provides a direct and easily accessible method to treat Desch–Schappacher perturbations in the non-autonomous setting.

Another more indirect way to tackle non-autonomous problems is given by means of so-called evolution semigroups. These have been used for non-autonomous perturbations of bounded and Miyadera–Voigt type, see [39] and [40]. However, we will not consider this method in the present paper.

Let us outline the paper. The first two sections have a preliminary character. We start by recalling the most important facts about autonomous abstract Cauchy problems as well as their associated extrapolation and Favard spaces. In Sect. 3, we revise non-autonomous abstract Cauchy problems and CD-systems. In Sect. 4, we state and prove the Desch–Schappacher perturbation result in our context, where we first consider extrapolation of CD-systems and then we prove our main result Theorem 4.10. The final section consists of an example discussing non-autonomous uniformly strongly elliptic differential operators on L^p-spaces, i.e. a so-called parabolic case.

2 Autonomous Systems and C_0-Semigroups

Let X be a Banach space.

2.1 C_0-Semigroups

The study of semigroups of operators started in the 1940s, see e.g. the monograph by Hille and Phillips [20], and has since then attracted a lot of attention in the literature, see for example the manuscripts by Engel and Nagel [13], Goldstein [16] or Pazy [38], just to mention a few.

Definition 2.1 A family $(T(t))_{t\geq 0}$ of bounded linear operators on X is called a C_0*-semigroup* if the following properties are satisfied:

(i) $T(0) = \mathrm{I}$, and $T(t+s) = T(t)T(s)$ for all $t, s \geq 0$.
(ii) $\lim_{t\to 0} \|T(t)x - x\| = 0$ for all $x \in X$.

To each operator semigroup we can assign a closed operator $(A, \mathrm{D}(A))$ called its (infinitesimal) generator.

Definition 2.2 Let $(T(t))_{t\geq 0}$ be a C_0-semigroup on X. The *(infinitesimal) generator* $(A, \mathrm{D}(A))$ of $(T(t))_{t\geq 0}$ is defined by

$$Ax := \lim_{t\to 0} \frac{T(t)x - x}{t}, \qquad \mathrm{D}(A) := \left\{ x \in X : \lim_{t\to 0} \frac{T(t)x - x}{t} \text{ exists} \right\}. \tag{2.1}$$

We notice that C_0-semigroups as well as their generators enjoy good properties, cf. [13, Chapter II, Lemma 1.3]. For example, if $(A, \mathrm{D}(A))$ is the generator of a C_0-semigroup $(T(t))_{t\geq 0}$ then by [13, Chapter II, Thm. 1.10] the resolvent $R(\lambda, A) := (\lambda - A)^{-1}$, $\lambda \in \rho(A)$, can be expressed by means of the Laplace transform whenever $\mathrm{Re}(\lambda)$ is sufficiently large, i.e., one has

$$R(\lambda, A)x = \int_0^\infty \mathrm{e}^{-\lambda t} T(t)x \, \mathrm{d}t, \qquad x \in X, \tag{2.2}$$

provided $\mathrm{Re}(\lambda) > \omega_0$, where ω_0 denotes the growth bound of the semigroup $(T(t))_{t\geq 0}$, cf. [13, Chapter I, Def. 5.6]. The celebrated Hille–Yosida theorem, cf. [13, Chapter II, Thm. 3.8], characterises generators of C_0-semigroups. Moreover, C_0-semigroups exactly provide the solution concept for autonomous abstract Cauchy problems

$$\begin{aligned} u'(t) &= Au(t), \quad t > 0, \\ u(0) &= x, \end{aligned} \tag{2.3}$$

where A is a linear operator in X and $x \in X$. Then (2.3) is well-posed in the sense of Hadamard if and only if A is the generator of a C_0-semigroup $(T(t))_{t\geq 0}$, and then $u := T(\cdot)x$ yields the solution, see e.g. [13, Corollary II.6.9].

2.2 Extrapolation Spaces

An important tool on our way towards a non-autonomous Desch–Schappacher theorem are the so-called extrapolation spaces. They are also important in several applications, see for example boundary perturbations [17, 19] or maximal regularity [11, 36], just to mention a few. Let us recall the basic construction of extrapolation spaces, cf. [13, Chapter II, Def. 5.4] or [34, Def. 1.4]. Notice, that if $(A, \mathrm{D}(A))$ is the generator of a C_0-semigroup then one may assume without loss of generality that $0 \in \rho(A)$, i.e., A is invertible with bounded inverse.

Definition 2.3 Let $(A, \mathrm{D}(A))$ be the generator of a C_0-semigroup $(T(t))_{t\geq 0}$ on X. With $\|\cdot\|_{-1}$ we denote the norm on X defined by

$$\|x\|_{-1} := \left\|A^{-1}x\right\|, \quad x \in X. \tag{2.4}$$

The completion of $(X, \|\cdot\|_{-1})$ is called the *(first) extrapolation space* and is denoted by X_{-1}. If we want to emphasize that the extrapolation is with respect to the semigroup $(T(t))_{t\geq 0}$ with generator $(A, \mathrm{D}(A))$ we write $X_{-1}(T)$ or $X_{-1}(A)$, respectively. Furthermore, we denote the continuous extension of the operators $T(t)$ to the extrapolation space X_{-1} by $T_{-1}(t)$.

The following result shows that in fact the family of operators $(T_{-1}(t))_{t\geq 0}$ yields again a C_0-semigroup but now on the extrapolated space X_{-1}, cf. [13, Chapter II, Thm. 5.5] or [34, Thm. 1.5].

Proposition 2.4

(a) *X_{-1} is a Banach space containing X as a dense subspace.*
(b) *$(T_{-1}(t))_{t\geq 0}$ is a C_0-semigroup on X_{-1}.*
(c) *The generator A_{-1} of $(T_{-1}(t))_{t\geq 0}$ has domain $\mathrm{D}(A_{-1}) = X$ and is the unique extension of $A : \mathrm{D}(A) \to X$ to an isometry from X to X_{-1}.*

Notice, that we only discussed the construction of the first extrapolation space since this is sufficient for our purpose. Furthermore, it follows from Proposition 2.4 that $\|T_{-1}(t)\| = \|T(t)\|$ for all $t \geq 0$, see also [35]. It is worth to mention that in general one can construct also extrapolation spaces of higher order, cf. [13, Chapter II, Sect. 5a] or [34, Sect. 1].

Remark 2.5 Observe, that whenever X is a separable Banach space, then X_{-1} is also separable. This follows directly from the construction of extrapolation spaces.

2.3 The Favard Class

An important concept in the context of Desch–Schappacher perturbations is the so-called Favard class.

Definition 2.6 Let $(T(t))_{t\geq 0}$ be a C_0-semigroup on X with generator $(A, \mathrm{D}(A))$ satisfying $[0,\infty) \subseteq \rho(A)$ and $\|\lambda R(\lambda, A)\| \leq M$ for all $\lambda > 0$ and some constant $M \geq 1$. The *Favard space* $F_1(A)$ corresponding to A and $(T(t))_{t\geq 0}$ is defined by

$$F_1(A) := \left\{ x \in X : \sup_{t>0} \frac{1}{t} \|T(t)x - x\| < \infty \right\}, \tag{2.5}$$

equipped with the norm $\|\cdot\|_{F_1(A)}$ defined by

$$\|x\|_{F_1(A)} := \sup_{t>0} \frac{1}{t} \|T(t)x - x\|, \quad x \in F_1(A). \tag{2.6}$$

By [13, Chapter II, Thm. 5.15] the space $F_1(A)$ is a Banach space. If X is reflexive then one has $F_1(A) = \mathrm{D}(A)$.

Remark 2.7 Let $(T(t))_{t\geq 0}$ be a C_0-semigroup with generator $(A, \mathrm{D}(A))$ satisfying $[0,\infty) \subseteq \rho(A)$ and $\|\lambda R(\lambda, A)\| \leq M$ for all $\lambda > 0$ and some constant $M \geq 1$. Then $F_1(A) = (X, D(A))_{1,\infty,K}$ (with equivalence of norms), see [8, Prop. 23] or [31, Prop. 5.7].

By Proposition 2.4 the extrapolated semigroup of a C_0-semigroup is a C_0-semigroup and hence we can construct the corresponding Favard space, which is then called the Favard class.

Definition 2.8 Let $(T(t))_{t\geq 0}$ be a C_0-semigroup on X with generator $(A, \mathrm{D}(A))$ satisfying $[0,\infty) \subseteq \rho(A)$ and $\|\lambda R(\lambda, A)\| \leq M$ for all $\lambda > 0$ and some constant $M \geq 1$. The *Favard class* $F_0(A)$ is defined to be the Favard space corresponding to the extrapolated semigroup $(T_{-1}(t))_{t\geq 0}$ and its generator A_{-1}, i.e.,

$$F_0(A) := \left\{ x \in X_{-1} : \sup_{t>0} \frac{1}{t} \|T_{-1}(t)x - x\|_{-1} < \infty \right\}, \tag{2.7}$$

equipped with the norm $\|\cdot\|_{F_0(A)}$ defined by

$$\|x\|_{F_0(A)} := \sup_{t>0} \frac{1}{t} \|T(t)x - x\|_{-1}, \quad x \in F_0(A). \tag{2.8}$$

Note that $F_0(A) = F_1(A_{-1})$. In view of Proposition 2.4(c) and Remark 2.7, we have $F_0(A) = (X_{-1}, X)_{1,\infty,K}$.

3 Non-autonomous Abstract Cauchy Problems

Let X be a separable Banach space.

3.1 Evolution Families

In what follows we consider, for fixed $T > 0$ and a given family of (unbounded) operators $\big((A(t), \mathrm{D}(A(t)))\big)_{t\in[0,T]}$ in X, the following non-autonomous abstract Cauchy problem

$$
\begin{aligned}
u'(t) &= A(t)u(t), \quad T \geq t \geq s \geq 0,\\
u(s) &= x,
\end{aligned}
\tag{nACP}
$$

where $x \in X$. For the sake of completeness we recall what we mean by a (classical) solution of (nACP).

Definition 3.1 Let $(A(t), \mathrm{D}(A(t)))$, $t \in [0, T]$, be linear operators in X and take $s \in [0, T]$ and $x \in \mathrm{D}(A(s))$. Then a *(classical) solution* of (nACP) is a function $u := u(\cdot; s, x) \in \mathrm{C}^1([s, T], X)$ such that $u(t) \in \mathrm{D}(A(t))$ and u satisfies (nACP) for $t \geq s$. The non-autonomous abstract Cauchy problem (nACP) is called *well-posed (on spaces Y_t)* if there are dense subspaces $Y_s \subseteq \mathrm{D}(A(s))$, $s \in [0, T]$, of X such that for $s \in [0, T]$ and $x \in Y_s$ there is a unique solution $t \mapsto u(t; s, x) \in Y_t$ of (nACP). In addition, for $s_n \to s$ and $Y_{s_n} \ni x_n \to x$, we have $\widetilde{u}(t; s_n, x_n) \to \widetilde{u}(t; s, x)$ uniformly for t in compact intervals in $\mathbb{R}$, where we set

$$
\widetilde{u}(t; s, x) := \begin{cases} u(t; s, x) & T \geq t \geq s \geq 0,\\ x & t < s,\\ u(T; s, x) & t > T. \end{cases}
$$

Remark 3.2 Note that there are different notions of solutions in the literature. For example, [15] use so-called strong solutions with weaker regularity properties.

The solution concept of strongly continuous semigroups of linear operators, in the autonomous case, is now replaced by evolution families in the non-autonomous situation.

Let $\Delta_T := \{(t, s) \in [0, T]^2 : t \geq s\}$.

Definition 3.3 A family of bounded linear operators $(U(t, s))_{t,s\in[0,T],t\geq s}$ on X is called a *(strongly continuous) evolution family* if

(i) $U(s, s) = \mathrm{I}$ and $U(t, r) = U(t, s)U(s, r)$ for all $t, s, r \in [0, T]$ with $T \geq t \geq s \geq r \geq 0$, and
(ii) the mapping $\Delta_T \ni (t, s) \mapsto U(t, s)$ is strongly continuous.

We say that $(U(t, s))_{t\geq s}$ *solves the non-autonomous abstract Cauchy problem* (nACP) *(on spaces Y_t)* if there are dense subspaces Y_s, $s \in [0, T]$, of X such that $U(t, s)Y_s \subseteq Y_t \subseteq \mathrm{D}(A(t))$ for $T \geq t \geq s \geq 0$ and the function $t \mapsto U(t, s)x$ is a solution of (nACP) for $s \in [0, T]$ and $x \in Y_s$.

That evolution families are the right concept in the non-autonomous situation is justified by the following fact.

Proposition 3.4 *The non-autonomous abstract Cauchy problem* (nACP) *is well-posed on* Y_t *if and only if there is an evolution family solving* (nACP) *on* Y_t.

Here and in the sequel we assume that (nACP) is well-posed in the sense of [13, Chapter VI, Def. 9.1]. Especially, there exists an evolution family $(U(t,s))_{t\geq s}$ solving (nACP). Observe that some authors also call $(U(t,s))_{t\geq s}$ a *parabolic fundamental solution*. The existence of such a fundamental solution is a priori not clear and is treated by Acquistapace and Terreni [1, 3] or Kato and Tanabe [24, 28, 45, 46] and many others.

3.2 CD-*Systems*

The main ingredient for the non-autonomous Desch–Schappacher perturbation theory will be the notion of CD-systems. Before introducing such systems, we recall the notion of Kato stability which will be important for our purpose.

Definition 3.5 (Kato Stability) Let $T > 0$. A family $\big((A(t), \mathrm{D}(A(t)))\big)_{t\in[0,T]}$ of linear operators on a Banach space X is called *stable* if there exist $M \geq 1$ and $\omega \in \mathbb{R}$ such that $(\omega, \infty) \subseteq \rho(A(t))$ for each $t \in [0, T]$ and

$$\left\| \prod_{j=1}^{k} R(\lambda, A(t_j)) \right\| = \| R(\lambda, A(t_k)) \cdots R(\lambda, A(t_1)) \| \leq \frac{M}{(\lambda - \omega)^k}, \tag{3.1}$$

for all $\lambda > \omega$ and every partition $0 \leq t_1 \leq t_2 \leq \ldots \leq t_k \leq T$, $k \in \mathbb{N}$.

Remark 3.6 It has been shown by Kato [25] that (3.1) is equivalent to the existence of $M \geq 1$ and $\omega \in \mathbb{R}$ such that

$$\left\| \prod_{j=1}^{k} \mathrm{e}^{\tau_j A(s_j)} \right\| = \left\| \mathrm{e}^{\tau_k A(s_k)} \cdots \mathrm{e}^{\tau_1 A(s_1)} \right\| \leq M \mathrm{e}^{\omega \sum_{j=1}^{k} s_j}, \tag{3.2}$$

for all partitions $0 \leq s_1 \leq s_2 \leq \ldots \leq s_k \leq T$ and all $\tau_k \geq 0$, $k \in \mathbb{N}$. Here, $(\mathrm{e}^{\tau A(s)})_{\tau\geq 0}$ denotes the C_0-semigroup generated by the operator $(A(s), \mathrm{D}(A(s)))$ for fixed $s \in [0, T]$.

Remark 3.7 Note that X_{-1} is separable since X is separable. As already mentioned by Pazy [38, Sect. 5.2], the order of the resolvent operators in (3.1) is important since in general the resolvent operators do not commute. There it is also noticed, that if each operator $(A(t), \mathrm{D}(A(t)))$, $t \in [0, T]$ is the generator of a quasi-contractive C_0-semigroup with a uniform growth bound then the family of operators

$\big((A(t), \mathrm{D}(A(t)))\big)_{t\in[0,T]}$ is stable in the sense of Definition 3.5. In particular, any family $\big((A(t), \mathrm{D}(A(t)))\big)_{t\in[0,T]}$ of generators of contractive C_0-semigroups is stable.

Let us now come to the definition of CD-systems.

Definition 3.8 (CD-System) Let $T > 0$, $\big((A(t), \mathrm{D}(A(t)))\big)_{t\in[0,T]}$ a family of generators of C_0-semigroups on X. Let D be a Banach space. Then the triplet $((A(t))_{t\in[0,T]}, X, \mathrm{D})$ is called a CD-*system* if the following properties are satisfied:

(a) The domain $\mathrm{D}(A(t)) = \mathrm{D}$ is independent of $t \in [0, T]$, where D is continuously and densely embedded in X.
(b) The family $\big((A(t), \mathrm{D})\big)_{t\in[0,T]}$ is stable in the sense of Definition 3.5.
(c) $A(\cdot) \in \mathrm{Lip}\left([0, T], \mathscr{L}_\mathrm{s}(\mathrm{D}, X)\right)$, i.e., $A(\cdot)$ is strongly Lipschitz continuous.

As a matter of fact, CD-systems automatically yield solutions of the associated non-autonomous abstract Cauchy problem (nACP), see e.g. [25, 26] or [27, Sect. 1.2].

Theorem 3.9 *Let $T > 0$, $((A(t))_{t\in[0,T]}, X, \mathrm{D})$ a* CD-*system. Then there exists a unique evolution family $(U(t, s))_{0\le s\le t\le T}$ with the following properties:*

(i) $U(t, r)U(r, s) = U(t, s)$ *and* $U(s, s) = \mathrm{I}$ *for all* $t, r, s \in [0, T]$ *with* $t \ge r \ge s$.
(ii) $U(\cdot, \cdot) \in \mathrm{C}\left(\Delta_T, \mathscr{L}_\mathrm{s}(X)\right) \cap \mathrm{C}\left(\Delta_T, \mathscr{L}_\mathrm{s}(\mathrm{D})\right)$.
(iii) $\frac{\partial}{\partial t}U(t, s)x = A(t)U(t, s)x$ *for all* $x \in \mathrm{D}$ *and* $t, s \in [0, T]$ *with* $t \ge s$.
(iv) $\frac{\partial}{\partial s}U(t, s)x = -U(t, s)A(s)x$ *for all* $x \in \mathrm{D}$ *and* $t, s \in [0, T]$ *with* $t \ge s$.
(v) *There exists* $M \ge 1$ *and* $\omega \in \mathbb{R}$ *such that* $\|U(t, s)\| \le M\mathrm{e}^{\omega(t-s)}$ *for all* $t, s \in [0, T]$ *with* $t \ge s$.

Examples 3.10 Let $T > 0$ and $g, \mu\colon [0, T] \times \mathbb{R}_{\ge 0} \to \mathbb{R}_{>0}$ such that μ is bounded away from zero, $g(t, \cdot) \in \mathrm{W}^{1,\infty}(\mathbb{R}_{\ge 0})$ and $\frac{1}{g(t,\cdot)}, \mu(t, \cdot) \in \mathrm{L}^\infty(\mathbb{R}_{\ge 0})$ for all $t \in [0, T]$, and $t \mapsto g(t, x)$, $t \mapsto \frac{\partial}{\partial x}g(t, x)$ and $t \mapsto \mu(t, x)$ are Lipschitz continuous uniformly in a.e. $x \in \mathbb{R}_{\ge 0}$. Let $X := \mathrm{L}^1\left(\mathbb{R}_{\ge 0}\right)$ and define $\big((A(t), \mathrm{D}(A(t)))\big)_{t\in[0,T]}$ by

$$(A(t)f)(x) := -\frac{\partial}{\partial x}g(t, x)f(x) - \mu(t, x)f(x), \quad t \in [0, T], \tag{3.3}$$

on the constant domain

$$\mathrm{D} := \mathrm{D}(A(t)) := \left\{f \in \mathrm{W}^{1,1}\left(\mathbb{R}_{\ge 0}\right) : \ f(0) = 0\right\}.$$

Then [8, Prop. 19] shows that $((A(t))_{t\in[0,T]}, X, \mathrm{D})$ is a CD-system.

4 Non-autonomous Desch–Schappacher Perturbations

Let X be a separable Banach space and $T > 0$.

4.1 *Perturbations of* CD-*Systems*

The following result by Bertoni [9, Prop. 6] regarding perturbations of CD-systems will be crucial for our non-autonomous Desch–Schappacher perturbation result. Note that given a CD-system $((A(t))_{t\in[0,T]}, X, \mathrm{D})$ we have that

$$F_1(A(t)) = (X; \mathrm{D})_{1,\infty,K} = F_1(A(s)), \quad 0 \leq s, t \leq T,$$

so the Favard spaces are constant. Let $F_1 := F_1(A(0)) = F_1(A(t))$ for all $t \in [0, T]$.

Proposition 4.1 *Let* $((A(t))_{t\in[0,T]}, X, \mathrm{D})$ *be a* CD-*system. If*

$$B(\cdot) \in \mathrm{L}^\infty([0, T], \mathscr{L}_\mathrm{s}(\mathrm{D}, F_1)) \cap \mathrm{Lip}([0, T], \mathscr{L}_\mathrm{s}(\mathrm{D}, X))$$

then $((A(t) + B(t))_{t\in[0,T]}, X, \mathrm{D})$ *is again a* CD-*system.*

In other words, Proposition 4.1 tells us that whenever we perturb a CD-system with a sufficiently regular family of operators then we obtain again a CD-system and hence by Theorem 3.9 an evolution family solving the associated (nACP).

4.2 CD-*Systems in Extrapolation Spaces*

Let $\big((A(t), \mathrm{D}(A(t)))\big)_{t\in[0,T]}$ be a family of closed and densely defined operators in X. Under suitable assumptions we are going to show that the family of operators $((A(t)_{-1})_{t\in[0,T]}, X_{-1}, X)$ is CD-system. To get this result, we first state and comment the on the assumptions we use.

Assumption A There exist $\theta \in \left(\frac{\pi}{2}, \pi\right)$ and $M \geq 1$ such that

(i) $\Sigma_{\theta,0} \subseteq \rho(A(t))$ for all $t \in [0, T]$ and
(ii) $\|R(\lambda, A(t))\| \leq \frac{M}{|\lambda|}$ for all $\lambda \in \Sigma_{\theta,0} \setminus \{0\}$ and $\left\|A(t)^{-1}\right\| \leq M$ for all $t \in [0, T]$,

where $\Sigma_{\theta,0} := \{z \in \mathbb{C} : |\arg(z)| \leq \theta\} \cup \{0\}$ is a sector in the complex plane.

Remark 4.2 Note that Assumption A guarantees that each operator $(A(t), \mathrm{D}(A(t)))$, $t \in [0, T]$ generates a strongly continuous and analytic semigroup on the Banach space X. In particular, one can construct the extrapolation spaces $X_{-1}(A(t))$ and extrapolated operators $A_{-1}(t) := A(t)_{-1}$ for each $t \in [0, T]$. We notice that by construction of extrapolation spaces all operators $A_{-1}(t)$ have the same domain, namely $\mathrm{D}(A_{-1}(t)) = X$ for all $t \in [0, T]$. Furthermore, by [6, Chapter V, Prop. 1.3.1], it holds that $\rho(A(t)) = \rho(A_{-1}(t))$ for all $t \in [0, T]$.

Lemma 4.3 *Let* $\big((A(t), \mathrm{D}(A(t)))\big)_{t\in[0,T]}$ *satisfy Assumption A and assume that* $\mathrm{D}(A(t)) = \mathrm{D}(A(s))$ *for all* $s, t \in [0, T]$.

(a) *The following are equivalent.*

  (i) *There exists* $C \geq 0$ *such that* $\left\|\mathrm{e}^{\tau A(t)} - \mathrm{e}^{\tau A(s)}\right\| \leq C\,|t-s|$ *for all* $t, s \in [0, T]$, $\tau \geq 0$.

  (ii) *There exists* $C \geq 0$ *such that* $\|R(\lambda, A(t)) - R(\lambda, A(s))\| \leq \frac{C}{|\lambda|}\,|t-s|$ *for all* $t, s \in [0, T]$, $\lambda \in \Sigma_{\theta,0} \setminus \{0\}$.

(b) *Assume there exists* $C \geq 0$ *such that* $\left\|\mathrm{I} - A(t)A(s)^{-1}\right\| \leq C\,|t-s|$ *for all* $t, s \in [0, T]$. *Then there exists* $C \geq 0$ *such that*

$$\|R(\lambda, A(t)) - R(\lambda, A(s))\| \leq \frac{C}{|\lambda|}\,|t-s|$$

for all $t, s \in [0, T]$, $\lambda \in \Sigma_{\theta,0} \setminus \{0\}$.

Proof

(a) The implication (i)⇒(ii) follows directly from the fact that the resolvent can be expressed in terms of the Laplace transformation (2.2) (at first for $\lambda > 0$ which then implies the statement for all $\lambda \in \Sigma_{\theta,0} \setminus \{0\}$). The converse (ii)⇒(i) is [43, Cor. 2.6] and uses that

$$\mathrm{e}^{\tau A(s)} = \frac{1}{2\pi i}\int_\Gamma \mathrm{e}^{\lambda\tau} R(\lambda, A(s))\,\mathrm{d}\lambda,$$

see also [13, Chapter II, Def. 4.2].

(b) This follows from [2, Lemma 2.2] by observing that

$$R(\lambda, A(t)) - R(\lambda, A(s)) = R(\lambda, A(t))(A(t)A(s)^{-1} - \mathrm{I})A(s)R(\lambda, A(s)).$$

□

Assumption B Let $\big((A(t), \mathrm{D}(A(t)))\big)_{t\in\mathbb{R}}$ satisfy Assumption A and denote by $X_{-1}(A(t))$, $t \in [0, T]$, the extrapolation space with respect to $(A(t), \mathrm{D}(A(t)))$, $t \in [0, T]$. We assume that

$$X_{-1}(A(t)) \cong X_{-1}(A(0)) =: X_{-1},$$

for all $t \in [0, T]$ and such that there exists $\kappa > 0$ with

$$\frac{1}{\kappa}\,\|x\|_{X_{-1}} \leq \|x\|_{X_{-1}(A(t))} \leq \kappa\,\|x\|_{X_{-1}}, \qquad x \in X_{-1},\ t \in [0, T]. \tag{4.1}$$

Remark 4.4 Assumption B appears to be natural in the context of extrapolation of evolution families. The problem is that, in general, for $s, t \in [0, T]$, $s \neq t$ the corresponding extrapolation spaces $X_{-1}(A(s))$ and $X_{-1}(A(t))$ may be completely different. However, Assumption B allows to extrapolate a given evolution family, cf. [5, Thm. 7.2]. Moreover, it allows us to transport the uniform resolvent estimate from Assumption A to the extrapolation space, i.e., we have that

$$\|R(\lambda, A_{-1}(t))\| \leq \frac{M}{|\lambda|}, \quad \lambda \in \Sigma_{\theta,0} \setminus \{0\} \quad \text{and} \quad \left\|A_{-1}(t)^{-1}\right\| \leq M, \quad t \in [0, T].$$

To see this, observe that for all $x \in X$ we have by (4.1) that

$$\|R(\lambda, A_{-1}(t))x\|_{X_{-1}} \leq \kappa \|R(\lambda, A_{-1}(t))x\|_{X_{-1}(A(t))} = \kappa \left\|R(\lambda, A(t))A(t)^{-1}x\right\|$$
$$\leq \|R(\lambda, A(t))\| \cdot \kappa \|x\|_{X_{-1}(A(t))} \leq \|R(\lambda, A(t))\| \cdot \kappa^2 \|x\|_{X_{-1}}.$$

Thus, by combining Assumptions A and B we observe that Assumption A also holds for the extrapolated family $\left((A_{-1}(t), X)\right)_{t\in[0,T]}$ (see Remark 4.2 for (i) and Remark 4.4 for (ii)). In particular, each operator $(A_{-1}(t), X)$, $t \in [0, T]$, generates a C_0-semigroup.

Examples 4.5 It was shown by Bertoni [8, Rem. 9] that the CD-system we mentioned in Example 3.10 has constant extrapolation spaces and that

$$X_{-1} = \left\{\varphi \in \mathscr{D}' : \exists g \in \mathrm{L}^1(\mathbb{R}_{\geq 0}) : \varphi = g'\right\}.$$

In view of Remarks 4.2 and 4.4, one might suspect that $((A_{-1}(t))_{t\in[0,T]}, X_{-1}, X)$ is a CD-system. In order to show this, we need another additional assumption which is inspired by [2, Hypo. II].

Assumption C There exists $L > 0$ such that $\left\|\mathrm{I} - A_{-1}(t)A_{-1}(s)^{-1}\right\| \leq L\,|t - s|$ for all $t, s \in [0, T]$.

Remark 4.6 Observe that in assumption [2, Hypo. II], Acquistapace and Terreni make use of a slightly weaker regularity assumption than Assumption C. In fact, they assume Hölder instead of Lipschitz continuity, i.e., there exist $L > 0$ and $\alpha \in (0, 1)$ such that $\left\|\mathrm{I} - A_{-1}(t)A_{-1}(s)^{-1}\right\| \leq L\,|t - s|^{\alpha}$ for all $t, s \in [0, T]$.

We can now show that $((A_{-1}(t))_{t\in[0,T]}, X_{-1}, X)$ is a CD-system.

Theorem 4.7 *Let the family* $\left((A(t), \mathrm{D}(A(t)))\right)_{t\in[0,T]}$ *satisfy Assumptions A–C. Then* $((A_{-1}(t))_{t\in[0,T]}, X_{-1}, X)$ *is a* CD*-system.*

Proof Note that X_{-1} is separable since X is separable. First of all, we show that $\left((A_{-1}(t), X)\right)_{t\in[0,T]}$ is stable in the sense of Definition 3.5. By Remark 4.2 and Remark 4.4, we observe that the family of operators $\left((A_{-1}(t), X)\right)_{t\in[0,T]}$ has constant domain and satisfies Assumption A. From Lemma 4.3 we conclude that there exists $C \geq 0$ such that

$$\|R(\lambda, A_{-1}(t)) - R(\lambda, A_{-1}(s))\| \leq \frac{C}{|\lambda|} |t - s|, \qquad (4.2)$$

for all $s, t \in [0, T]$ and each $\lambda \in \Sigma_{\theta,0} \setminus \{0\}$ or equivalently that there exists $C \geq 0$ such that

$$\left\| e^{\tau A_{-1}(t)} - e^{\tau A_{-1}(s)} \right\| \leq C |t - s|,$$

for all $s, t \in [0, T]$ and $\tau \geq 0$, which by [42, Prop. 4.9] implies that the family of operators $\big((A_{-1}(t), X)\big)_{t\in[0,T]}$ is stable. In order to show that the family $((A_{-1}(t))_{t\in[0,T]}, X_{-1}, X)$ is a CD-system, it remains to show that $A_{-1}(\cdot)$ is strongly Lipschitz continuous, which directly follows from the Lipschitz continuity of the resolvents, see (4.2), and [7, Thm. 3(b)]. □

Remark 4.8 Theorem 4.7 yields that the Assumptions A–C imply that the family $((A_{-1}(t))_{t\in[0,T]}, X_{-1}, X)$ is a CD-system and hence by Theorem 3.9 we obtain a unique evolution family $(\widetilde{U}(t, s))_{t\geq s}$ corresponding to the family of operators $\big((A_{-1}(t), X)\big)_{t\in[0,T]}$. Moreover, by [5, Thm. 7.2], there exists an evolution family $(U_{-1}(t, s))_{t\geq s}$ associated with $\big((A_{-1}(t), X)\big)_{t\in[0,T]}$ such that $U_{-1}(t, s)_{|X} = U(t, s)$ (therefore also called extrapolated evolution family). Thus, uniqueness implies $\widetilde{U}(t, s) = U_{-1}(t, s)$ for all $t, s \in [0, T]$ with $t \geq s$.

Remark 4.9 By following the proof of Theorem 4.7 it suffices that (4.2) holds true (which by Lemma 4.3 follows from Assumption C). Since also Assumption B holds and $R(\lambda, A_{-1}(t))_{|X} = R(\lambda, A(t))$ we obtain that (4.2) is already satisfied if we assume that the resolvents of $(A(t))_{t\in[0,T]}$ on X are commuting and Lipschitz continuous; that is,

$$R(\lambda, A(t))R(\mu, A(s)) = R(\mu, A(s))R(\lambda, A(t)), \quad s, t \in [0, T], \ \lambda, \mu \in \Sigma_{\theta,0},$$

and that there exists $C' \geq 0$ such that

$$\|R(\lambda, A(t)) - R(\lambda, A(s))\| \leq \frac{C'}{|\lambda|} |t - s|, \qquad (4.3)$$

for all $s, t \in [0, T]$ and each $\lambda \in \Sigma_{\theta,0} \setminus \{0\}$. Indeed, we then observe for $x \in X$ that

$$\begin{aligned}
&\|\big(R(\lambda, A_{-1}(t)) - R(\lambda, A_{-1}(s))\big)x\|_{X_{-1}} \\
&\leq \kappa \|A(t)^{-1}\big(R(\lambda, A(t)) - R(\lambda, A(s))\big)x\|_X \\
&= \kappa \|\big(R(\lambda, A(t)) - R(\lambda, A(s))\big)A(t)^{-1}x\|_X \\
&\leq \kappa^2 \frac{C'}{|\lambda|} |t - s| \, \|x\|_{X_{-1}}
\end{aligned}$$

for all $s, t \in [0, T]$ and $\lambda \in \Sigma_{\theta,0} \setminus \{0\}$, which implies (4.2) by density of X in X_{-1}. Note that commuting resolvents, sometimes used as definition of commutation of unbounded operators, have also been used in other contexts [12, 33].

In general, condition (4.3) can be checked easier than Assumption C due to the fact that one does not need the knowledge of the extrapolation spaces and the extrapolated operators.

4.3 A Non-autonomous Desch–Schappacher Perturbation

We are now able to state and prove our main result, a non-autonomous Desch–Schappacher perturbation type result. Let $\big((A(t), \mathrm{D}(A(t)))\big)_{t\in[0,T]}$ be a family of closed and densely defined operators in X satisfying Assumption A. Due to the fact that $\mathrm{D}(A_{-1}(t)) = X$ for all $t \in [0, T]$ we conclude that the Favard classes $F_0(A(t))$ are also constant in time; that is, $F_0 := F_0(A(0)) = F_0(A(t))$ for all $t \in [0, T]$, since $F_0(A(t)) = (X_{-1}, X)_{1,\infty,K} = F_0(A(0))$, see the last sentence of Sect. 2.

Theorem 4.10 *Let* $\big((A(t), \mathrm{D}(A(t)))\big)_{t\in[0,T]}$ *satisfy Assumption A–C and assume that the corresponding* (nACP) *yields a solution* $(U(t, s))_{t\geq s}$. *If*

$$B(\cdot) \in \mathrm{L}^{\infty}\left([0, T], \mathscr{L}_{\mathrm{s}}(X, F_0)\right) \cap \mathrm{Lip}\left([0, T], \mathscr{L}_{\mathrm{s}}(X, X_{-1})\right),$$

then there exists an evolution family $(V(t, s))_{t\geq s}$ *on* X *satisfying the variation of constant formula*

$$V(t, s)x = U(t, s)x + \int_s^t U_{-1}(t, \sigma)B(\sigma)V(\sigma, s)x \,\mathrm{d}\sigma, \quad t \geq s, x \in X.$$

Proof By Theorem 4.7 we know that $((A_{-1}(t))_{t\in[0,T]}, X_{-1}, X)$ is a CD-system, yielding the solution $(U_{-1}(t, s))_{t\geq s}$, see Remark 4.8. Due to the assumed regularity of $B(\cdot)$ we conclude by Proposition 4.1 that $((A_{-1}(t)+B(t))_{t\in[0,T]}, X_{-1}, X)$ is also a CD-system. Therefore, Theorem 3.9 ensures the existence of a unique solution of the corresponding (nACP) which we call $(\widetilde{V}(t, s))_{t\geq s}$. This family of operators satisfies the variation of constants formula on X_{-1}, i.e.,

$$\widetilde{V}(t, s)x = U_{-1}(t, s)x + \int_s^t U_{-1}(t, \sigma)B(\sigma)\widetilde{V}(\sigma, s)x \,\mathrm{d}\sigma, \quad t \geq s, \ x \in X_{-1}.$$

By Theorem 3.9 we have that $\widetilde{V} \in \mathrm{C}(\Delta_T, \mathscr{L}_{\mathrm{s}}(X_{-1}, X_{-1})) \cap \mathrm{C}(\Delta_T, \mathscr{L}_{\mathrm{s}}(X, X))$. Thus, we know that the restriction $(V(t, s))_{t\geq s}$ of $(\widetilde{V}(t, s))_{t\geq s}$ to X yields a continuous family which then satisfies

$$V(t, s)x = U(t, s)x + \int_s^t U_{-1}(t, \sigma)B(\sigma)V(\sigma, s)x \,\mathrm{d}\sigma, \quad t \geq s, \ x \in X.$$

□

Remark 4.11 Let us relate Theorem 4.10 to the existing literature.

(a) In [41, Thm. 2.3], non-autonomous perturbations of autonomous abstract Cauchy problems were considered, i.e. there exists a generator A_0 in X such that $A(t) = A_0$ for all $t \in [0, T]$, and B was only assumed to be continuous (with values in $\mathscr{L}_s(X, F_0(A))$). We need Lipschitz continuity (with values in $\mathscr{L}_s(X, X_1)$) due to the fact that we have non-autonomous $A(\cdot)$.
(b) In [48, Cor. 3.7], also non-autonomous perturbations of autonomous abstract Cauchy problems were considered, i.e. there exists a generator A_0 in X such that $A(t) = A_0$ for all $t \in [0, T]$, where B also only needs to be continuous (with values in $\mathscr{L}_s(X)$) and bounded (with values in $\mathscr{L}_s(\mathrm{D}(A_0), F_1(A_0))$). We need a corresponding boundedness assumption on B (with values in $\mathscr{L}_s(X, F_0(A_0))$, which is the same assumption as in [48] considered in the extrapolation space), but Lipschitz continuity (with values in $\mathscr{L}_s(X, X_1)$).

Remark 4.12 It is worth to mention that even if (nACP) is well-posed the perturbed non-autonomous abstract Cauchy problem does not have to be well-posed in general. In fact, the evolution family $(V(t, s))_{t \geq s}$ one obtains from Theorem 4.10 can be interpreted as mild solution of the corresponding Cauchy problem. We also refer to [13, Chapter VI, Ex. 9.21].

In view of Remark 4.9, we can replace Assumption C by (4.3) in Theorems 4.7 and 4.10 if the resolvents of $(A(t))_{t\in[0,T]}$ are commuting. We formulate this as a corollary.

Corollary 4.13 *Let* $\big((A(t), \mathrm{D}(A(t)))\big)_{t\in[0,T]}$ *satisfy Assumption A–B and let the resolvents of* $(A(t))_{t\in[0,T]}$ *be commuting and satisfy* (4.3). *Then* $((A_{-1}(t))_{t\in[0,T]}, X_{-1}, X)$ *is a* CD*-system. Moreover, if we assume that the corresponding* (nACP) *yields a solution* $(U(t, s))_{t\geq s}$ *and if*

$$B(\cdot) \in \mathrm{L}^\infty\left([0, T], \mathscr{L}_s(X, F_0)\right) \cap \mathrm{Lip}\left([0, T], \mathscr{L}_s(X, X_{-1})\right),$$

then there exists an evolution family $(V(t, s))_{t\geq s}$ *on* X *satisfying the variation of constants formula*

$$V(t, s)x = U(t, s)x + \int_s^t U_{-1}(t, \sigma)B(\sigma)V(\sigma, s)x \,\mathrm{d}\sigma, \qquad t \geq s, x \in X.$$

5 Example: Non-autonomous Uniformly Strongly Elliptic Differential Operators

Let $T > 0$. Let $m \in \mathbb{N}$, and for $\alpha \in \mathbb{N}_0^d$ with $|\alpha| \leq m$ let $a_\alpha\colon [0, T] \to \mathbb{C}$ be Lipschitz continuous. Let $a\colon [0, T] \times \mathbb{R}^d \to \mathbb{C}^d$ be defined by

$$a(t, \xi) := \sum_{|\alpha|\leq m} a_\alpha(t)(i\xi)^\alpha.$$

We assume that a is uniformly strongly elliptic, i.e. for the principal part $a_m\colon [0,T]\times\mathbb{R}^d \to \mathbb{C}^d$, $a_m(t,\xi) := \sum_{|\alpha|=m} a_\alpha(t)(i\xi)^\alpha$ of a there exists $c>0$ such that

$$\operatorname{Re} a_m(t,\xi) \geq c\,|\xi|^m, \quad t\in[0,T], \xi\in\mathbb{R}^d.$$

Additionally, let us further assume that there exists $\omega>0$ such that $\operatorname{Re} a(t,\xi)\geq \omega$ for all $t\in[0,T]$ and $\xi\in\mathbb{R}^d$ (which we can achieve by adding a suitable constant to a_0).

Let $p\in(1,\infty)$. Then a gives rise to a family $\big((A_p(t),\mathrm{D}(A_p(t)))\big)_{t\in[0,T]}$ of negatives of m-sectorial differential operators in $X:=\mathrm{L}^p(\mathbb{R}^d)$ as

$$\begin{aligned}\mathrm{D}(A_p(t)) &:= \mathrm{W}^{p,m}(\mathbb{R}^d),\\ A_p(t)u &:= -\sum_{|\alpha|\leq m} a_\alpha(t)\partial^\alpha u\end{aligned}$$

for $t\in[0,T]$, see, e.g., [18, Chapter 8], [10]. Note that by uniform strong ellipticity, $\|\cdot\|_{\mathrm{W}^{p,m}(\mathbb{R}^d)}$ and the graph norm of $A_p(s)$ are equivalent (with uniform constants for $s\in[0,T]$).

By [18, Theorem 8.2.1], Assumption A is satisfied. Moreover, it is easy to see that the family of operators $\big((A_p(t),\mathrm{W}^{p,m}(\mathbb{R}^d))\big)_{t\in[0,T]}$ satisfies Assumption B and that $X_{-1}\cong \mathrm{W}^{p,m}(\mathbb{R}^d)' = \mathrm{W}^{p',-m}(\mathbb{R}^d)$, which follows from a duality argument and [4, Chapter 3, 3.14]. Here, $p'\in(1,\infty)$ is the Hölder dual exponent, i.e. $\frac{1}{p}+\frac{1}{p'}=1$.

Moreover, since $A_p(s)$ and $A_p(t)$ commute on $W^{p,2m}(\mathbb{R}^d)$ for all $s,t\in[0,T]$ (as the coefficients are only depending on time and not on space), we observe that the resolvents of $(A_p(t))_{t\in[0,T]}$ commute. In order to show that (4.3) is satisfied, we make use of Lemma 4.3(b). Let $f\in\mathrm{L}^p(\mathbb{R}^d)$, $u(t):=A_p(t)^{-1}f$ for $t\in[0,T]$. By the Lipschitz continuity of the coefficient functions a_α there exists $C\geq 0$ such that

$$\|(\mathrm{I}-A_p(t)A_p(s)^{-1})f\| = \|A_p(s)u(s)-A_p(t)u(s)\| \leq C\,|t-s| \sum_{|\alpha|\leq m}\|\partial^\alpha u(s)\|$$

for $t,s\in[0,T]$. Thus, there exists $\widetilde{C}_p\geq 0$ such that

$$\|(\mathrm{I}-A_p(t)A_p(s)^{-1})f\| \leq \widetilde{C}_p\,|t-s|\,\|u(s)\|_{\mathrm{W}^{p,m}(\mathbb{R}^d)}, \quad t,s\in[0,T].$$

Hence, the uniform bound for $\|A_p(s)^{-1}\|$ yields existence of $C_p\geq 0$ such that

$$\|(\mathrm{I}-A_p(t)A_p(s)^{-1})f\| \leq C_p\,|t-s|\,\|f\|, \quad f\in\mathrm{L}^p(\mathbb{R}^d).$$

By Lemma 4.3(b) we obtain that (4.3) is satisfied. Thus, the first part of Corollary 4.13 yields that $\big(((A_p)_{-1}(t))_{t\in[0,T]},\mathrm{W}^{p',-m}(\mathbb{R}^d),\mathrm{L}^p(\mathbb{R}^d)\big)$ is a CD-system. Moreover, by [30, Section 6.1], there exists a unique evolution family $(U_p(t,s))_{t\geq s}$

associated with $\big((A_p(t), \mathrm{W}^{p,m}(\mathbb{R}^d))\big)_{t\in[0,T]}$. Furthermore, since $\mathrm{W}^{p',-m}(\mathbb{R}^d)$ is reflexive (see [4, Theorem 3.3.12]), we obtain $F_0 = \mathrm{D}((A_p)_{-1}(t))$ and therefore $F_0 = \mathrm{L}^p(\mathbb{R}^d)$ for all $t \in [0, T]$ by [13, Corollary II.5.21]. Hence, the second part of Corollary 4.13 yields that if

$$B(\cdot) \in \mathrm{L}^\infty\big([0, T], \mathscr{L}_s(\mathrm{L}^p(\mathbb{R}^d), \mathrm{L}^p(\mathbb{R}^d))\big) \cap \mathrm{Lip}\big([0, T], \mathscr{L}_s(\mathrm{L}^p(\mathbb{R}^d), \mathrm{W}^{p',-m}(\mathbb{R}^d))\big)$$

then there exists an evolution system $(V_p(t, s))_{t\geq s}$ on X giving rise to so-called mild solutions for the perturbed non-autonomous Cauchy problem

$$\begin{aligned} u'(t) &= (A(t) + B(t))u(t), \quad t \in [s, T], \\ u(s) &= u_s \in \mathrm{L}^p(\mathbb{R}^d). \end{aligned}$$

Examples 5.1 As a particular example, for $j \in \{1, \ldots, d\}$ let $a_{2e_j} : [0, T] \to \mathbb{C}$ be Lipschitz continuous with $c := -\max_{j\in\{1,\ldots,d\}} \max_{t\in[0,T]} \operatorname{Re} a_{2e_j}(t) > 0$, and for $\alpha \in \mathbb{N}_0^d$ with $|\alpha| \leq 1$ let $a_\alpha : [0, T] \to \mathbb{C}$ be Lipschitz continuous. Let

$$a(t, \xi) := \sum_{j=1}^{d} a_{2e_j}(t) i^2 \xi_j^2 + \sum_{|\alpha|\leq 1} a_\alpha(t)(i\xi)^\alpha, \quad t \in [0, T], \xi \in \mathbb{R}^d,$$

and hence $A_p(t) = -\sum_{j=1}^d a_{2e_j}(t)\partial_j^2 - \sum_{|\alpha|\leq 1} a_\alpha(t)\partial^\alpha$ for $t \in [0, T]$. As above, let us assume that there exists $\omega > 0$ such that $\operatorname{Re} a(t, \xi) \geq \omega$ for all $t \in [0, T]$, $\xi \in \mathbb{R}^d$. Then Assumptions A–B are satisfied, the resolvents of $(A_p(t))_{t\in[0,T]}$ are commuting and (4.3) are satisfied. For $t \in [0, T]$ and $f \in \mathrm{L}^p(\mathbb{R}^d)$ let

$$B(t)f := \begin{cases} f & t = 0, \\ \frac{1}{(2t)^d} \mathbf{1}_{(-t,t)^d} * f & t > 0. \end{cases}$$

Then clearly $B(\cdot) \in \mathrm{C}\big([0, T], \mathscr{L}_s(\mathrm{L}^p(\mathbb{R}^d), \mathrm{L}^p(\mathbb{R}^d))\big)$ (however, $B(\cdot)$ is not Lipschitz continuous). Moreover, a short calculation reveals that

$$B(\cdot) \in \mathrm{Lip}\big([0, T], \mathscr{L}_s(\mathrm{L}^p(\mathbb{R}^d), \mathrm{W}^{p',-2}(\mathbb{R}^d))\big).$$

Thus, Corollary 4.13 is applicable.

Acknowledgments C.B. thanks the University of Wuppertal for the possibility of funding a stay at the University of Salerno within the Erasmus exchange program as well as A. Rhandi for fruitful discussions and valuable comments. Moreover, C.B. acknowledges funding by the Deutsche Forschungsgemeinschaft (DFG, German Research Foundation) - 468736785 and by the National Research Foundation (Grant number: SRUG220317136). It is acknowledged that opinions, findings and conclusions or recommendations expressed in any publication generated by this supported research is that of the author(s). The National Research Foundation accepts no liability whatsoever in this regard. Both authors want to thank the anonymous referee for the valuable feedback. The authors are also thankful for the fruitful comments of the anonymous referees which allowed us to improve the article.

References

1. P. Acquistapace, Evolution operators and strong solutions of abstract linear parabolic equations. Differential Integral Equations **1**(4), 433–457 (1988)
2. P. Acquistapace, B. Terreni, On the abstract nonautonomous parabolic Cauchy problem in the case of constant domains. Ann. Mat. Pura Appl. (4) **140**, 1–55 (1985). https://doi.org/10.1007/BF01776844
3. P. Acquistapace, B. Terreni, A unified approach to abstract linear nonautonomous parabolic equations. Rend. Sem. Mat. Univ. Padova **78**, 47–107 (1987). http://www.numdam.org/item?id=RSMUP_1987__78__47_0
4. R.A. Adams, J.J.F. Fournier, *Sobolev Spaces*. Pure and Applied Mathematics (Amsterdam), vol. 140, 2nd edn. (Elsevier/Academic, Amsterdam, 2003)
5. H. Amann, Parabolic evolution equations in interpolation and extrapolation spaces. J. Funct. Anal. **78**(2), 233–270 (1988). https://doi.org/10.1016/0022-1236(88)90120-6
6. H. Amann, *Linear and Quasilinear Parabolic Problems*, vol. I. Monographs in Mathematics, vol. 89 (Birkhäuser Boston, Inc., Boston, 1995). https://doi.org/10.1007/978-3-0348-9221-6. Abstract linear theory
7. S. Bertoni, Evolutionary co-processes and perturbations of Kato-stable families of operators. Dyn. Contin. Discrete Impuls. Syst. Ser. A Math. Anal. **9**(2), 225–235 (2002)
8. S. Bertoni, Multiplicative perturbations of constant-domain evolution equations. J. Evol. Equ. **5**(2), 291–316 (2005). https://doi.org/10.1007/s00028-005-0192-0
9. S. Bertoni, Stability of CD-systems under perturbations in the Favard class. Mediterr. J. Math. **11**(4), 1195–1204 (2014). https://doi.org/10.1007/s00009-013-0376-8
10. C. Bombach, F. Gabel, C. Seifert, M. Tautenhahn, Observability for non-autonomous systems. SIAM J. Control Optim. **61**(1), 313–339 (2023). https://doi.org/10.1137/22M1485139
11. G. Da Prato, P. Grisvard, Maximal regularity for evolution equations by interpolation and extrapolation. J. Funct. Anal. **58**(2), 107–124 (1984). https://doi.org/10.1016/0022-1236(84)90034-X
12. G. Dore, A. Venni, On the closedness of the sum of two closed operators. Math. Z. **196**(2), 189–201 (1987). https://doi.org/10.1007/BF01163654
13. K.J. Engel, R. Nagel, *One-Parameter Semigroups for Linear Evolution Equations*. Graduate Texts in Mathematics, vol. 194 (Springer, New York, 2000). With contributions by S. Brendle, M. Campiti, T. Hahn, G. Metafune, G. Nickel, D. Pallara, C. Perazzoli, A. Rhandi, S. Romanelli and R. Schnaubelt
14. D.E. Evans, Time dependent perturbations and scattering of strongly continuous groups on Banach spaces. Math. Ann. **221**(3), 275–290 (1976). https://doi.org/10.1007/BF01596393
15. C. Gallarati, M. Veraar, Maximal regularity for non-autonomous equations with measurable dependence on time. Potential Anal. **46**(3), 527–567 (2017). https://doi.org/10.1007/s11118-016-9593-7
16. J.A. Goldstein, *Semigroups of Linear Operators & Applications* (Dover Publications, Mineola, 2017). Second edition of [MR0790497]. Including transcriptions of five lectures from the 1989 workshop at Blaubeuren, Germany
17. G. Greiner, Perturbing the boundary conditions of a generator. Houston J. Math. **13**(2), 213–229 (1987)
18. M. Haase, The functional calculus for sectorial operators, in *Operator Theory: Advances and Applications*, vol. 169 (Birkhäuser Verlag, Basel, 2006). https://doi.org/10.1007/3-7643-7698-8

19. S. Hadd, R. Manzo, A. Rhandi, Unbounded perturbations of the generator domain. Discrete Contin. Dyn. Syst. **35**(2), 703–723 (2015). https://doi.org/10.3934/dcds.2015.35.703
20. E. Hille, R.S. Phillips, *Functional Analysis and Semi-groups*. American Mathematical Society Colloquium Publications, vol. 31, rev. edn. (American Mathematical Society, Providence, 1957)
21. D. Hinrichsen, A.J. Pritchard, Robust stability of linear evolution operators on Banach spaces. SIAM J. Control Optim. **32**(6), 1503–1541 (1994). https://doi.org/10.1137/S0363012992230404
22. B. Jacob, H. Laasri, Well-posedness of infinite-dimensional non-autonomous passive boundary control systems. Evol. Equ. Control Theory **10**(2), 385–409 (2021). https://doi.org/10.3934/eect.2020072
23. T. Kato, Integration of the equation of evolution in a Banach space. J. Math. Soc. Japan **5**, 208–234 (1953). https://doi.org/10.2969/jmsj/00520208
24. T. Kato, Abstract evolution equations of parabolic type in Banach and Hilbert spaces. Nagoya Math. J. **19**, 93–125 (1961). http://projecteuclid.org/euclid.nmj/1118800864
25. T. Kato, Linear evolution equations of "hyperbolic" type. J. Fac. Sci. Univ. Tokyo Sect. I **17**, 241–258 (1970)
26. T. Kato, Linear evolution equations of "hyperbolic" type. II. J. Math. Soc. Japan **25**, 648–666 (1973). https://doi.org/10.2969/jmsj/02540648
27. T. Kato, Abstract differential equations and nonlinear mixed problems. Lezioni Fermiane. [Fermi Lectures]. Scuola Normale Superiore, Pisa; Accademia Nazionale dei Lincei, Rome (1985)
28. T. Kato, H. Tanabe, On the abstract evolution equation. Osaka Math. J. **14**, 107–133 (1962)
29. P.E. Kloeden, C. Pötzsche (eds.), *Nonautonomous Dynamical Systems in the Life Sciences*. Lecture Notes in Mathematics, vol. 2102 (Springer, Cham, 2013). https://doi.org/10.1007/978-3-319-03080-7. Mathematical Biosciences Subseries
30. A. Lunardi, *Analytic Semigroups and Optimal Regularity in Parabolic Problems*. Modern Birkhäuser Classics (Birkhäuser/Springer Basel AG, Basel, 1995) [2013 reprint of the 1995 original] [MR1329547]
31. A. Lunardi, Interpolation theory, *Appunti. Scuola Normale Superiore di Pisa (Nuova Serie) [Lecture Notes. Scuola Normale Superiore di Pisa (New Series)]*, vol. 16. Edizioni della Normale, Pisa (2018). https://doi.org/10.1007/978-88-7642-638-4. Third edition [of MR2523200]
32. L. Maniar, R. Schnaubelt, Robustness of Fredholm properties of parabolic evolution equations under boundary perturbations. J. Lond. Math. Soc. (2) **77**(3), 558–580 (2008). https://doi.org/10.1112/jlms/jdn001
33. S. Monniaux, J. Prüss, A theorem of the Dore-Venni type for noncommuting operators. Trans. Am. Math. Soc. **349**(12), 4787–4814 (1997). https://doi.org/10.1090/S0002-9947-97-01997-1
34. R. Nagel, Extrapolation spaces for semigroups (nonlinear evolution equations and applications). RIMS Kôkyûroku **1009**, 181–191 (1997). http://hdl.handle.net/2433/61496
35. R. Nagel, E. Sinestrari, Inhomogeneous Volterra integrodifferential equations for Hille-Yosida operators, in *Functional Analysis (Essen, 1991)*. Lecture Notes in Pure and Appl. Math., vol. 150 (Dekker, New York, 1994), pp. 51–70
36. R. Nagel, E. Sinestrari, Extrapolation spaces and minimal regularity for evolution equations. J. Evol. Equ. **6**(2), 287–303 (2006). https://doi.org/10.1007/s00028-006-0246-y
37. G. Nickel, On evolution semigroups and wellposedness of nonautonomous Cauchy problems. Ph.D. thesis, Eberhard-Karls-Universität Tübingen (1996)
38. A. Pazy, *Semigroups of Linear Operators and Applications to Partial Differential Equations*. Applied Mathematical Sciences, vol. 44 (Springer, New York, 1983). https://doi.org/10.1007/978-1-4612-5561-1
39. F. Räbiger, A. Rhandi, R. Schnaubelt, Perturbation and an abstract characterization of evolution semigroups. J. Math. Anal. Appl. **198**(2), 516–533 (1996). https://doi.org/10.1006/jmaa.1996.0096

40. F. Räbiger, R. Schnaubelt, A. Rhandi, J. Voigt, Non-autonomous Miyadera perturbations. Differential Integral Equations **13**(1–3), 341–368 (2000)
41. A. Rhandi, Extrapolation methods to solve non-autonomous retarded partial differential equations. Studia Math. **126**(3), 219–233 (1997). https://doi.org/10.4064/sm-126-3-219-233
42. R. Schnaubelt, Exponential bounds and hyperbolicity of evolution families. Ph.D. thesis, Eberhard–Karls–Universität Tübingen (1996)
43. R. Schnaubelt, Sufficient conditions for exponential stability and dichotomy of evolution equations. Forum Math. **11**(5), 543–566 (1999). https://doi.org/10.1515/form.1999.013
44. R. Schnaubelt, Feedbacks for nonautonomous regular linear systems. SIAM J. Control Optim. **41**(4), 1141–1165 (2002). https://doi.org/10.1137/S036301290139169X
45. H. Tanabe, On the equations of evolution in a Banach space. Osaka Math. J. **12**, 363–376 (1960)
46. H. Tanabe, Evolutional equations of parabolic type. Proc. Japan Acad. **37**, 610–613 (1961). http://projecteuclid.org/euclid.pja/1195523536
47. H. Tanabe, *Equations of Evolution*. Monographs and Studies in Mathematics, vol. 6 (Pitman (Advanced Publishing Program), Boston, 1979). Translated from the Japanese by N. Mugibayashi and H. Haneda
48. T.J. Xiao, J. Liang, J. van Casteren, Time dependent Desch-Schappacher type perturbations of Volterra integral equations. Integr. Equ. Oper. Theory **44**(4), 494–506 (2002). https://doi.org/10.1007/BF01193674

Operator Algebras Associated with Graphs and Categories of Paths: A Survey

Juliana Bukoski and Sushil Singla

Abstract Many interesting examples of operator algebras, both self-adjoint and non-self-adjoint, can be constructed from directed graphs. In this survey, we overview the construction of C^*-algebras from directed graphs and from two generalizations of graphs: higher rank graphs and categories of paths. We also look at free semigroupoid algebras generated from graphs and higher rank graphs, with an emphasis on the left regular free semigroupoid algebra. We give examples of specific graphs and the algebras they generate, and we discuss properties such as semisimplicity and reflexivity. Finally, we propose a new construction: applying the left regular free semigroupoid construction to categories of paths.

Keywords Graph C^*-algebras · Free semigroupoid algebras · Higher rank graphs · Category of paths

1 Introduction

A directed graph is a set of vertices along with a set of edges, where each edge has a source vertex and a range vertex. Such a graph can be represented by a collection of operators on a Hilbert space $\mathcal{H}$; each vertex is associated to a projection, and each edge is associated to a partial isometry that maps between the subspaces corresponding to its source and range vertices. These projections and partial isometries are used to construct a C^*-algebra called the *graph C^*-algebra* of

J. Bukoski
Department of Mathematics, Physics, and Computer Science, Georgetown College, Georgetown, KY, USA
e-mail: juliana_bukoski@georgetowncollege.edu

S. Singla (✉)
Department of Mathematics, Shiv Nadar University Delhi NCR, Tehsil Dadri, Greater Noida, India
e-mail: ss774@snu.edu.in

Y. Choi et al. (eds.), *Operators, Semigroups, Algebras and Function Theory*, Operator Theory: Advances and Applications 292,
https://doi.org/10.1007/978-3-031-38020-4_5

the directed graph. There are many examples of common C^*-algebras which can be realized as graph algebras, and many properties of graph algebras are determined by structural properties of the graph.

Free semigroupoid algebras generated by directed graphs are a class of non-self-adjoint operator algebras introduced by Kribs and Power [46]. The construction of these algebras from a graph is similar to the graph C^*-algebra construction in that vertices are represented by projections and edges by partial isometries. However, a free semigroupoid algebra is closed in the weak operator topology and does not (necessarily) include adjoints. As in the graph C^*-algebra case, many previously-studied non-self-adjoint operator algebras can be expressed as free semigroupoid algebras for some directed graph, and many properties of the algebra correspond to properties of the graph. In fact, this relationship is in some sense stronger than the self-adjoint case; while it is possible to find two non-isomorphic graphs that produce the same graph C^*-algebra, Kribs and Power [46] showed that two free semigroupoid algebras from graphs are unitarily equivalent if and only if their corresponding graphs are isomorphic.

Both the C^*-algebra and the free semigroupoid algebra construction have been extended to higher rank graphs, which are a generalization of graphs where edges have length in $\mathbb{N}^k$ instead of $\mathbb{N}$; higher rank graphs can be thought of as graphs where certain paths are identified, according to a factorization property. Categories of paths are another generalization of graphs, introduced by Spielberg [75], which allow identifications under conditions less restrictive than the higher rank graph factorization property. Spielberg has defined C^*-algebras from categories of paths and considered some properties of these algebras [75].

In Sect. 2 of this paper, we give some basic definitions. In Sect. 3, we outline the graph C^*-algebra and free semigroupoid algebra constructions and discuss some of the properties of these operator algebras. In Sect. 4, we overview the higher rank graphs of Kumjian and Pask and the analogous operator algebras associated to them. Finally, in Sect. 5, we discuss Spielberg's C^*-algebra construction and results for categories of paths, and we define and propose some results for free semigroupoid algebras from categories of paths. The main goal of this survey article is to bring all work related to operator algebras associated with graphs and categories of paths under one roof, as well as introduce free semigroupoid algebras from categories of paths. Throughout the paper, we give examples of specific graphs, higher rank graphs, and categories of paths, along with the operator algebras they generate.

2 Definitions

Let $G = (G^0, G^1, r, s)$ be a directed graph consisting of countable sets G^0 and G^1 and functions $r, s : G^1 \to G^0$. The elements of G^0 are called *vertices*, and the elements of G^1 are called *edges*. For each edge $e \in G^1$, $s(e)$ and $r(e)$ are called the *source* and *range* of e, respectively. A *source* is a vertex that receives no edges.

A *path of length n* in G is a sequence of edges $\mu = \mu_1\mu_2\ldots\mu_n$ such that $s(\mu_i) = r(\mu_{i+1})$ for all $1 \leq i \leq n-1$. We write $|\mu| = n$. Notice that we concatenate paths from right to left, to be consistent with composition of the operators that will be associated with these edges. Likewise, we define the range and source of a path $\mu = \mu_1\mu_2\ldots\mu_n$ by $r(\mu) = r(\mu_1)$ and $s(\mu) = s(\mu_n)$ for $|\mu| > 1$, and $r(v) = v = s(v)$ for all $v \in G^0$.

We often represent graphs with diagrams where either a dot or the name of the vertex represents each vertex, and an arrow (possibly with a label) represents each edge. For example, the graph with two vertices v_1 and v_2, and two edges, e with $s(e) = r(e) = v_1$ and f with $s(f) = v_1$ and $r(f) = v_2$, can be represented as:

$$e \circlearrowleft v_1 \xrightarrow{\quad f \quad} v_2$$

We will not generally require graphs to be finite; for example, the graph represented by

$$v_1 \longrightarrow v_2 \longrightarrow \cdots \longrightarrow v_n \longrightarrow \cdots$$

has an infinite number of both vertices and edges. However, we will often be interested in graphs where each vertex receives at most finitely many edges; that is, $\{e \in G^1 : r(e) = v\}$ is finite for all $v \in G^0$. Such a graph is called *row-finite*.

In general, the process for defining an operator algebra from a graph starts with assigning projections to each vertex and partial isometries to each edge. Let $\mathcal{H}$ be a Hilbert space and $\mathcal{B}(\mathcal{H})$ the space of bounded linear operators on $\mathcal{H}$.

Definition 2.1 Let G be a row-finite directed graph. A *Cuntz-Krieger G-family* in $\mathcal{H}$ is a family $\{S, P\}$ of partial isometries $\{S_e : e \in G^1\}$ and pairwise orthogonal projections $\{P_v : v \in G^0\}$ on $\mathcal{H}$ satisfying

(I) $\quad S_e^* S_e = P_{s(e)}$ for all $e \in G^1$

(CK) $\quad P_v = \sum_{e \in G^1 : r(e) = v} S_e S_e^*$ for each $v \in G^0$ that is not a source.

Note that the sum in (CK) is finite because the graph is assumed to be row-finite.

A *Cuntz-Krieger-Toeplitz (CKT) G-family* satisfies (I) as well as

(CKT) $\quad P_v < \sum_{e \in G^1 : r(e) = v} S_e S_e^*$ for each $v \in G^0$ that is not a source.

Conditions (I) and (CK) are called the *Cuntz-Krieger relations*, and Conditions (I) and (CKT) are called the *Cuntz-Krieger-Toeplitz relations*. Note that Condition (I) implies that S_e is a partial isometry with initial space $P_{s(e)}\mathcal{H}$. Additionally, (CK) and (CKT) imply that the partial isometries S_e associated to the edges e with $r(e) = v$ have mutually orthogonal ranges which are closed subspaces of $P_{r(e)}\mathcal{H}$.

In Sect. 3, we will use these families of isometries to define the graph C^*-algebra and the free semigroupoid algebra of a graph. First, however, we define some properties of operator algebras that we will explore for the algebras discussed in this survey.

An operator $T \in \mathcal{B}(\mathcal{H})$ is called *nilpotent* if $T^n = 0$. We say that T is *quasi-nilpotent* if the spectrum of T is 0, or, equivalently, if $\lim_{n\to\infty} \|T^n\|^{1/n} = 0$. The *Jacobson radical* $\mathrm{rad}(\mathcal{A})$ of a Banach algebra $\mathcal{A}$ is the intersection of the kernels of all algebraically irreducible representations. An algebra $\mathcal{A}$ is called *semisimple* if $\mathrm{rad}(\mathcal{A}) = 0$. It is a well-known fact (for example, Theorem 2.3.5(ii) in [70]), that the Jacobson radical of an algebra of operators is the largest quasinilpotent ideal in the algebra.

The *commutant* $\mathcal{A}'$ of an algebra $\mathcal{A}$ is the set of all operators in $\mathcal{B}(\mathcal{H})$ that commute with all operators in $\mathcal{A}$. The *bicommutant* of $\mathcal{A}$ is the commutant of $\mathcal{A}'$. We will look at when the bicommutant of an algebra is equal to the algebra itself. The famous von Neumann Bicommutant Theorem states that if we have an algebra $\mathcal{A}$ consisting of bounded operators on a Hilbert space that contains the identity operator and is closed under taking adjoints, then the closures of $\mathcal{A}$ in the weak operator topology and the strong operator topology are equal, and are in turn equal to the bicommutant $\mathcal{A}''$ of $\mathcal{A}$ (see Corollary 3.3 of [76]). Although the free semigroupoid algebras that we will be examining may not be von Neumann algebras, and will just be weak operator topology (WOT)-closed algebras, they still satisfy that the bicommutant is equal to itself, as we will see in Sect. 3.

Finally, for the non-self-adjoint algebras, we will be interested in whether or not the algebra is reflexive and hyper-reflexive. Roughly speaking, a reflexive algebra is one that can be characterized by its invariant subspaces. A subspace M of a Hilbert space $\mathcal{H}$ is *invariant* for an operator $A \in \mathcal{B}(\mathcal{H})$ if $A(M) \subseteq M$. For a subalgebra $\mathcal{A}$ of $\mathcal{B}(\mathcal{H})$, the set of all subspaces that are invariant for all operators in $\mathcal{A}$ forms a lattice, written $\mathrm{Lat}(\mathcal{A})$. The set of all operators in $\mathcal{B}(\mathcal{H})$ for which all subspaces in $\mathrm{Lat}(\mathcal{A})$ are invariant forms an algebra, written $\mathrm{Alg\,Lat}(\mathcal{A})$. It is immediate that $\mathcal{A} \subseteq \mathrm{Alg\,Lat}(\mathcal{A})$. When the opposite containment holds, $\mathcal{A}$ is called *reflexive*. The notion of reflexivity was introduced by Radjavi and Rosenthal in [63], and terminology was suggested by Halmos in [33, 34].

All von Neumann algebras are reflexive algebras (see Theorem 9.17 of [64]). Since all reflexive operator algebras are weakly closed subalgebras containing the identity, it follows that von Neumann algebras are precisely the self-adjoint reflexive operator algebras. Another important class of examples of reflexive algebras are nest algebras, which in finite dimensions are algebras of upper triangular matrices. These were introduced by Ringrose in [71] as an example of reflexive operator algebras. For more examples of reflexive algebras, see [1, 35, 52, 53, 59, 73].

Hyper-reflexivity, defined by Arveson in [2], is a stronger condition than reflexivity and is defined as follows. Let $\mathcal{L} = \mathrm{Lat}(\mathcal{A})$. Then $\mathcal{L}$ determines a seminorm on $\mathcal{B}(\mathcal{H})$ by

$$\beta_{\mathcal{L}}(T) = \sup_{L\in\mathcal{L}} \|P_L^{\perp} T P_L\|,$$

where P_L is the projection onto the subspace L. Then $\beta_{\mathcal{L}}(T) = 0$ if $T \in \mathrm{Alg}(\mathcal{L})$. Thus, $\mathcal{A}$ is reflexive if and only if

$$\mathcal{A} = \{T \in \mathcal{B}(\mathcal{H}) : \beta_{\mathcal{L}}(T) = 0\}.$$

Note that this is equivalent to: for every $T \in \mathcal{B}(\mathcal{H})$, there exists a constant C_T such that $\mathrm{dist}(T, \mathrm{Alg}(\mathcal{L})) \leq C_T \beta_{\mathcal{L}}(T)$. Moreover, we have

$$\beta_{\mathcal{L}}(T) \leq \mathrm{dist}(T, \mathrm{Alg}(\mathcal{L}))$$

for all $T \in \mathcal{B}(\mathcal{H})$. The algebra is said to be *hyper-reflexive* if these norms are comparable, and the constant of reflexivity is the smallest value of C such that

$$\mathrm{dist}(T, \mathrm{Alg}(\mathcal{L})) \leq C\beta_{\mathcal{L}}(T)$$

Clearly, all hyper-reflexive operator algebras are reflexive operator algebras. But converse may not be true, see [21, 42]. Nest algebras are also hyper-reflexive, with distance constant one (see Theorem 1.1 of [2]). For more about nest algebras and hyper-reflexivity, we refer readers to [3, 15, 43]. For more examples of hyper-reflexive algebras, see [10, 14, 32, 41, 72].

3 Graph C^*-Algebras and Free Semigroupoid Algebras

We begin this section by looking at C^*-algebras generated from graphs. The C^*-algebra generated by a Cuntz-Krieger family $\{S, P\}$ is written $C^*(S, P)$. The *graph C^*-algebra* of a graph G, written as $C^*(G)$, is defined as the universal C^*-algebra generated by families of isometries satisfying the Cuntz-Krieger relations. The graph C^*-algebra $C^*(G)$ always exists and is unique up to isomorphism sending generators to generators (see Proposition 1.21 and Corollary 1.22 of [65]). A good introduction to graph C^*-algebras is Raeburn's book [65].

We mention here an important result called the Cuntz-Krieger Uniqueness Theorem that is useful in finding graph C^*-algebras. A *cycle* is a path $\mu = \mu_1\mu_2 \ldots \mu_n$ with $n \geq 1$, $s(\mu_n) = r(\mu_1)$, and $s(\mu_i) \neq s(\mu_j)$ for all $i \neq j$. An edge e is an *entry* to the cycle μ if there exists i such that $r(e) = r(\mu_i)$ and $e \neq \mu_i$.

Theorem 3.1 ([4], Theorem 3.1) *Suppose G is a row-finite directed graph in which every cycle has an entry, and $\{T, Q\}$ is a Cuntz-Krieger G-family in a Hilbert space $\mathcal{H}$ such that $Q_v \neq 0$ for every $v \in G^0$. Then $C^*(T, Q)$ is isomorphic to $C^*(G)$.*

As stated in Remark 2.17 of [65], the above theorem for finite graphs follows as a special case of Theorem 2.13 of [13]. Thus, the above theorem is essentially due to Cuntz and Krieger. The condition 'every cycle has an entry' was introduced by Kumjian, Pask, and Raeburn in [51] and the version stated above was first proved as Theorem 3.1 of [4]. Another important uniqueness result is the Gauge Invariant Uniqueness Theorem (see Theorem 2.1 of [4]).

We now give a few examples of directed graphs and the graph C^*-algebras they generate.

Examples 3.2 ([65], Example 1.23) Let G be the graph with one vertex x and one edge e. A Cuntz-Krieger G-family $\{S, P\}$ on $\mathcal{H}$ must satisfy $S_e^* S_e = P_v = S_e S_e^*$. Thus, any Cuntz-Krieger G-family consists of a unitary and the identity. So $C^*(G)$ is the universal C^*-algebra generated by a unitary, and so is isomorphic to $C(\mathbb{T})$, the complex-valued continuous functions on the unit circle $\mathbb{T}$ in $\mathbb{R}^2$.

Examples 3.3 If G is the graph with one vertex and $n \geq 2$ edges, then any Cuntz-Krieger G-family $\{P, S\}$ satisfies $P = I$ and $\sum_{i=1}^{n} S_i = I$, and so $C^*(G)$ is the Cuntz algebra O_n. For a detailed study of O_n, see [12].

Examples 3.4 Consider the graph G with n vertices $x_1, \ldots, x_n$ and $n - 1$ edges $e_1, \ldots, e_{n-1}$ satisfying $s(e_j) = x_j$ and $r(e_j) = x_{j+1}$:

$$x_1 \xrightarrow{e_1} x_2 \xrightarrow{e_2} \cdots \xrightarrow{e_{n-1}} x_n$$

Let $H = \mathbb{C}^n$ with basis $\{b_1, \ldots, b_n\}$. Let P_{e_j} be the matrix unit E_{jj}, and let S_{e_j} be the matrix unit $E_{j,j+1}$. Then $\{S, P\}$ is a Cuntz-Krieger G-family with $C^*(S, P) = M_n(\mathbb{C})$. Since G has no cycles, the Cuntz-Krieger Uniqueness Theorem applies, giving us that $C^*(G)$ is isomorphic to $M_n(\mathbb{C})$.

Examples 3.5 Consider the graph G with infinitely many vertices $\{x_n\}_{n\in\mathbb{N}}$ and infinitely many edges e_n satisfying $s(e_n) = x_n$ and $r(e_n) = x_{n+1}$:

$$x_1 \xrightarrow{e_1} x_2 \xrightarrow{e_2} \cdots \xrightarrow{e_{n-1}} x_n \xrightarrow{e_n} \cdots$$

Let $\mathcal{H} = \ell^2$ with basis $\{b_n\}_{n\in\mathbb{N}}$. Let P_{x_n} be the projection onto $\mathrm{span}\{b_n\}$ and let S_{e_n} be given by

$$S_{e_n}(b_m) = \begin{cases} b_{m+1} & \text{if } m = n \\ 0 & \text{else} \end{cases}.$$

Then $\{S, P\}$ is a Cuntz-Krieger G-family with $C^*(S, P)$ equal to the compact operators on separable Hilbert space $\mathcal{K}(\mathcal{H})$. Since G has no cycles, the Cuntz-Krieger Uniqueness Theorem applies, giving us that $C^*(G)$ is isomorphic to $\mathcal{K}(\mathcal{H})$.

Note that a similar argument shows that the graph

$$x_1 \xleftarrow[e_1]{} x_2 \xleftarrow[e_2]{} \cdots \xleftarrow[e_{n-1]{} x_n \xleftarrow[e_n]{} \cdots$$

also has graph C^*-algebra isomorphic to $\mathcal{K}(\mathcal{H})$.

Examples 3.6 Consider the graph G with two vertices v_1 and v_2, and two edges, e with $s(e) = r(e) = v_1$ and f with $s(f) = v_1$ and $r(f) = v_2$:

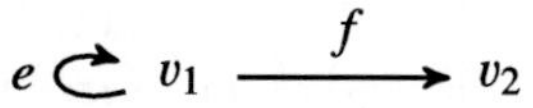

A Cuntz-Krieger G family $\{S, P\}$ must satisfy $S_e^* S_e = P_{v_1} = S_f^* S_f$, $P_{v_1} = S_e S_e^*$, and $P_{v_2} = S_f S_f^*$. Note that $(S_e + S_f)(S_e + S_f)^* = I$, $(S_e + S_f)^*(S_e + S_f) = 2P_{v_1}$ and $I - P_{v_1} = P_{v_2}$. So any Cuntz-Krieger G-family is generated by $S_e + S_f$, which is a non-unitary isometry when $P_{v_2} \neq 0$. Thus, $C^*(G)$ is unitarily equivalent to the Toeplitz algebra.

Examples 3.7 Let C_n be the graph with n vertices $v_1, \ldots, v_n$ and n edges $e_1, \ldots, e_n$ satisfying $r(e_j) = v_{j+1} = s(e_{j+1})$ for $j < n$ and $r(e_n) = v_1 = s(e_1)$. The graph C_4 is pictured below:

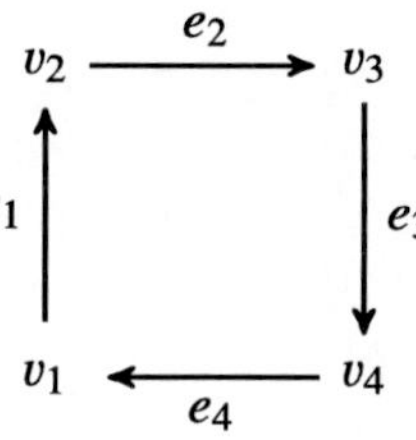

Then $C^*(C_n)$ is isomorphic to $C(\mathbb{T}, M_n(\mathbb{C}))$. For more details, see Lemma 2.4 of [36] and Theorem 2.2 of [27].

In addition to graph C^*-algebras, there are also non-self-adjoint operator algebras associated to graphs. Given a graph G, a free semigroupoid algebra is the unital WOT-closed algebra generated by a CKT G-family. The most common such algebra is the *left regular free semigroupoid algebra*, which is generated by the left regular representation on the *path space* $\mathbb{F}^+(G)$ of G, which is the set of all paths in G. More specifically, the left regular free semigroupoid algebra of a graph G is defined as follows. Let $\mathcal{H}_G$ be a Hilbert space with orthonormal basis $\{\xi_w\}_{w \in \mathbb{F}^+(G)}$, indexed by the path space of G. This is called a *Fock space*. For each $w \in \mathbb{F}^+(G)$, we can define a linear operator $L_w \in \mathcal{B}(\mathcal{H}_G)$ as follows. For $\mu \in \mathbb{F}^+(G)$, let

$$L_w(\xi_\mu) = \begin{cases} \xi_{w\mu} & \text{if } s(w) = r(\mu) \\ 0 & \text{else} \end{cases} .$$

Then L_w is a partial isometry on the Fock space, sometimes called a *partial creation operator*.

Notice that if e and f are edges, then L_e and L_f have orthogonal ranges. Also, for any vertex $x \in G^0$, L_x is a projection:

$$L_x \xi_\nu = \begin{cases} \xi_\nu & \text{if } r(\nu) = x \\ 0 & \text{else} \end{cases} .$$

For a path $\mu = e_1 e_2 \dots e_n$, note that $L_\mu = L_{e_1} L_{e_2} \dots L_{e_n}$. Then $\{L_\mu\}_{\mu \in \mathbb{F}^+(G)}$ is a CKT G-family as defined in Sect. 2.

Definition 3.8 ([46], Definition 3.2) Let $\mathfrak{L}_G$ be the WOT-closed algebra generated by $\{L_w\}_{w \in \mathbb{F}^+(G)}$. This is called the *(left regular) free semigroupoid algebra.*

Remark The norm-closed algebra generated by $\{L_\mu\}_{\mu \in \mathbb{F}^+(G)}$ for a finite graph G is called a *quiver algebra*. For more on quiver algebras, see [39, 55–57].

We now include a few examples of left regular free semigroupoid algebras generated by graphs.

Examples 3.9 ([46], Example 6.1) Consider the graph G with a single vertex x and a single edge e, as in Example 3.2. The Hilbert space $\mathcal{H}_G$ is isomorphic to the Hardy space H^2, and L_e and L_x are isomorphic to the unilateral shift and the identity operator, respectively. Thus, $\mathfrak{L}_G$ is isomorphic to the WOT-closed algebra generated by those two operators, which is H^∞.

Examples 3.10 If G has only a single vertex, then the unital WOT-closed algebra generated by a CKT family is called a *free semigroup algebra*, and can also be said to be generated by a family of n isometries with orthogonal ranges. The left regular free semigroup algebra, also called the non-commutative analytic Toeplitz algebra, is written $\mathfrak{L}_n$. It was introduced by Popescu [60–62] and has been studied extensively by Arias and Popescu [1], and Davidson and Pitts [20].

Free semigroup algebras other than the left regular free semigroup algebra have also been studied. Davidson, Katsoulis, and Pitts prove a structure theorem for all free semigroup algebras in [23]. Read gives an example of a CKT family for which the free semigroup algebra is $\mathcal{B}(\mathcal{H})$ ([68]; see also [17]). Davidson gives other examples in [16]. For more work on free semigroup algebras, see [18, 19, 22, 24, 25, 44].

Examples 3.11 ([46], Example 6.3) Consider the graph G given in Example 3.6. Then $\mathfrak{L}_G$ is generated by L_e, L_f, L_x, and L_y. If we make the identifications $\mathcal{H}_G = L_x \mathcal{H}_G \oplus L_y \mathcal{H}_G \cong H^2 \oplus H^2$, where H^2 is the Hardy space, then

$$L_e \cong \begin{bmatrix} S & 0 \\ 0 & 0 \end{bmatrix}; \; L_f \cong \begin{bmatrix} 0 & 0 \\ S & 0 \end{bmatrix}; \; L_e \cong \begin{bmatrix} I & 0 \\ 0 & 0 \end{bmatrix}; \; L_e \cong \begin{bmatrix} 0 & 0 \\ 0 & I \end{bmatrix},$$

where S is the unilateral shift. Thus, $\mathfrak{L}_G$ is unitarily equivalent to

$$\mathfrak{L}_G \cong \begin{bmatrix} H^\infty & 0 \\ H_0^\infty & \mathbb{C}I \end{bmatrix}.$$

We now consider some properties of free semigroupoid algebras. Analogous to the left regular free semigroupoid algebra, we can use the right regular representation on $\mathbb{F}^+(G)$ to define a right analogue to $\mathfrak{L}_G$. Given $\mu \in \mathbb{F}^+(G)$, we define an operator $R_{\tilde{\mu}}$ by

$$R_{\tilde{\mu}}\xi_\nu = \begin{cases} \xi_{\nu\mu} & \text{if } r(\mu) = s(\nu) \\ 0 & \text{else} \end{cases},$$

where $\tilde{\mu}$ is μ with the edges in the reverse order. Let $\mathfrak{R}_G$ be the WOT-closed algebra generated by $\{R_{\tilde{\mu}}\}_{\mu \in \mathbb{F}^+(G)}$.

If $A \in \mathfrak{L}_G$, then clearly $A \in \mathfrak{R}'_G$, since for any $\mu, \nu, w \in \mathbb{F}^+(G)$,

$$L_\nu R_{\tilde{\mu}}\xi_w = L_\nu \xi_{w\mu} = \xi_{\nu w\mu} = R_{\tilde{\mu}}\xi_{\nu w} = R_\mu L_\nu \xi_x.$$

The equality $\mathfrak{L}_G = \mathfrak{R}'_G$ was shown by Davidson and Pitts for the one vertex case in [20], and by Kribs and Power in the general case in [46]. These authors also show that $\mathfrak{R}_G = \mathfrak{L}'_G$, so it follows that $\mathfrak{L}_G$ is its own double commutant, $\mathfrak{L}_G = \mathfrak{L}''_G$. The proof of these results involves a "Fourier expansion" of A which is worth outlining below.

Let A be in $\mathfrak{L}_G$. For each vertex $x \in G^0$, there are constants $\{a_w\}_{s(w)=x}$ such that

$$A\xi_x = AL_x\xi_x = R_x(AL_x)\xi_x = \sum_{s(w)=x} a_w \xi_w.$$

So for $v = xv \in \mathbb{F}^+(G)$,

$$A\xi_v = R_v A\xi_x = \sum_{s(w)=x} a_w \xi_{wv}.$$

We would like to conclude that A is equal to the (possibly infinite) sum $\sum_{w \in \mathbb{F}^+(G)} a_w L_w$ in some sense. Davidson and Pitts [20] prove that the Cesàro partial sums of $\sum_{w \in \mathbb{F}^+(G)} a_w L_w$ converge SOT to A by relating them to the following Cesàro partial sums for A:

Definition 3.12 For a graph G, let E_i be the projection onto $\operatorname{span}\{\xi_\mu : \mu \in \mathbb{F}^+(G),\ |\mu| = i\}$. The *Cesàro sums* of $A \in \mathcal{B}(\mathcal{H}_G)$ are given by

$$\Sigma_k(A) = \sum_{|j|<k} \left(1 - \frac{|j|}{k}\right) \Phi_j(A),$$

where $\Phi_j(A) = \sum\limits_{\ell \geq \max\{0,-j\}} E_\ell A E_{\ell+j}$ is the ith diagonal of A in the matrix form associated to the partition $I = E_0 + E_1 + E_2 + \dots$.

The fact that these Cesàro sums of A converge in the strong operator topology (SOT) to A is a consequence of the following proposition, which is referenced but not explicitly proven in [20]. We outline a proof here.

Proposition 3.13 *Let $A \in \mathcal{B}(\mathcal{H})$. Let $\{P_j\}$ be a sequence of finite rank projections such that $\sum_{j=1}^{\infty} P_j = I$. The Cesàro sums of A, given by*

$$\Sigma_k(A) = \sum_{|j|<k} \left(1 - \frac{|j|}{k}\right) \Phi_j(A),$$

where $\Phi_j(A) = \sum_{\ell \geq \max\{0,-j\}} P_\ell A P_{\ell+j}$, converge SOT to A.

Proof Define a function $f_h : \mathbb{T} \to \mathcal{H}$ as follows. Let n_j be the dimension of $P_j \mathcal{H} P_j$. For $\lambda \in \mathbb{T}$, let U_λ be the following matrix:

$$U_\lambda = \begin{bmatrix} \lambda^{-1}(I_{n_1}) & 0 & 0 & \dots \\ 0 & \lambda^{-2}(I_{n_2}) & 0 & \dots \\ 0 & 0 & \lambda^{-3}(I_{n_3}) & \dots \\ \vdots & \vdots & \vdots & \ddots \end{bmatrix},$$

where I_{n_j} is the $n_j \times n_j$ identity matrix. Note U_λ is a unitary, and for $A \in \mathcal{B}(\mathcal{H})$:

$$U_\lambda A U_\lambda^{-1} = \begin{bmatrix} 1(A_{1,1}) & \lambda(A_{1,2}) & \lambda^2(A_{1,3}) & \dots \\ \lambda^{-1}(A_{2,1}) & 1(A_{2,2}) & \lambda(A_{2,3}) & \dots \\ \lambda^{-2}(A_{3,1}) & \lambda^{-1}(A_{3,2}) & 1(A_{3,3}) & \dots \\ \vdots & \vdots & \vdots & \ddots \end{bmatrix}.$$

In other words, conjugating A by U_λ^{-1} results in multiplying the (i, j)th block component of A by λ^{j-i}.

Now let $A \in \mathcal{B}(\mathcal{H})$. Fix $h \in \mathcal{H}$. Define the function $f_h : \mathbb{T} \to \mathcal{H}$ by $f_h(\lambda) = U_\lambda A U_{\lambda^{-1}} h$. It can be shown that $f_h : \mathbb{T} \to \mathcal{H}$ is a continuous function, which implies that $g_h : [0, 2\pi] \to \mathcal{H}$ given by $g_h(t) = f_h(e^{it})$ is also continuous. Furthermore,

$$g_h(t) = \sum_{j=-\infty}^{\infty} e^{ijt} \Phi_j(A)h.$$

It can then be shown that the Fourier coefficients of $g_h(t)$ are given by

$$z_n = \frac{1}{2\pi} \int_0^{2\pi} e^{-ins} g_h(s) \, ds = \Phi_n(A)h.$$

This means that the Cesàro partial sums $\sigma_k g_h(t)$ of $g_h(t)$ are given by:

$$\sigma_k g_h(t) = \sum_{|j|<k} e^{ijt}\left(\frac{k-|j|}{k}\right)\Phi_j(A)h = U_{e^{it}}\Sigma_k(A)U_{e^{it}}^{-1}h.$$

Here we can apply a vector-valued version of Fejér's theorem (stated as Theorem 1.1 in [11], for example) to conclude that for $t \in (0, 2\pi)$, $\|\sigma_k g_h(t) - g_h(t)\| \to 0$ as $k \to \infty$. That is, $\|U_{e^{it}}\Sigma_k(A)U_{e^{-it}}h - U_{e^{it}}AU_{e^{-it}}h\|$ converges to 0. This holds for all $h \in \mathcal{H}$. So for $t \in (0, 2\pi)$, we have that $U_{e^{it}}\Sigma_k(A)U_{e^{-it}}$ converges SOT to $U_{e^{it}}AU_{e^{-it}}$. Since SOT convergence is preserved by unitary equivalence, this means that $\Sigma_k(A)$ converges SOT to A as desired. □

It is a well-known fact that all C^*-algebras are semisimple (see II.1.6.4 of [6]) and therefore all graph C^*-algebras are semisimple. Davidson and Pitts showed that $\mathfrak{L}_n$ contains no non-trivial idempotents or non-zero quasinilpotent elements, implying that $\mathfrak{L}_n$ is also semisimple ([20], Corollary 1.9; see Remark 1.10 in [20] for a possible earlier reference). For an arbitrary graph G, Kribs and Power showed that $\mathfrak{L}_G$ is semisimple if and only if every path in $\mathbb{F}^+(G)$ lies on a cycle ([46], Theorem 5.1).

In Theorem 4.1 of [1], Arias and Popescu prove that $\mathfrak{L}_n$ is a reflexive algebra. Kribs and Power showed that for any graph G, $\mathfrak{L}_G$ is reflexive [46]. For the more generalized free semigroupoid algebras beyond that induced by the left regular representation, Kennedy showed that any free semigroup algebra is reflexive [40] and Davidson, Dor-on, and Li showed that any free semigroupoid algebra is reflexive [26].

Concerning the stronger condition of hyper-reflexivity, Davidson [14] showed that $\mathfrak{L}_1$ is hyper-reflexive with constant at most 19. In [20], Davidson and Pitts go further and prove that $\mathfrak{L}_n$ is hyper-reflexive for $n \geq 2$, with constant at most 51. Davidson notes in [16] that this bound can be lowered to 3 by applying Bercovici's hyper-reflexivity theorem (Theorem 3.1, [5]). Kennedy showed that a certain class of free semigroup algebras is hyper-reflexive [40], and Jaëck and Power showed that if G is any finite graph, then $\mathfrak{L}_G$ is hyper-reflexive [37]. Fuller and Kennedy strengthened these results when they proved in Theorem 1.1 of [32] that all free semigroup algebras are hyper-reflexive.

Kribs and Power also showed that unitarily equivalent free semigroupoid algebras have isomorphic directed graphs (Theorem 9.1, [46]). This interesting fact is not true for graph C^*-algebras; see Example 3.5. For more discussion of isomorphic graphs and associated algebras, see [39]. For more on the structure of free semigroupoid algebras, see [38]. For a study of weighted versions of free semigroupoid algebras, see [45] and [48].

4 Operator Algebras from Higher Rank Graphs

Higher rank graphs were introduced by Kumjian and Pask in [49], and are defined using the language of categories.

Definition 4.1 A *k-graph* is a countable category Λ with range and source maps r and s respectively, and a functor $d : \Lambda \to \mathbb{N}^k$ satisfying the following factorization property: for every $\lambda \in \Lambda$ and $m, n \in \mathbb{N}^k$ with $d(\lambda) = m + n$, there are unique elements $\mu, \nu \in \Lambda$ such that $\lambda = \mu\nu$ and $d(\mu) = m$ and $d(\nu) = n$.

For a path with degree $n = (n_1, n_2, \dots, n_k) \in \mathbb{N}^k$, we say its *length* is $|n| = n_1 + n_2 + \cdots + n_k$. Let Λ^0 be the set of paths with length 0 (the vertices of Λ) and let Λ^1 be the set of paths of length 1.

Examples 4.2 Any directed graph (G^0, G^1, r, s) gives a category where Λ is the collection of all paths and r, s are the range and source maps. This is a 1-graph, satisfying the factorization property trivially since each path can be written as a concatenation of edges in one unique way. In fact, any 1-graph Λ comes from a directed graph.

As with graphs, we represent higher rank graphs in diagrams by using a dot or label for each element of Λ^0, and using arrows between the dots to represent elements of Λ^1. We represent the degree of each edge with a color. In the diagrams that follow, unbroken, dotted, and dashed arrows will represent three different colors.

Examples 4.3 Consider the following 2-graph, where the edges b_1 and b_2 have degree $(1, 0)$, the edges r_1 and r_2 have degree $(0, 1)$. In order for this graph to satisfy the factorization property, we must have the identification $b_2 r_1 = r_2 b_1$.

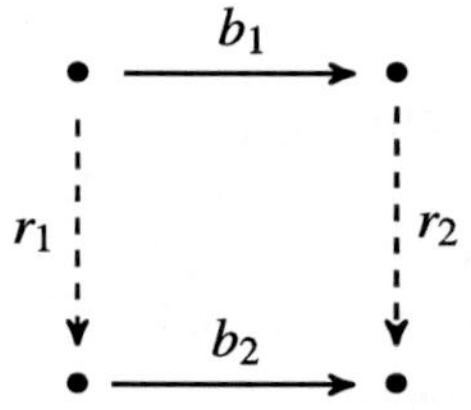

Such a diagram which indicates the vertices, edges, and degrees of the edges is called a 1-skeleton, and does not necessarily determine the k-graph, as shown in the next example.

Examples 4.4 ([65], Example 10.5) The following 1-skeleton corresponds to two different 2-graphs:

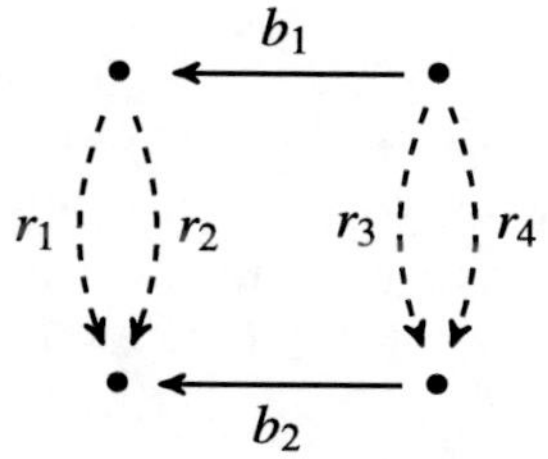

The path r_1b_1 must be identified with either b_2r_3 or b_2r_4 in order for the factorization property to be satisfied. If $r_1b_1 = b_2r_3$, then it must be true that $r_2b_1 = b_2r_4$. Likewise, if $r_1b_1 = b_2r_4$, then $r_2b_1 = b_2r_3$.

Examples 4.5 Consider the 3-graph with one edge e of degree $(1, 0, 0)$, one edge f of degree $(0, 1, 0)$ and one edge g of degree $(0, 0, 1)$.

The factorization property implies that for each $(n, m, k) \in \mathbb{N}^3$, there is one unique path of degree (n, m, k). So, for example, $ffgfe = egf^3$.

Next, we mention some definitions and results for higher rank graph C^*-algebras analogous to graph C^*-algebras.

Definition 4.6 For a k-graph Λ with degree map d, we define $d^{-1}(n)$ to be paths of degree n for all $n \in \mathbb{N}^k$. We say Λ is *row-finite* if $r^{-1}(v) \cap d^{-1}(n)$ is finite for all $v \in \Lambda^0$ and $n \in \mathbb{N}^k$.

In Definition 2.1 of a Cuntz-Krieger G-family for a row-finite directed graph G, we saw that the orthogonal projection P_v was the sum of $S_eS_e^*$ for all $e \in G^1$ with $r(e) = v$, for each $v \in G^0$ that is not a source. So to define a Cuntz-Krieger family for higher rank graphs, we first make this as a basic assumption that motivates the following definition.

Definition 4.7 A k-graph Λ is said to *have no source* if for every $v \in \Lambda^0$ and $n \in \mathbb{N}^k$, there exists $\mu \in \Lambda$ such that $d(\mu) = n$ and $r(\mu) = v$.

Definition 4.8 Let Λ be a k-graph that is row-finite and has no source. A *Cuntz-Krieger Λ-family* in $\mathcal{H}$ is a collection of partial isometries $S = \{S_\mu : \mu \in \Lambda\}$ satisfying the following:

(a) $\{S_v : v \in \Lambda^0\}$ are mutually orthogonal projections,
(b) $S_{\mu_1}S_{\mu_2} = S_{\mu_1\mu_2}$ when $s(\mu_1) = r(\mu_2)$,

(c) $S_\mu^* S_\mu = S_{s(\mu)}$ for all $\mu \in \Lambda$,

(d) $S_v = \sum_{\mu\in\Lambda, d(\mu)=n, r(\mu)=v} S_\mu S_\mu^*$ for all $v \in \Lambda^0$ and $n \in \mathbb{N}^k$.

Let Λ be a k-graph that is row-finite and has no sources. Let S be a Cuntz-Krieger Λ-family. Then the partial isometries associated to paths of the same degree have orthogonal ranges; that is, $\{S_\mu S_\mu^* : d(\mu) = n\}$ is a mutually orthogonal family of projections for each $n \in \mathbb{N}^k$. It is natural to ask if there is a universal C^*-algebra $C^*(\Lambda)$ generated by a Cuntz-Krieger Λ-family with the property that if $\{S_\mu : \mu \in \Lambda\}$ is any Cuntz-Krieger Λ-family, then there is a $*$-homomorphism from $C^*(\Lambda)$ to $C^*\left(\{S_\mu : \mu \in \Lambda\}\right)$.

For any k-graph Λ, $C^*(\Lambda)$ always exists (see Proposition 10.9 of [65]) and every partial isometry in the generating family of the universal Cuntz-Krieger Λ-family is non-zero (see Proposition 2.11 of [49] or Corollary 10.13 of [65]). An appropriate analogue for the Gauge Invariant Uniqueness Theorem for k-graphs, can be found in Theorem 3.4 of [49]. The most important application of the Gauge Invariant Uniqueness Theorem for k-graphs is that for a k-graph Λ, there is an isomorphism between $C^*(\Lambda)$ and the C^*-algebra of the locally compact groupoid with unit space equal to the infinite path space of Λ. For a detailed study of groupoids and C^*-algebras generated by them, see [69]. For the details of the proof of isomorphism in the case of directed graphs, see [50], and for k-graphs, see Corollary 3.5(i) of [49].

It would be interesting to find appropriate version(s) of the Cuntz-Krieger Uniqueness Theorem for k-graphs. For directed graphs, in the statement of the Cuntz-Krieger Uniqueness Theorem, we had the condition that every cycle has an entry. We have a version of the Cuntz-Krieger Uniqueness Theorem for k-graphs under an *aperiodicity condition*; see Section 4 of [49].

Below we give a few examples of C^*-algebras generated by higher rank graphs.

Examples 4.9 If Λ is a 1-graph associated to a row-finite directed graph G with no sources, then by restricting the Cuntz-Krieger Λ-family to G^0 and G^1, we obtain a Cuntz-Krieger G-family in the sense of Definition 2.1. Thus for a directed graph G, we have $C^*(\Lambda) = C^*(G)$.

Examples 4.10 Let T be the semigroup $\mathbb{N}^k$ viewed as a small category and $d : T \to \mathbb{N}^k$ be the identity map. Then (T, d) is a k-graph. We have $C^*(T) = C(\mathbb{T}^k)$, where $\mathbb{T}^k$ denotes the k-torus (see Example 1.7(iii) of [49]).

Examples 4.11 For $k \geq 1$, let Ω_k be the small category with $\Omega_k^0 = \mathbb{N}^k$ and

$$\Omega_j^* = \{(p, q) \in \mathbb{N}^k \times \mathbb{N}^k : p \leq q \text{ if and only if } p_i \leq q_i \text{ for all } 1 \leq i \leq k\}.$$

Let the source and range maps be given by $r(m, n) = m$ and $s(m, n) = n$ with $d : \Omega_k \to \mathbb{N}^k$ defined as $d(m, n) = n - m$. Then Ω_k is a k-graph with $C^*(\Omega_k) = \mathcal{K}(l^2(\mathbb{N}^k))$ (see Example 1.7(ii) of [49]).

Below we give the 1-skeleton for the infinite graph Ω_2.

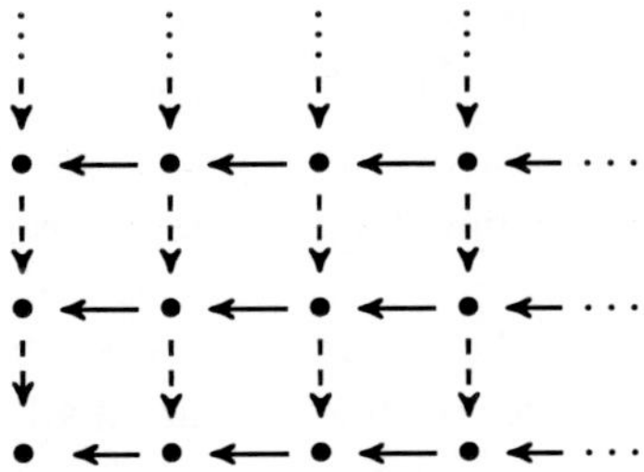

We also mention that 2-graphs have been of special interest; see [58].

Before moving to free semigroupoid algebras for higher rank graphs, we note that row-finiteness and having no sources have been important conditions on a k-graph Λ to define $C^*(\Lambda)$. Authors have tried to relax these conditions. The procedure for dealing with 1-graphs (or directed graphs) with sources can be done using Lemma 2.1 of [4]. However, when $k > 1$, there are many different kinds of sources, and there is as yet no analogous procedure for dealing with them. For some particular classes of row-finite k-graphs with sources with an extra local convexity condition, see [66].

But the question remains : Can we analyze k-graphs that are not row-finite? If a vertex emits infinitely many edges, then the Cuntz-Krieger relation

$$S_v = \sum_{\mu \in \Lambda, d(\mu)=n, r(\mu)=v} S_\mu S_\mu^* \tag{4.1}$$

does not make sense. Fowler and Raeburn noticed that even if vertices emit infinitely many edges in a 1-graph (or directed graph), one can work with graph algebras if we merely insist that equality in (4.1) occurs when v emits finitely many edges; see Corollaries 4.2 and 4.5 of [30]. For a more detailed study of this, see [31]. For an alternative approach to studying the graph C^*-algebra of a directed graph, see [74].

In [67], the theory of Cuntz-Krieger families and graph algebras has been generalized to the class of finitely-aligned k-graphs. This class contains in particular all row-finite k-graphs. A Gauge-Invariant Uniqueness Theorem and a Cuntz-Krieger Uniqueness Theorem have also been proved for finitely-aligned k-graphs in [67]. For the study of general k-graphs, the work of Spielberg in the more general setting of categories of paths, which we discuss in Sect. 5, is useful.

Kribs and Power extended the free semigroupoid algebra construction to higher rank graphs in [47]. For a higher rank graph Λ, define a Fock space Hilbert space $\mathcal{H}_\Lambda$ with orthonormal basis $\{\xi_\mu\}_{\mu \in \Lambda}$, indexed by the elements of Λ. We can then define linear operators $L_\mu \in \mathcal{B}(\mathcal{H}_\Lambda)$ as follows. For $\nu \in \Lambda$, define:

$$L_\mu \xi_\nu = \begin{cases} \xi_{\mu\nu} & \text{if } s(\mu) = r(\nu) \\ 0 & \text{else} \end{cases}$$

As before, if x is a vertex of Λ, then L_x is a projection.

Definition 4.12 ([47], Definition 3.1) The WOT-closed algebra generated by $\{L_\mu\}_{\mu\in\Lambda}$ is called the *(left regular) free semigroupoid algebra* for Λ and is written $\mathfrak{L}_\Lambda$.

The right regular free semigroupoid algebra $\mathfrak{R}_\Lambda$ can be defined analogously as in Sect. 3. Kribs and Power showed that, as with free semigroupoid algebras from graphs, the commutant of $\mathfrak{L}_\Lambda$ is $\mathfrak{R}_\Lambda$, and $\mathfrak{L}_\Lambda$ is its own double commutant [47].

We now include a few examples of free semigroupoid algebras from higher rank graphs.

Examples 4.13 The 2-graph in Example 4.3 has nine elements in its path space: four vertices, two edges with degree $(1, 0)$, two edges with degree $(0, 1)$, and a single edge of degree $(1, 1)$. So $\mathcal{H}_\Lambda$ is 9-dimensional, and $\mathfrak{L}_\Lambda$ can be identified with a matrix algebra on this Hilbert space.

Examples 4.14 ([47], Example 4.1) Let Λ be the 3-graph with one vertex and three loops e, f, and g, as in Example 4.5. In this higher rank graph, a path μ is uniquely determined by the number of times e, f, and g appear in any decomposition of μ into edges. The Hilbert space $\mathcal{H}_\Lambda$ can be identified with $\mathcal{H}_G \otimes \mathcal{H}_G \otimes \mathcal{H}_G$, where G is the graph with one vertex and one edge. Then $\mathfrak{L}_\Lambda$ is unitarily equivalent to $H^\infty \otimes H^\infty \otimes H^\infty$.

Examples 4.15 ([47], Example 4.3) Define $C_n^{(k)}$ to be the k-graph with n vertices $v_1, \ldots, v_n$ such that for each $i \in \{1, \ldots, n-1\}$, there is one edge of each color with source v_i and range v_{i+1}, and in addition, there is one edge of each color with source v_n and range v_1. This is called a higher rank cyclic graph. The 1-skeleton for $C_3^{(2)}$ is:

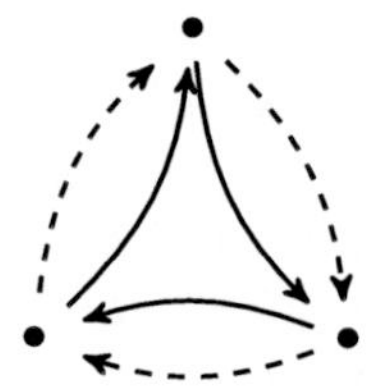

Let $H^2(z, w)$ be the Hardy space for the torus $\mathbb{T}^2 = \{(z, w) : |z| = |w| = 1\}$, with its basis $\{z^p w^p : p, q \in \mathbb{N}\}$. Kribs and Power show that $\mathfrak{L}_{C_3^{(2)}}$ is unitarily equivalent to the matrix function algebra

$$\begin{bmatrix} H^\infty_{(3,0)}(z, w) & H^\infty_{(3,2)}(z, w) & H^\infty_{(3,1)}(z, w) \\ H^\infty_{(3,1)}(z, w) & H^\infty_{(3,0)}(z, w) & H^\infty_{(3,2)}(z, w) \\ H^\infty_{(3,2)}(z, w) & H^\infty_{(3,1)}(z, w) & H^\infty_{(3,0)}(z, w) \end{bmatrix},$$

where $H^\infty_{3,i}(z, w)$ is the closed span of the basis elements $\{z^p w^q : p+q \equiv i \bmod 3\}$ for $i = 1, 2$.

Kribs and Power proved that, as with free semigroupoid algebras from graphs, a higher rank free semigroupoid algebra is isomorphic to its bicommutant ([47], Corollary 3.5). Also analogous to the graph case, $\mathfrak{L}_\Lambda$ is semisimple if and only if every edge in Λ lines in a cycle (Theorem 7.2, [47]). In Corollary 6.3 of the same paper [47], Kribs and Power apply Bercovici's hyper-reflexivity theorem to show that every single-vertex higher rank free semigroupoid algebra is reflexive. They then use this to prove reflexivity for a large class of free semigroupoid algebras. Specifically:

Theorem 4.16 ([47], Theorem 6.4) *Let Λ be a higher rank graph such that no vertex v satisfies all three of the following properties:*

(a) for each $\lambda \in \Lambda$ with degree one, $r(v) = \lambda$ implies $s(v) = \lambda$;
(b) there is at most one loop edge of each color at v; and
(c) there are loop edges $\mu \neq \mu'$ and paths $\lambda, \lambda' \in \Lambda$ with $s(\lambda) = v = s(\lambda')$ that immediately leave v such that $\lambda\mu = \lambda'\mu'$.

Then $\mathfrak{L}_\Lambda$ is reflexive.

In the next section, we will move on to a further generalization of graphs, and see the corresponding constructions and properties in that setting.

5 Operator Algebras from Categories of Paths

Categories of paths were introduced by Spielberg in [75] as a generalization of graphs and higher rank graphs.

Definition 5.1 ([75], Definition 2.1) A small category Λ with source map s and range map r is called a *category of paths* if, for $\alpha, \beta, \gamma \in \Lambda$,

(a) $\alpha\beta = \alpha\gamma$ implies $\beta = \gamma$ (left cancellation)
(b) $\beta\alpha = \gamma\alpha$ implies $\beta = \gamma$ (right cancellation)
(c) $\alpha\beta = s(\beta)$ implies $\alpha = \beta = s(\beta)$ (no inverses)

Examples 5.2 Graphs and higher rank graphs are examples of categories of paths. Note that the factorization property implies that left and right cancellation hold in a higher rank graph.

Examples 5.3 The p-graphs introduced by Brownlowe, Sims, and Vittadello in [7] are an example of a category of paths. A p-graph is a generalization of a directed graph where paths have a degree in a semigroup P rather than a length in $\mathbb{N}$. See [7] for more details.

Examples 5.4 Consider the graph $x_1 \underset{b_1}{\overset{a_1}{\rightrightarrows}} x_2 \underset{b_2}{\overset{a_2}{\rightrightarrows}} x_3$ along with the identifications $b_2a_1 = a_2b_1$ and $a_2a_1 = b_2b_1$. This forms a category of paths with three vertices (x_1, x_2, x_3) and six non-vertex paths $(a_1, b_1, a_2, b_2, a_2a_1, a_2b_1)$. This is not a higher rank graph, since there is no way to assign degrees to the edges in such a way that the factorization property would be satisfied. For more on this example, see Example 2.9 of [54].

Examples 5.5 Consider the category of paths defined by the graph

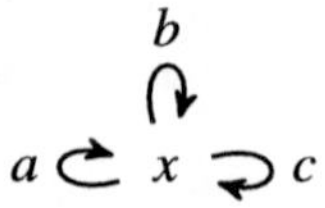

along with the identifications $ab = bc = ca$, $ac = cb = ba$, and $a^2 = b^2 = c^2$. For this category of paths, we can define the length of a path in the normal sense. The identifications imply that for every $k \geq 1$, there are exactly three paths of length k. For $k \geq 2$, they can be written in the forms: a^k, ba^{k-1}, ca^{k-1}.

Our construction of $C^*(\Lambda)$ for a k-graph or directed graph Λ was heavily based on the degree functor (that was used in the definition of a Cuntz-Krieger Λ-family). Indeed, for a category of paths, many different degrees can exist, giving different decompositions of the algebra (see Section 9 of [75]). Exel [29] proposed an idea that a degree functor is not at all necessary for defining the C^*-algebra.

For a semigroupoid $(\Lambda, \Lambda^{(2)}, \cdot)$ (see Definition 2.1 of [29]), a representation f of Λ in a unital C^*-algebra B was defined in Definition 4.1 of [29]. Using the notion of tight representations (see Definition 4.5 of [29]), the universal representation $\tilde{O}(\Lambda)$ of a semigroupoid Λ was defined in Definition 4.6 of [29]. The most interesting thing is that any small category is a sort of semigroupoid in several ways, and it turns out that the universal representation $\tilde{O}(\Lambda)$ of a k-graph, viewed as a semigroupoid, is isomorphic to $C^*(\Lambda)$ as defined in last section (see Theorem 8.7 of [29]). We saw in Sect. 4 that for a k-graph Λ, as an application of the Gauge Invariant Uniqueness Theorem for k-graphs, there is isomorphism between $C^*(\Lambda)$ and the C^*-algebra of a locally compact groupoid with unit space equal to the infinite path space of Λ. Exel [28] also observed that any semigroupoid C^*-algebra can be described as a groupoid C^*-algebra.

The work of Exel motivated Spielberg to define the C^*-algebra of a category of paths. The idea of Spielberg's construction was based on an analogy from Cuntz and Krieger's works: taking ultrafilters on a Boolean ring generated by a set of paths extending a particular path and defining the C^*-algebras via a groupoid (this produces the usual Cuntz-Krieger algebra). Since this process just uses the operations of concatenation and cancellation, this construction can be translated to categories of paths. Spielberg not only found a generalization of $C^*(\Lambda)$ for a k-graph Λ to categories of paths, he also found generalization of gauge actions and aperiodicity for categories of paths in Sections 9 and 10 of [75]. We will refer readers to [75] for the technical details and we instead discuss here the extension of free semigroupoid algebras to categories of paths and our proposal to extend results of free semigroupoid algebras associated with k-graphs to those of categories of paths. The following material was done in PhD work of the first author and can be accessed from [9] and is under preparation in paper form in [8].

The same construction used for the left regular free semigroupoid algebra from a graph can be easily applied to a category of paths. For a category of paths Λ, define a Hilbert space $\mathcal{H}_\Lambda$ with basis $\{\xi_\mu\}_{\mu\in\Lambda}$ indexed by the path space of the category of paths, and for each $\mu \in \Lambda$, define partial isometries L_μ by

$$L_\mu(\xi_\nu) = \begin{cases} \xi_{\mu\nu} & \text{if } s(\mu) = r(\nu) \\ 0 & \text{else} \end{cases},$$

for each $\nu \in \Lambda$.

Definition 5.6 The *left regular free semigroupoid algebra* $\mathfrak{L}_\Lambda$ is the WOT-closed algebra generated by the operators $\{L_\mu\}_{\mu\in\Lambda}$.

Examples 5.7 ([9], Lemma 3.4.4) Let Λ be the category of paths from Example 5.5. As mentioned, this category of paths has exactly 3 paths of length k for every $k \geq 1$. For $k \geq 2$, they can be written in the forms: a^k, ba^{k-1}, ca^{k-1}. Consider the Hilbert spaces $\{H_k\}_{k\geq 0}$ where $H_0 = \{x\}$, $H_1 = \{a, b, c\}$, and $H_k = \{a^k, ba^{k-1}, ca^{k-1}\}$ for $k \geq 2$. Then $I = \sum_{k=0}^{\infty} P_k$, where P_k is the projection onto H_k.

In this matrix decomposition, L_a, L_b, and L_c are represented by

$$L_a = \begin{bmatrix} 0 & 0 & 0 & \cdots \\ A_1 & 0 & 0 & \cdots \\ 0 & A & 0 & \cdots \\ 0 & 0 & A & \cdots \\ \vdots & \vdots & \vdots & \ddots \end{bmatrix}, \quad L_b = \begin{bmatrix} 0 & 0 & 0 & \cdots \\ B_1 & 0 & 0 & \cdots \\ 0 & B & 0 & \cdots \\ 0 & 0 & B & \cdots \\ \vdots & \vdots & \vdots & \ddots \end{bmatrix}, \quad L_c = \begin{bmatrix} 0 & 0 & 0 & \cdots \\ C_1 & 0 & 0 & \cdots \\ 0 & C & 0 & \cdots \\ 0 & 0 & C & \cdots \\ \vdots & \vdots & \vdots & \ddots \end{bmatrix}$$

where $A_1 = \begin{bmatrix} 1 \\ 0 \\ 0 \end{bmatrix}$, $B_1 = \begin{bmatrix} 0 \\ 1 \\ 0 \end{bmatrix}$, $C_1 = \begin{bmatrix} 0 \\ 0 \\ 1 \end{bmatrix}$, and A, B, and C are defined as

$$A = \begin{bmatrix} 1 & 0 & 0 \\ 0 & 0 & 1 \\ 0 & 1 & 0 \end{bmatrix}, \quad B = \begin{bmatrix} 0 & 1 & 0 \\ 1 & 0 & 0 \\ 0 & 0 & 1 \end{bmatrix}, \quad C = \begin{bmatrix} 0 & 0 & 1 \\ 0 & 1 & 0 \\ 1 & 0 & 0 \end{bmatrix}.$$

Notice that this free semigroupoid algebra has a non-zero nilpotent element $T = L_a + \omega L_b + \omega^2 L_c$, where ω is a primitive third root of unity. This differs from the higher rank graph case, since left regular free semigroupoid algebras from single-vertex higher rank graphs have no non-zero nilpotents ([9], Proposition 3.4.2).

To study semisimplicity in the category of paths case, it will be helpful to employ the definition of a degree functor from [75] (restricted here to functions mapping into $\mathbb{N}^k$):

Definition 5.8 A *degree functor* on Λ is a function φ from Λ to $\mathbb{N}^k$ such that for all $\mu, \nu \in \Lambda$ satisfying $s(\mu) = r(\nu)$:

$$\varphi(\mu\nu) = \varphi(\mu) + \varphi(\nu).$$

We say the degree functor is *non-degenerate* if $\varphi(\alpha) \neq 0$ when $\alpha \notin \Lambda^0$. Let $|\mu| = |\varphi(\mu)|$.

Moreover, we propose the following definitions:

Definition 5.9 Let Λ be a category of paths with a degree functor. A path $\mu \in \Lambda$ is called *minimal* if for $\nu, \eta \in \Lambda$, $\mu = \nu\eta$ implies $\mu = \nu$ or $\mu = \eta$.

Definition 5.10 We say that a category of paths Λ *satisfies property (P)* if:

(a) For each vertex $v \in \Lambda^0$, the set of minimal paths in $v\Lambda$ is finite; and
(b) If $A \neq 0$ and $A = a_1 L_{w_1} + a_2 L_{w_2} + \cdots + a_k L_{w_k}$ where $|w_1| = |w_2| = \cdots = |w_k|$, then there is some $\mu \in \Lambda$ such that $L_\mu A$ is not nilpotent.

This condition has two parts: the first is similar to row-finiteness in a graph; the second is a restriction on which elements of the algebra can be nilpotent, which is similar to, but more general than, the requirement that all paths lie on a cycle. The following can be shown:

Theorem 5.11 ([9], Theorem 4.1.9) *Let Λ be a category of paths with a non-degenerate degree functor. Then:*

(a) If Λ satisfies (P), then $\mathfrak{L}_\Lambda$ is semisimple.
(b) If $\mathfrak{L}_\Lambda$ is semisimple, then each path in Λ lies on a cycle.

An example of a category of paths that satisfies (P) is Example 5.7 ([9], Corollary 4.2.4). Therefore, $\mathfrak{L}_\Lambda$ is semisimple for this example.

As for reflexivity, the proof of Theorem 6.4 from [47] also proves the following proposition. We say v is a *radiating vertex* if for all $\lambda \in \Lambda$, $r(\lambda) = v$ implies $s(\lambda) = v$.

Proposition 5.12 *Suppose that Λ is a category of paths with a non-degenerate degree functor such that each radiating vertex v satisfies*

(a) for the category of paths Λ' consisting of v and all paths $\mu \in \Lambda$ with $r(\mu) = s(\mu) = v$, we have $\mathfrak{L}_{\Lambda'}$ is reflexive; and
(b) if μ_1 and μ_2 are loops at v and w_1 and w_2 are paths with source v, then $w_1\mu_1 \neq w_2\mu_2$.

Then $\mathfrak{L}_\Lambda$ is reflexive.

This suggests that the place to start with reflexivity is to determine which single-vertex categories of paths have reflexive free semigroupoid algebras. Then we can use that information to analyze the reflexivity of multiple-vertex categories of paths.

Acknowledgments We acknowledge very helpful comments by the editors and referee that led to significant changes in the first draft of this article.

References

1. A. Arias, G. Popescu, Factorization and reflexivity on Fock spaces. Integr. Equ. Oper. Theory **23**(3), 268–286 (1995). https://doi.org/10.1007/BF01198485
2. W. Arveson, Interpolation problems in nest algebras. J. Funct. Anal. **20**(3), 208–233 (1975). https://doi.org/10.1016/0022-1236(75)90041-5
3. W. Arveson, Ten lectures on operator algebras, in *CBMS Regional Conference Series in Mathematics*, vol. 55. Published for the Conference Board of the Mathematical Sciences, Washington, DC; by the American Mathematical Society, Providence, RI (1984). https://doi.org/10.1090/cbms/055
4. T. Bates, D. Pask, I. Raeburn, W. Szymański, The C^*-algebras of row-finite graphs. New York J. Math. **6**, 307–324 (2000). http://nyjm.albany.edu:8000/j/2000/6_307.html
5. H. Bercovici, Hyper-reflexivity and the factorization of linear functionals. J. Funct. Anal. **158**(1), 242–252 (1998). https://doi.org/10.1006/jfan.1998.3288
6. B. Blackadar, Operator algebras, in *Encyclopaedia of Mathematical Sciences*, vol. 122 (Springer, Berlin, 2006). Theory of C^*-algebras and von Neumann algebras, Operator Algebras and Non-commutative Geometry, III. https://doi.org/10.1007/3-540-28517-2.
7. N. Brownlowe, A. Sims, S.T. Vittadello, Co-universal C^*-algebras associated to generalised graphs. Isr. J. Math. **193**(1), 399–440 (2013). https://doi.org/10.1007/s11856-012-0106-0
8. J. Bukoski, Free Semigroupoid Algebras from Categories of Paths. Submitted
9. J. Bukoski, Free Semigroupoid Algebras from Categories of Paths. Ph.D. Thesis, The University of Nebraska - Lincoln, ProQuest LLC, Ann Arbor, MI, 2021. http://gateway.proquest.com/openurl?url_ver=Z39.88-2004&rft_val_fmt=info:ofi/fmt:kev:mtx:dissertation&res_dat=xri:pqm&rft_dat=xri:pqdiss:28489973
10. E. Christensen, Extension of derivations. J. Funct. Anal. **27**(2), 234–247 (1978). https://doi.org/10.1016/0022-1236(78)90029-0
11. I. Ciorănescu, C. Lizama, Some applications of Fejér's theorem to operator cosine functions in Banach spaces. Proc. Am. Math. Soc. **125**(8), 2353–2362 (1997). https://doi.org/10.1090/S0002-9939-97-03837-9
12. J. Cuntz, Simple C^*-algebras generated by isometries. Commun. Math. Phys. **57**(2), 173–185 (1977). http://projecteuclid.org/euclid.cmp/1103901288
13. J. Cuntz, W. Krieger, A class of C^*-algebras and topological Markov chains. Invent. Math. **56**(3), 251–268 (1980). https://doi.org/10.1007/BF01390048
14. K.R. Davidson, The distance to the analytic Toeplitz operators. Ill. J. Math. **31**(2), 265–273 (1987). http://projecteuclid.org/euclid.ijm/1255989320
15. K.R. Davidson, Nest algebras, *Pitman Research Notes in Mathematics Series*, vol. 191 (Longman Scientific & Technical, Harlow, 1988). Copublished in the United States with John Wiley & Sons, Inc., New York (1988). Triangular forms for operator algebras on Hilbert space
16. K.R. Davidson, Free semigroup algebras: a survey, in *Systems, Approximation, Singular Integral Operators, and Related Topics (Bordeaux, 2000)*. Operator Theory: Advances and Applications, vol. 129 (Birkhäuser, Basel, 2001), pp. 209–240
17. K.R. Davidson, $\mathcal{B}(\mathcal{H})$ is a free semigroup algebra. Proc. Am. Math. Soc. **134**(6), 1753–1757 (2006). https://doi.org/10.1090/S0002-9939-05-08147-5
18. K.R. Davidson, D.R. Pitts, The algebraic structure of non-commutative analytic Toeplitz algebras. Math. Ann. **311**(2), 275–303 (1998). https://doi.org/10.1007/s002080050188
19. K.R. Davidson, D.R. Pitts, Nevanlinna-Pick interpolation for non-commutative analytic Toeplitz algebras. Integr. Equ. Oper. Theory **31**(3), 321–337 (1998). https://doi.org/10.1007/BF01195123

20. K.R. Davidson, D.R. Pitts, Invariant subspaces and hyper-reflexivity for free semigroup algebras. Proc. Lond. Math. Soc. (3) **78**(2), 401–430 (1999). https://doi.org/10.1112/S002461159900180X
21. K.R. Davidson, S.C. Power, Failure of the distance formula. J. Lond. Math. Soc. (2) **32**(1), 157–165 (1985). https://doi.org/10.1112/jlms/s2-32.1.157
22. K.R. Davidson, A. Wright, Operator algebras with unique preduals. Can. Math. Bull. **54**(3), 411–421 (2011). https://doi.org/10.4153/CMB-2011-036-0
23. K.R. Davidson, E. Katsoulis, D.R. Pitts, The structure of free semigroup algebras. J. Reine Angew. Math. **533**, 99–125 (2001). https://doi.org/10.1515/crll.2001.028
24. K.R. Davidson, D.W. Kribs, M.E. Shpigel, Isometric dilations of non-commuting finite rank n-tuples. Can. J. Math. **53**(3), 506–545 (2001). https://doi.org/10.4153/CJM-2001-022-0
25. K.R. Davidson, J. Li, D.R. Pitts, Absolutely continuous representations and a Kaplansky density theorem for free semigroup algebras. J. Funct. Anal. **224**(1), 160–191 (2005). https://doi.org/10.1016/j.jfa.2004.08.005
26. K.R. Davidson, A. Dor-On, B. Li, Structure of free semigroupoid algebras. J. Funct. Anal. **277**(9), 3283–3350 (2019). https://doi.org/10.1016/j.jfa.2019.06.002
27. D.E. Evans, Gauge actions on $\mathcal{O}_A$. J. Oper. Theory **7**(1), 79–100 (1982)
28. R. Exel, Inverse semigroups and combinatorial C^*-algebras. Bull. Braz. Math. Soc. (N.S.) **39**(2), 191–313 (2008). https://doi.org/10.1007/s00574-008-0080-7
29. R. Exel, Semigroupoid C^*-algebras. J. Math. Anal. Appl. **377**(1), 303–318 (2011). https://doi.org/10.1016/j.jmaa.2010.10.061
30. N.J. Fowler, I. Raeburn, The Toeplitz algebra of a Hilbert bimodule. Indiana Univ. Math. J. **48**(1), 155–181 (1999). https://doi.org/10.1512/iumj.1999.48.1639
31. N.J. Fowler, M. Laca, I. Raeburn, The C^*-algebras of infinite graphs. Proc. Am. Math. Soc. **128**(8), 2319–2327 (2000). https://doi.org/10.1090/S0002-9939-99-05378-2
32. A.H. Fuller, M. Kennedy, Isometric tuples are hyperreflexive. Indiana Univ. Math. J. **62**(5), 1679–1689 (2013). https://doi.org/10.1512/iumj.2013.62.5144
33. P.R. Halmos, Ten problems in Hilbert space. Bull. Am. Math. Soc. **76**, 887–933 (1970). https://doi.org/10.1090/S0002-9904-1970-12502-2
34. P.R. Halmos, Reflexive lattices of subspaces. J. Lond. Math. Soc. (2) **4**, 257–263 (1971). https://doi.org/10.1112/jlms/s2-4.2.257
35. L. Helmer, Reflexivity of non-commutative Hardy algebras. J. Funct. Anal. **272**(7), 2752–2794 (2017). https://doi.org/10.1016/j.jfa.2016.12.004
36. A. Huef, I. Raeburn, The ideal structure of Cuntz-Krieger algebras. Ergodic Theory Dyn. Syst. **17**(3), 611–624 (1997). https://doi.org/10.1017/S0143385797079200
37. F. Jaëck, S.C. Power, Hyper-reflexivity of free semigroupoid algebras. Proc. Am. Math. Soc. **134**(7), 2027–2035 (2006). https://doi.org/10.1090/S0002-9939-05-08209-2
38. M.T. Jury, D.W. Kribs, Ideal structure in free semigroupoid algebras from directed graphs. J. Oper. Theory **53**(2), 273–302 (2005)
39. E. Katsoulis, D.W. Kribs, Isomorphisms of algebras associated with directed graphs. Math. Ann. **330**(4), 709–728 (2004). https://doi.org/10.1007/s00208-004-0566-6
40. M. Kennedy, Wandering vectors and the reflexivity of free semigroup algebras. J. Reine Angew. Math. **653**, 47–73 (2011). https://doi.org/10.1515/CRELLE.2011.019
41. K. Kliś, M. Ptak, Quasinormal operators are hyperreflexive, in *Topological Algebras, Their Applications, and Related Topics*. Banach Center Publications, vol. 67 (Institute of Mathematics of the Polish Academy of Sciences, Warsaw, 2005), pp. 241–244. https://doi.org/10.4064/bc67-0-19
42. J. Kraus, D.R. Larson, Some applications of a technique for constructing reflexive operator algebras. J. Oper. Theory **13**(2), 227–236 (1985)
43. J. Kraus, D.R. Larson, Reflexivity and distance formulae. Proc. Lond. Math. Soc. (3) **53**(2), 340–356 (1986). https://doi.org/10.1112/plms/s3-53.2.340
44. D.W. Kribs, Factoring in non-commutative analytic Toeplitz algebras. J. Oper. Theory **45**(1), 175–193 (2001)

45. D.W. Kribs, Non-selfadjoint operator algebras generated by weighted shifts on Fock space. Houston J. Math. **30**(1), 147–169 (2004)
46. D.W. Kribs, S.C. Power, Free semigroupoid algebras. J. Ramanujan Math. Soc. **19**(2), 117–159 (2004)
47. D.W. Kribs, S.C. Power, The analytic algebras of higher rank graphs. Math. Proc. R. Ir. Acad. **106A**(2), 199–218 (2006). https://doi.org/10.3318/PRIA.2006.106.2.199
48. D.W. Kribs, R.H. Levene, S.C. Power, Commutants of weighted shift directed graph operator algebras. Proc. Am. Math. Soc. **145**(8), 3465–3480 (2017). https://doi.org/10.1090/proc/13477
49. A. Kumjian, D. Pask, Higher rank graph C^*-algebras. New York J. Math. **6**, 1–20 (2000). http://nyjm.albany.edu:8000/j/2000/6_1.html
50. A. Kumjian, D. Pask, I. Raeburn, J. Renault, Graphs, groupoids, and Cuntz-Krieger algebras. J. Funct. Anal. **144**(2), 505–541 (1997). https://doi.org/10.1006/jfan.1996.3001
51. A. Kumjian, D. Pask, I. Raeburn, Cuntz-Krieger algebras of directed graphs. Pacific J. Math. **184**(1), 161–174 (1998). https://doi.org/10.2140/pjm.1998.184.161
52. Z. Liu, Reflexivity of Murray–von Neumann algebras, in *Operator Algebras and Their Applications*. Contemporary Mathematics, vol. 671 (American Mathematical Society, Providence, 2016), pp. 175–184. https://doi.org/10.1090/conm/671/13509
53. F. Merlevède, C. Peligrad, M. Peligrad, Reflexive operator algebras on Banach spaces. Pac. J. Math. **267**(2), 451–464 (2014). https://doi.org/10.2140/pjm.2014.267.451
54. I. Mitscher, J. Spielberg, AF C^*-algebras from non-AF groupoids. Trans. Am. Math. Soc. Article electronically published. https://doi.org/10.1090/trans/8723
55. P.S. Muhly, A finite-dimensional introduction to operator algebra, in *Operator Algebras and Applications (Samos, 1996)*. Nato Science Series C: Mathematical and Physical Sciences, vol. 495 (Kluwer Academic Publishers, Dordrecht, 1997), pp. 313–354
56. P.S. Muhly, B. Solel, Tensor algebras over C^*-correspondences: representations, dilations, and C^*-envelopes. J. Funct. Anal. **158**(2), 389–457 (1998). https://doi.org/10.1006/jfan.1998.3294
57. P.S. Muhly, B. Solel, Tensor algebras, induced representations, and the Wold decomposition. Can. J. Math. **51**(4), 850–880 (1999). https://doi.org/10.4153/CJM-1999-037-8
58. D. Pask, I. Raeburn, M. Rordam, A. Sims, Rank-two graphs whose C^*-algebras are direct limits of circle algebras. J. Funct. Anal. **239**(1), 137–178 (2006). https://doi.org/10.1016/j.jfa.2006.04.003
59. C. Peligrad, Reflexive operator algebras on noncommutative Hardy spaces. Math. Ann. **253**(2), 165–175 (1980). https://doi.org/10.1007/BF01578912
60. G. Popescu, Functional calculus for noncommuting operators. Mich. Math. J. **42**(2), 345–356 (1995). https://doi.org/10.1307/mmj/1029005232
61. G. Popescu, Multi-analytic operators on Fock spaces. Math. Ann. **303**(1), 31–46 (1995). https://doi.org/10.1007/BF01460977
62. G. Popescu, Non-commutative disc algebras and their representations. Proc. Am. Math. Soc. **124**(7), 2137–2148 (1996). https://doi.org/10.1090/S0002-9939-96-03514-9
63. H. Radjavi, P. Rosenthal, On invariant subspaces and reflexive algebras. Am. J. Math. **91**, 683–692 (1969). https://doi.org/10.2307/2373347
64. H. Radjavi, P. Rosenthal, *Invariant Subspaces*, 2nd edn. (Dover Publications, Mineola, 2003)
65. I. Raeburn, Graph algebras, in *CBMS Regional Conference Series in Mathematics*, vol. 103 (American Mathematical Society, Providence, 2005). https://doi.org/10.1090/cbms/103. Published for the Conference Board of the Mathematical Sciences, Washington, DC
66. I. Raeburn, A. Sims, T. Yeend, Higher-rank graphs and their C^*-algebras. Proc. Edinb. Math. Soc. (2) **46**(1), 99–115 (2003). https://doi.org/10.1017/S0013091501000645
67. I. Raeburn, A. Sims, T. Yeend, The C^*-algebras of finitely aligned higher-rank graphs. J. Funct. Anal. **213**(1), 206–240 (2004). https://doi.org/10.1016/j.jfa.2003.10.014
68. C.J. Read, A large weak operator closure for the algebra generated by two isometries. J. Oper. Theory **54**(2), 305–316 (2005)
69. J. Renault, *A Groupoid Approach to C^*-algebras*. Lecture Notes in Mathematics, vol. 793 (Springer, Berlin, 1980)

70. C.E. Rickart, *General Theory of Banach Algebras*. The University Series in Higher Mathematics (D. Van Nostrand Co., Princeton/Toronto/London/New York, 1960)
71. J.R. Ringrose, On some algebras of operators. Proc. Lond. Math. Soc. (3) **15**, 61–83 (1965). https://doi.org/10.1112/plms/s3-15.1.61
72. S. Rosenoer, Distance estimates for von Neumann algebras. Proc. Am. Math. Soc. **86**(2), 248–252 (1982). https://doi.org/10.2307/2043390
73. D. Sarason, Invariant subspaces and unstarred operator algebras. Pac. J. Math. **17**, 511–517 (1966). http://projecteuclid.org/euclid.pjm/1102994515
74. J. Spielberg, A functorial approach to the C^*-algebras of a graph. Int. J. Math. **13**(3), 245–277 (2002). https://doi.org/10.1142/S0129167X02001319
75. J. Spielberg, Groupoids and C^*-algebras for categories of paths. Trans. Am. Math. Soc. **366**(11), 5771–5819 (2014). https://doi.org/10.1090/S0002-9947-2014-06008-X
76. C. Strătilă, L. Zsidó, *Lectures on von Neumann Algebras* (Editura Academiei/Abacus Press, Bucharest/Tunbridge Wells, 1979). Revision of the 1975 original, Translated from the Romanian by Silviu Teleman

Finite Sections of Periodic Schrödinger Operators

Fabian Gabel, Dennis Gallaun, Julian Großmann, Marko Lindner, and Riko Ukena

Abstract We study discrete Schrödinger operators H with periodic potentials as they are typically used to approximate aperiodic Schrödinger operators like the Fibonacci Hamiltonian. We prove an efficient test for applicability of the finite section method, a procedure that approximates H by growing finite square submatrices H_n. For integer-valued potentials, we show that the finite section method is applicable as soon as H is invertible. This statement remains true for $\{0, \lambda\}$-valued potentials with fixed rational λ and period less than nine as well as for arbitrary real-valued potentials of period two.

1 Introduction

Discrete Schrödinger Operators

If one discretises the one-dimensional Schrödinger operator $\Delta + a\cdot$ with potential a by finite differences, one derives a so-called *discrete Schrödinger operator H*, acting on a two-sided infinite sequence $x : \mathbb{Z} \to \mathbb{C}$ via

$$(Hx)_k = x_{k-1} + x_{k+1} + v(k)x_k\,, \qquad k \in \mathbb{Z}\,. \tag{1.1}$$

Typically, x is an element of $\ell^p(\mathbb{Z})$ with some $p \in [1, \infty]$, where v, again called the *potential* of H, is essentially formed by samples of a. We will assume that a is bounded on $\mathbb{R}$, whence v is bounded on $\mathbb{Z}$, making H a bounded linear operator on every space $\ell^p(\mathbb{Z})$. Identifying x with a two-sided infinite vector, H acts as a two-sided infinite tridiagonal matrix $(H_{ij})_{i,j\in\mathbb{Z}}$ with the sequence v on the main

F. Gabel · D. Gallaun · J. Großmann (✉) · M. Lindner · R. Ukena
Hamburg University of Technology, Institute of Mathematics, Hamburg, Germany
e-mail: fabian.gabel@tuhh.de; dennis.gallaun@tuhh.de; julian.grossmann@jp-g.de; lindner@tuhh.de; riko.ukena@tuhh.de

Y. Choi et al. (eds.), *Operators, Semigroups, Algebras and Function Theory*, Operator Theory: Advances and Applications 292,
https://doi.org/10.1007/978-3-031-38020-4_6

diagonal and ones on the diagonals immediately below and above. Note that H is *self-adjoint* on $\ell^2(\mathbb{Z})$ if v is real-valued. Otherwise, H still is *normal.*

If v is a periodic function, H is called a *periodic Schrödinger operator.* In theoretical solid-state physics, periodic Schrödinger operators represent a standard model for periodic crystals. The class of periodic discrete Schrödinger operators is very well studied and several methods have been developed to understand its spectral properties. This includes *Floquet–Bloch theory*, see, e.g., [35, Chapter 7] and [29, Chapter XIII.16], the *transfer matrix formalism*, see, e.g., [9, Section 3] and [19], and more recently and generally the application of the dynamical systems formalism, see [8] and the references therein. In this article, we study the spectrum and the approximation of periodic Schrödinger operators via the method of limit operators and the finite section method.

The Finite Section Method

If H is invertible, to find the unique solution of the system $Hx = b$, one often uses a truncation technique that replaces the original (infinite-dimensional) system with a sequence of finite linear systems: the *finite section method (FSM).*

Let H_n correspond to the finite square submatrix $(H_{ij})_{i,j=-n}^{n}$, the so-called *n-th finite section* of H. We say that the FSM is *applicable* to H if, for sufficiently large n, the inverses H_n^{-1} exist and approximate the inverse of H. By *approximation* we mean that we first embed H_n^{-1} into a two-sided infinite matrix with the same shape as H and that $H_n^{-1}e_j \to H^{-1}e_j$ holds for all canonical unit vectors $e_j \in \ell^p(\mathbb{Z})$. For $p < \infty$ this implies strong convergence $H_n^{-1}x \to H^{-1}x$ for all $x \in \ell^p(\mathbb{Z})$. In particular, the solution of $Hx = b$ can be approximated by the solutions of corresponding finite-dimensional equations.

Some models also use half-line Schrödinger operators H_+ on $\ell^p(\mathbb{Z}_+)$, with $\mathbb{Z}_+ := \{0, 1, 2, \dots\}$, which also can be seen as a restriction of H from (1.1) with a Dirichlet boundary condition, given by

$$(H_+x)_k = x_{k-1} + x_{k+1} + v(k)x_k\,, \quad k \in \mathbb{Z}_+\,, \quad \text{where } x_{-1} = 0\,. \tag{1.2}$$

This operator is then represented by a one-sided infinite matrix $(H_{ij})_{i,j=0}^{\infty}$. In this case, the *n-th finite section* H_n of H_+ is given by the matrix $(H_{ij})_{i,j=0}^{n}$.

A standard result, see, e.g., [20, 27], shows that:

$$\textit{FSM is applicable to } H \quad \Longleftrightarrow \quad \begin{cases} (a)\ H \textit{ is invertible} \\ (b)\ \textit{all but finitely many } H_n \textit{ are invertible} \\ (c)\ \textit{with uniformly bounded inverses } (H_n^{-1}) \end{cases}$$

Note that the equivalence above also analogously holds for one-sided operators H_+ instead of H.

For some classes of operators, it turns out that the invertibility of H (or H_+) alone is already sufficient for the applicability of the FSM, as it already implies the two other conditions. Of course, this simplifies FSM matters a lot. Let us call such an operator *FSM-simple*—either it has an applicable FSM, or it is not invertible. Consequently, if an FSM-simple operator is invertible, then the FSM is applicable.

Example: Fibonacci Hamiltonian
The two-sided infinite discrete Schrödinger operator with potential

$$v(k) = \chi_{[1-\alpha,1)}(k\alpha \bmod 1), \qquad k \in \mathbb{Z},$$

where $\alpha = \frac{1}{2}(\sqrt{5} - 1)$ is the golden ratio, is called the *Fibonacci Hamiltonian*. As [24] shows, this operator is FSM-simple.

The Fibonacci potential v is a combinatorically very interesting mix of zeros and ones. Since α is irrational, it is not periodic but a so-called quasiperiodic sequence. The entries $v(k)$ with $k = -1, \ldots, 55$ look like this:

$$101101011011010110101101101011011010110101101101011010110101$$

Certain patterns are impossible, others reoccur in interesting ways that are not quite periodic. The patterns 101 and 10 repeat like the ones and zeros on the original level: we observe self-similarity. The Fibonacci potential is, of course, well-studied and well-understood, see, e.g., [5–7, 13, 34, 37]. The proof in [24] that shows that H is FSM-simple exploits a lot of this complexity and beauty of the Fibonacci Hamiltonian. At the same time, the proof heavily depends on the explicit structure of the Fibonacci Hamiltonian which raises the question of whether every discrete Schrödinger operator with a potential of zeros and ones is in fact FSM-simple.

Our Results
In this paper, we prove sufficient conditions for a periodic Schrödinger operator to be FSM-simple. In particular, all periodic Schrödinger operators with $\{0, 1\}$-valued potential are FSM-simple.

Theorem 1.1 *Let $p \in [1, \infty]$ and H be a periodic discrete Schrödinger operator on $\ell^p(\mathbb{Z})$ given by* (1.1)*, where v is the potential.*

(a) If v is integer-valued, then H is FSM-simple.
(b) If the period is less than nine and v only takes values in $\{0, \lambda\}$ with a fixed rational λ, then H is FSM-simple.
(c) If the period is two and the potential is real-valued, then H is FSM-simple.

If H_+ is a periodic discrete Schrödinger operator on $\ell^p(\mathbb{Z}_+)$ given by (1.2)*, then statements (a)–(c) hold analogously for H_+ instead of H.*

Remark 1.2

(a) The bound on the period length in Theorem 1.1(b) is optimal. The operator H with the 9-periodic potential given by repeating the vector $\frac{1}{2}(1, 1, 0, 1, 0, 1, 0, 1, 1)$ is not FSM-simple, see Example 4.5 below. The 5-periodic potential given by repeating $\frac{1}{\sqrt{2}}(1, 1, 0, 1, 0)$ shows that Theorem 1.1(b) fails if we drop rationality of λ, see Example 4.4 below.

(b) The 3-periodic potential given by repeating $(2, \frac{1}{2}, \frac{1}{2})$ shows that Theorem 1.1(c) cannot include period three, see Example 4.2 below.

For K-periodic Schrödinger operators H on $\ell^p(\mathbb{Z})$ with real-valued potential v, the invertibility is particularly easy to check, see, e.g., [25]: H is invertible if and only if any of the following matrices has its trace outside $[-2, 2]$:

$$M^{(j)} = \begin{pmatrix} -v(K-1+j) & -1 \\ 1 & 0 \end{pmatrix} \cdots \begin{pmatrix} -v(1+j) & -1 \\ 1 & 0 \end{pmatrix} \begin{pmatrix} -v(j) & -1 \\ 1 & 0 \end{pmatrix} \tag{1.3}$$

$$\widetilde{M}^{(j)} = \begin{pmatrix} -v(j) & -1 \\ 1 & 0 \end{pmatrix} \begin{pmatrix} -v(1+j) & -1 \\ 1 & 0 \end{pmatrix} \cdots \begin{pmatrix} -v(K-1+j) & -1 \\ 1 & 0 \end{pmatrix} \tag{1.4}$$

with $j \in \mathbb{Z}$. If H is FSM-simple, then that already settles the applicability of the FSM. However, for periodic Schrödinger operators that are possibly not FSM-simple, in addition to the aforementioned trace condition, one has to check whether

$$M_{2,1} \neq 0 \qquad \text{or} \qquad |M_{1,1}| > 1\,, \tag{1.5}$$

where M runs through the set of matrices (1.3) and (1.4).

Theorem 1.3 *Let $p \in [1, \infty]$, H be a K-periodic discrete Schrödinger operator on $\ell^p(\mathbb{Z})$ with real-valued potential v defined by* (1.1)*, and H_+ be the restriction on $\ell^p(\mathbb{Z}_+)$ defined by* (1.2)*. In addition, let $M^{(j)}$ and $\widetilde{M}^{(j)}$ be given by* (1.3) *and* (1.4)*. Then the following holds:*

(a) The FSM is applicable to H if and only if $|\operatorname{tr}(M^{(0)})| > 2$ and

$$M^{(0)}, \ldots, M^{(K-1)} \text{ and } \widetilde{M}^{(0)}, \ldots, \widetilde{M}^{(K-1)} \text{are subject to (1.5)}.$$

(b) The FSM is applicable to H_+ if and only if $|\operatorname{tr}(M^{(0)})| > 2$ and

$$M^{(0)} \text{ and } \widetilde{M}^{(0)}, \ldots, \widetilde{M}^{(K-1)} \text{are subject to (1.5)}.$$

In particular, the FSM is applicable to H_+ if it is applicable to its periodic extension H on $\ell^p(\mathbb{Z})$. Moreover, H_+ is FSM-simple if H is FSM-simple.

Remark 1.4 The 9-periodic Schrödinger operator H with potential given by repeating $\frac{1}{\sqrt{2}}(1, 1, 1, 0, 1, 1, 0, 1, 0)$ is an example where the FSM is applicable to H_+ but not to H, see Example 4.6 below.

2 The FSM: Operators and Tools

2.1 Band Operators and the FSM

A discrete Schrödinger operator H can analogously be described by a two-sided infinite tridiagonal matrix $A = (a_{ij})_{i,j\in\mathbb{Z}}$ which is defined by

$$A = \begin{pmatrix} \ddots & \ddots & & & \\ \ddots & v(0) & 1 & & \\ & 1 & v(1) & 1 & \\ & & 1 & v(2) & \ddots \\ & & & \ddots & \ddots \end{pmatrix}. \tag{2.1}$$

The so-called *finite section method (FSM) for H or A* considers the sequence of finite submatrices

$$A_n = (a_{ij})_{i,j=-n}^{n}, \qquad n \in \mathbb{N}, \tag{2.2}$$

and asks the following:

> *Are the matrices A and A_n invertible for all sufficiently large n and are their inverses (after embedding them into a two-sided infinite matrix) strongly convergent to the inverse of A?*

In the case of a positive answer, we call the FSM (for A) *applicable*. This approximation of A^{-1} can be used for solving equations $Ax = b$ approximately via the solutions of finite systems corresponding to A_n. Rigorous treatments of this natural approximation can be found in [1, 14]. In case the FSM is not applicable in the above sense, it may still be possible to establish an applicability result by passing to suitable modifications $A_n = (a_{ij})_{i,j=l_n}^{r_n}$ with well-chosen sequences $l_n \to -\infty$ and $r_n \to \infty$, see [22, 28].

Neither the FSM nor the following tools and results in Sect. 2 require self-adjointness (unless we explicitly say so).

Operators like the Schrödinger operator H whose infinite matrix representation only exhibits finitely many non-zero diagonals are a well-known subject of investigation regarding the applicability of the FSM. Operators of this type are summarised in the class of *band operators*, which we introduce next.

Definition 2.1 (Band-Width and Band Operator) A finite sum

$$A = \sum_{k=-\omega}^{\omega} M_{a^{(k)}} S^k$$

of products of multiplication operators $(M_{a^{(k)}}x)_n = a_n^{(k)} x_n$ with $a^{(k)} \in \ell^\infty(\mathbb{Z})$ and powers of the shift operator, $(Sx)_n = x_{n-1}$, is called a *band operator* with *band-width* $\omega \in \mathbb{N}$.

By definition, a band operator A acts as a bounded operator on every $\ell^p(\mathbb{Z})$ with $p \in [1, \infty]$. In the following, we will identify an operator A on $\ell^p(\mathbb{Z})$ with its usual *matrix representation* $(a_{ij})_{i,j\in\mathbb{Z}}$ with respect to the canonical basis in $\ell^p(\mathbb{Z})$. For band operators with band-width ω, this matrix is only supported on the k-th diagonals with $-\omega \leq k \leq \omega$.

Remark 2.2 Whenever necessary, we consider for some index set $\mathbb{I} \subset \mathbb{Z}$ the space $\ell^p(\mathbb{I})$ as a subspace of $\ell^p(\mathbb{Z})$. Then the terms in Definition 2.1 naturally carry over to operators A on $\ell^p(\mathbb{I})$ by associating them with the restriction onto the subspace $\ell^p(\mathbb{I})$ of their canonical extension by zero.

Here is one fundamental fact about band operators:

Proposition 2.3 *Let B be a band operator on $\ell^p(\mathbb{I})$ for some $\mathbb{I} \subset \mathbb{Z}$ and $p \in [1, \infty]$. If B is invertible, then B is also invertible as an operator on $\ell^q(\mathbb{I})$ for all $q \in [1, \infty]$.*

Proof First, we identify the Banach space $\ell^p(\mathbb{Z})$ with the p-direct sum $\ell^p(\mathbb{J}) \oplus \ell^p(\mathbb{I})$ where $\mathbb{J} := \mathbb{Z} \setminus \mathbb{I}$ and the norm is given by

$$\|x \oplus y\| := \begin{cases} \left(\|x\|_{\ell^p(\mathbb{J})}^p + \|y\|_{\ell^p(\mathbb{I})}^p\right)^{\frac{1}{p}} & \text{if } p < \infty\,, \\ \max\left\{\|x\|_{\ell^\infty(\mathbb{J})}, \|y\|_{\ell^\infty(\mathbb{I})}\right\} & \text{if } p = \infty\,. \end{cases}$$

Now, assume that B is invertible on $\ell^p(\mathbb{I})$. We extend the operator B to an invertible operator A on $\ell^p(\mathbb{Z})$ via

$$A := \begin{pmatrix} \mathbf{1}_{\mathbb{J}} & 0 \\ 0 & B \end{pmatrix}$$

with respect to the direct decomposition $\ell^p(\mathbb{Z}) = \ell^p(\mathbb{J}) \oplus \ell^p(\mathbb{I})$. Here, $\mathbf{1}_{\mathbb{J}}$ denotes the identity on $\ell^p(\mathbb{J})$. Note that, as a band operator, A is an element of the so-called *Wiener algebra* $\mathcal{W}$, see, e.g., [21, Section 3.7.3]. Due to [18, 5.2.10], the algebra $\mathcal{W}$ is closed under taking inverses; a concise proof of this fact can also be found in [27, Corollary 2.5.4]. As an element of $\mathcal{W}$, the inverse A^{-1} acts boundedly on every space $\ell^q(\mathbb{Z})$, $q \in [1, \infty]$. Consequently, the operator A is invertible on all spaces $\ell^q(\mathbb{Z})$, $q \in [1, \infty]$. Moreover, we have

$$A^{-1} = \begin{pmatrix} \mathbf{1}_{\mathbb{J}} & 0 \\ 0 & B^{-1} \end{pmatrix}.$$

Finally, we conclude that B^{-1} is a bounded operator on $\ell^q(\mathbb{I})$ for all $q \in [1, \infty]$. □

Going back to the FSM: assuming invertibility of A on $\ell^p(\mathbb{Z})$, the applicability of the FSM is equivalent to the uniform boundedness of the inverses A_n^{-1}, a concept that is also known as *stability*.

Definition 2.4 (Stability) A sequence of operators $(A_n)_{n\in\mathbb{N}}$ defined on $\ell^p(\mathbb{Z})$ or $\ell^p(\mathbb{Z}_+)$ for $p \in [1,\infty]$ is called *stable* if there exists $n_0 \in \mathbb{N}$ such that for all $n \geq n_0$, the operators A_n are invertible and their inverses are uniformly bounded.

The basic result connecting the notions of applicability and stability is known as *Polski's theorem*, cf. [15, Theorem 1.4], and it says that (A_n) is applicable to A if and only if it is stable and A is invertible.

Now, stability, and hence applicability, of the FSM sequence (A_n) in (2.2) is closely connected to the following entrywise limits

$$(a_{i+l_n,j+l_n})_{i,j=0}^{\infty} \xrightarrow{n\to\infty} L_+ \quad \text{and} \quad (a_{i+r_n,j+r_n})_{i,j=-\infty}^{0} \xrightarrow{n\to\infty} R_- \tag{2.3}$$

of one-sided infinite submatrices of A, where we consider sequences (l_n) and (r_n) with $\lim_{n\to\infty} l_n = -\infty$ and $\lim_{n\to\infty} r_n = \infty$ such that the limits in (2.3) exist. The following result summarises the aforementioned connections and is a standard outcome regarding the applicability of the FSM for band operators.

Lemma 2.5 ([4, Lemma 1.2], [28, Theorem 2.3])

(a) Let A be a band operator on $\ell^p(\mathbb{Z})$ for $p \in [1,\infty]$. Then the following are equivalent:

(i) The FSM is applicable to A.
(ii) The sequence (A_n) with $A_n = (a_{ij})_{i,j=-n}^{n}$ is stable.
(iii) The operator A and the limits L_+ and R_- from (2.3) *are invertible for all suitable sequences (l_n) and (r_n).*

(b) Let A_+ be a band operator on $\ell^p(\mathbb{Z}_+)$ for $p \in [1,\infty]$, where $\mathbb{Z}_+ := \{k \in \mathbb{Z} : k \geq 0\}$. Then the following are equivalent:

(iv) The FSM is applicable to A_+.
(v) The sequence (A_n) with $A_n = (a_{ij})_{i,j=0}^{n}$ is stable.
(vi) The operator A_+ and the limits R_- from (2.3) *are invertible for all suitable sequences (r_n).*

2.2 Limit Operators

We now focus on an operator-theoretical tool in order to reformulate conditions (a)(iii) and (b)(vi) of Lemma 2.5: the so-called *limit operators*, cf. [20, 26, 27]. In the following, let $\mathbb{Z}_- := -\mathbb{Z}_+$. Note that both sets, $\mathbb{Z}_-$ and $\mathbb{Z}_+$, include zero.

Definition 2.6 (Limit Operators and Compressions) Let A be a band operator on $\ell^p(\mathbb{I})$ for $p \in [1, \infty]$ and $\mathbb{I} \in \{\mathbb{Z}_+, \mathbb{Z}_-, \mathbb{Z}\}$. An operator $B \in \ell^p(\mathbb{Z})$ with matrix representation $(b_{ij})_{i,j\in\mathbb{Z}}$ is called *a limit operator of* A if there is a sequence $h = (h_n)_{n\in\mathbb{N}} \subset \mathbb{Z}$ with $\lim_{n\to\infty} |h_n| = \infty$ and

$$a_{i+h_n, j+h_n} \xrightarrow{n\to\infty} b_{ij}$$

for all $i, j \in \mathbb{Z}$. In this case, we also write $A_h := B$ and say that h is the *corresponding sequence to* B. We denote the set of all limit operators of A by $\mathrm{Lim}(A)$.

For $A_h \in \mathrm{Lim}(A)$, we write $A_h \in \mathrm{Lim}_+(A)$ or $A_h \in \mathrm{Lim}_-(A)$ if the corresponding sequence h tends to $+\infty$ or $-\infty$, respectively. Moreover, if $\mathbb{I} = \mathbb{Z}$, we call both operators

$$A_+ := (a_{ij})_{i,j=0}^{\infty} \quad \text{and} \quad A_- := (a_{ij})_{i,j=-\infty}^{0}$$

(one-sided) compressions of A. Then we have the identities $\mathrm{Lim}_\pm(A) = \mathrm{Lim}(A_\pm)$.

In fact, the operators L_+ and R_- from (2.3) correspond to the one-sided compressions of the limit operators $A_{(l_n)}$ and $A_{(r_n)}$. Evidently, a limit operator A_h of a band operator A is again a band operator with the same band-width.

We now reformulate Lemma 2.5 in the language of limit operators.

Proposition 2.7 ([4, Lemma 1.2], [28, Theorem 2.3]) *Let A be a band operator on $\ell^p(\mathbb{Z})$ for $p \in [1, \infty]$. Then the FSM for A is applicable if and only if*

(a) A is invertible on $\ell^p(\mathbb{Z})$,
(b) for all $R \in \mathrm{Lim}_+(A)$ the compressions R_- are invertible on $\ell^p(\mathbb{Z}_-)$, and
(c) for all $L \in \mathrm{Lim}_-(A)$ the compressions L_+ are invertible on $\ell^p(\mathbb{Z}_+)$.

Now let A_+ be a band operator on $\ell^p(\mathbb{Z}_+)$ for $p \in [1, \infty]$. Then the FSM for A_+ is applicable if and only if

(d) A_+ is invertible on $\ell^p(\mathbb{Z}_+)$ and
(e) for all $R \in \mathrm{Lim}_+(A)$ the compressions R_- are invertible on $\ell^p(\mathbb{Z}_-)$.

The restriction to a particular choice of p in Proposition 2.7 can be dropped, thanks to Proposition 2.3. In subsequent parts of this article, this result will allow us to fall back onto the Hilbert space case $p = 2$ whenever needed.

Recall that a bounded linear operator $A\colon X \to Y$ between Banach spaces X and Y is called a *Fredholm operator* if the kernel $\mathrm{Ker}(A)$ and the cokernel $Y/\mathrm{Ran}(A)$ are finite-dimensional. Furthermore, we define the *essential spectrum of* A by

$$\sigma_{\mathrm{ess}}(A) := \{E \in \mathbb{C} : A - E \text{ is not a Fredholm operator}\}.$$

The next lemma establishes the equivalence of the Fredholmness of a band operator A, the invertibility of its limit operators, see [23, 26], or, alternatively, their injectivity on the space $\ell^\infty(\mathbb{Z})$, see [2, 3]. The lemma previously appeared in this form in [24].

Lemma 2.8 ([24, Lemma 2.6]) *Let* $\mathbb{I} \in \{\mathbb{Z}, \mathbb{Z}_+, \mathbb{Z}_-, \mathbb{N}\}$ *and* $p \in [1, \infty]$. *For a band operator A, the following are equivalent:*

(i) A is a Fredholm operator on $\ell^p(\mathbb{I})$.
(ii) All limit operators of A are invertible on $\ell^p(\mathbb{Z})$.
(iii) All limit operators of A are injective on $\ell^\infty(\mathbb{Z})$.

Remark 2.9 Actually, it is Lemma 2.8, whose first versions (with proof) go back to [26] and are somehow finalised in [2, 23], where limit operators naturally enter the scene, and they only reappear in theorems on applicability and stability of operator sequences because the latter is equivalent to Fredholmness of an associated operator.

Here is a simple consequence of Lemma 2.8 about spectra and essential spectra of band operators and their limit operators, see also [23, Corollary 12]:

Proposition 2.10 *Let A be a band operator on* $\ell^p(\mathbb{I})$ *with* $\mathbb{I} \in \{\mathbb{Z}, \mathbb{Z}_+, \mathbb{Z}_-, \mathbb{N}\}$. *Then*

$$\sigma_{\text{ess}}(A) = \bigcup_{B \in \text{Lim}(A)} \sigma(B) .$$

In particular, we have:

(a) If $\sigma(B) = \sigma(B')$ *for all* $B, B' \in \text{Lim}(A)$, *then* $\sigma_{\text{ess}}(A) = \sigma(B)$.
(b) If $B \in \text{Lim}(A)$, *then* $\sigma(B) \subset \sigma_{\text{ess}}(A) \subset \sigma(A)$.
(c) If $\mathbb{I} = \mathbb{Z}$ *and* $A \in \text{Lim}(A)$, *then* $\sigma(A) = \sigma_{\text{ess}}(A)$.

Proof By definition, $E \in \sigma_{\text{ess}}(A)$ if and only if $A - E$ is not Fredholm. By the equivalence (i)⇔(ii) in Lemma 2.8, this is equivalent to $B - E$ not being invertible, i.e., $E \in \sigma(B)$, for some $B \in \text{Lim}(A)$. □

Lemma 2.8 is particularly useful when dealing with a band operator A that additionally has the property $A \in \text{Lim}(A)$. All subsequent examples of Schrödinger operators will have this property, which is sometimes referred to as *self-similarity*, cf. [2]. The combination of Proposition 2.3 with Lemma 2.8 leads to the following results that allow us to translate the invertibility problem of an operator into an injectivity problem of its limit operators.

Corollary 2.11 *Let A be a band operator with* $A \in \text{Lim}(A)$. *Then the following are equivalent:*

(i) All limit operators of A are injective on $\ell^\infty(\mathbb{Z})$.
(ii) A is invertible on $\ell^p(\mathbb{Z})$ *for all* $p \in [1, \infty]$.
(iii) A is invertible on $\ell^p(\mathbb{Z})$ *for some* $p \in [1, \infty]$.

Proof The equivalence (ii)⇔(iii) is immediately given by Proposition 2.3. Since $A \in \mathrm{Lim}(A)$, the implication (i)⇒(iii) follows from Lemma 2.8. Using that an invertible operator is a Fredholm operator, we also have (iii)⇒(i) by Lemma 2.8. □

Note that Corollary 2.11 only handles operators that are defined on $\ell^p(\mathbb{Z})$. However, Lemma 2.5 also relies on the invertibility of one-sided compressions. Therefore, the next corollary gives the corresponding result. For this, we consider only the case $p = 2$ because there we are able to use the *Hilbert space adjoint* A^* for an operator A on the Hilbert space $\ell^2(\mathbb{I})$ with $\mathbb{I} \subset \mathbb{Z}$.

Corollary 2.12 *Let B be a self-adjoint invertible band operator on $\ell^2(\mathbb{Z})$.*

(a) If the compression B_- is injective on $\ell^\infty(\mathbb{Z}_-)$, then B_- is invertible on $\ell^p(\mathbb{Z}_-)$ for all $p \in [1, \infty]$.
(b) If the compression B_+ is injective on $\ell^\infty(\mathbb{Z}_+)$, then B_+ is invertible on $\ell^p(\mathbb{Z}_+)$ for all $p \in [1, \infty]$.

Proof We only prove (a) since the proof of (b) works completely analogously. For the proof of (a), we note that, due to Proposition 2.3, it suffices to consider $p = 2$. As B_- is injective on $\ell^\infty(\mathbb{Z}_-)$, it is also injective on the subset $\ell^2(\mathbb{Z}_-) \subset \ell^\infty(\mathbb{Z}_-)$. We will show that B_- has a dense and closed range making B_- also surjective.

The range of B_- is dense in $\ell^2(\mathbb{Z}_-)$: Since B is self-adjoint, its compression B_- is also self-adjoint. Therefore, the adjoint $(B_-)^*$ is also injective on $\ell^2(\mathbb{Z}_-)$ which implies that the range of B_- is dense in $\ell^2(\mathbb{Z}_-)$.

The range of B_- is closed in $\ell^2(\mathbb{Z}_-)$: Since B is invertible, it is in particular Fredholm on $\ell^2(\mathbb{Z})$. Consequently, Lemma 2.8(i)⇒(ii) gives that all operators in $\mathrm{Lim}(B) \supset \mathrm{Lim}(B_-)$ are invertible on $\ell^2(\mathbb{Z})$. As all limit operators of B_- are invertible on $\ell^2(\mathbb{Z})$, the implication (ii)⇒(i) in Lemma 2.8 gives that B_- is Fredholm. In particular, B_- has a closed range. □

3 Periodic Schrödinger Operators

In this section, we will analyse periodic Schrödinger operators and always assume the hypothesis below.

Hypothesis 3.1 Let $p \in [1, \infty]$ and $H : \ell^p(\mathbb{Z}) \to \ell^p(\mathbb{Z})$ be given by

$$(Hx)_n = x_{n+1} + x_{n-1} + v(n)x_n\,, \quad n \in \mathbb{Z} \tag{3.1}$$

with the real-valued potential $v : \mathbb{Z} \to \mathbb{R}$. Assume in addition that v is periodic with period K, which means $K \in \mathbb{N}$ and $v(n + K) = v(n)$ for all $n \in \mathbb{Z}$.

The potential v as well as the operator H from above are simply called *K-periodic*. Many facts about such periodic Schrödinger operators are known, see, e.g., [11, 25, 35], so this section mainly focuses on the interpretation of these facts in the framework of limit operator theory.

3.1 Trace Condition for the Spectrum of Periodic Schrödinger Operators

If we choose an energy $E \in \mathbb{R}$ and a vector $x \in \ker(H - E)$, relation (3.1) yields the eigenvalue equation

$$0 = ((H - E)x)_n = x_{n+1} + x_{n-1} + (v(n) - E)x_n\,, \qquad n \in \mathbb{Z}\,.$$

This scalar three-term recurrence can be written as the vector-valued two-term recursion

$$\begin{pmatrix} x_{n+1} \\ x_n \end{pmatrix} = \begin{pmatrix} E - v(n) & -1 \\ 1 & 0 \end{pmatrix} \begin{pmatrix} x_n \\ x_{n-1} \end{pmatrix} \quad \text{for all } n \in \mathbb{Z}. \tag{3.2}$$

The 2×2 matrix in (3.2) is called the *transfer matrix* and denoted for each $n \in \mathbb{Z}$ by:

$$T(n, E) := \begin{pmatrix} E - v(n) & -1 \\ 1 & 0 \end{pmatrix}. \tag{3.3}$$

The transfer matrices lie in the *symplectic group*, which means that they satisfy the relation $T(n, E)^T \left(\begin{smallmatrix} 0 & -1 \\ 1 & 0 \end{smallmatrix}\right) T(n, E) = \left(\begin{smallmatrix} 0 & -1 \\ 1 & 0 \end{smallmatrix}\right)$. In particular their determinant is 1, since eigenvalues of a symplectic matrix always come in pairs $\{\lambda_i, \lambda_i^{-1}\}$. For a periodic potential with period K, we define the *monodromy matrix*

$$M(E) := T(K - 1, E) \cdots T(1, E)\, T(0, E)\,. \tag{3.4}$$

Obviously, $M(E)$ brings us from $\left(\begin{smallmatrix} x_0 \\ x_{-1} \end{smallmatrix}\right)$ to $\left(\begin{smallmatrix} x_K \\ x_{K-1} \end{smallmatrix}\right)$, and then again from $\left(\begin{smallmatrix} x_K \\ x_{K-1} \end{smallmatrix}\right)$ to $\left(\begin{smallmatrix} x_{2K} \\ x_{2K-1} \end{smallmatrix}\right)$, and so on. When it is clear from the context that we only consider the energy $E = 0$, we will write $T(n) := T(n, 0)$ and $M := M(0)$.

As we have seen in Sect. 2.2, limit operators provide a useful tool for studying Fredholmness and invertibility of operators. The next lemma gives a complete description of the limit operators for periodic Schrödinger operators.

Lemma 3.2 *Assume Hypothesis 3.1.*

(a) Every limit operator of H is again a K-periodic Schrödinger operator and

$$\operatorname{Lim}(H) = \{S^{-k} H S^k : k = 0, \ldots, K - 1\} = \operatorname{Lim}_+(H) = \operatorname{Lim}_-(H)\,,$$

where S denotes the right shift operator, i.e. $(Sx)_n = x_{n-1}$ for all $n \in \mathbb{Z}$.

(b) $\sigma(B) = \sigma(H)$ *for all* $B \in \mathrm{Lim}(H)$.
(c) If $B \in \mathrm{Lim}(H)$ *with monodromy matrix* $M_B(E)$*, then we have the identity* $\mathrm{tr}(M_B(E)) = \mathrm{tr}(M(E))$ *for all* $E \in \mathbb{R}$.

Proof (a) Let $B \in \mathrm{Lim}(H)$ with corresponding sequence (h_k). By Definition 2.6, the sequence of representation matrices $(H_k)_{k\in\mathbb{N}}$ with $H_k := S^{-h_k} H S^{h_k}$ converges entrywise to the representation matrix of B. As the diagonal of H only consists of the periodic continuation of $w := (v(0), \dots, v(K-1))$, the diagonal of the shifted matrix H_k only consists of a periodic continuation of a cyclic permutation of w. Hence, in order to converge, the sequence (H_k) has to become constant, eventually. This proves that $B = S^{-k} H S^k$ for some $k \in \{0, \dots, K-1\}$.

The other inclusion is straightforward: choosing $h = (k + n \cdot K)_n$, we get

$$S^{-k} H S^k = \lim_{n\to\infty} S^{-(k+n\cdot K)} H S^{k+n\cdot K} = A_h \in \mathrm{Lim}(H)\,.$$

(b) This follows from the fact that the shift operator is an isomorphism on $\ell^p(\mathbb{Z})$.

(c) From part (a) it follows that the monodromy matrix for B is given by

$$M_B(E) = T(\tau(K-1), E) \cdots T(\tau(1), E)\, T(\tau(0), E)\,,$$

where $(\tau(K-1), \dots, \tau(1), \tau(0))$ is just a cyclic permutation of $(K-1, \dots, 1, 0)$. Consequently $\mathrm{tr}(M_B(E)) = \mathrm{tr}(M(E))$. □

The complete description of the limit operators of a periodic Schrödinger operator from Lemma 3.2 now allows for a powerful characterisation of invertibility.

Lemma 3.3 *Assume Hypothesis 3.1. The periodic Schrödinger operator* H *is invertible on* $\ell^p(\mathbb{Z})$ *for any* $p \in [1, \infty]$ *if and only if* H *is injective on* $\ell^\infty(\mathbb{Z})$.

Proof Let H be injective on $\ell^\infty(\mathbb{Z})$. By Lemma 3.2(a), all limit operators of H are shifts of H and hence, also injective on $\ell^\infty(\mathbb{Z})$. In particular, by Lemma 3.2(a) we have $H \in \mathrm{Lim}(H)$. Now, Corollary 2.11 implies that H is invertible on $\ell^p(\mathbb{Z})$ for all $p \in [1, \infty]$. The only-if-part also follows from Corollary 2.11. □

The following proposition resembles a well-known result about the spectrum of periodic Schrödinger operators, which can be described by a trace condition of the monodromy matrix, see, e.g., [25]. We still give a proof here, because it employs limit operator techniques and is not very long.

Proposition 3.4 *Assume Hypothesis 3.1 and let* $E \in \mathbb{R}$*. Then* $E \in \sigma(H)$ *if and only if the so-called* trace condition, $|\mathrm{tr}(M(E))| \le 2$, *holds.*

Proof A real number E lies in the resolvent set if and only if $H - E$ is invertible on $\ell^p(\mathbb{Z})$. By Lemma 3.3, this is equivalent to $H - E$ being injective on $\ell^\infty(\mathbb{Z})$. However, using (3.2), this is equivalent to the claim that for all $(\alpha, \beta) \neq (0, 0)$, the two-sided sequence

$$\left(M(E)^n \begin{pmatrix}\alpha\\ \beta\end{pmatrix}\right)_{n\in\mathbb{Z}}$$

is unbounded. This is satisfied if and only if both eigenvalues λ_1, λ_2 of $M(E)$ fulfil $|\lambda_i| \neq 1$. Since $M(E)$ has determinant $\det(M(E)) = 1$, this is equivalent to the characteristic polynomial $p(\lambda) = \lambda^2 - \operatorname{tr}(M(E))\lambda + 1$ having two distinct real solutions. Note here that two identical solutions would be ± 1 and two proper complex solutions would be $\lambda_1, \overline{\lambda_1}$ with product 1, hence $|\lambda_i| = 1$. However, having two distinct real solutions is equivalent to the discriminant of p being positive, which yields the condition $\operatorname{tr}(M(E))^2 > 4$. □

In the following examples, we illustrate the use of the trace condition from Proposition 3.4 for determining the spectrum of periodic Schrödinger operators.

Examples 3.5

(a) If the potential v is constant with $v(0) \in \mathbb{R}$, which means $K = 1$, then

$$\operatorname{tr}(M(E)) = E - v(0) \text{ and } \sigma(H) = [-2 + v(0), 2 + v(0)].$$

(b) If the potential v has period 2, with $v(0) = -1$ and $v(1) = 1$, we get

$$\operatorname{tr}(M(E)) = E^2 - 3 \text{ and } \sigma(H) = [-\sqrt{5}, -1] \cup [1, \sqrt{5}]\,.$$

(c) More generally, for a 2-periodic potential, we have

$$\operatorname{tr}(M(E)) = -2 + (E - v(0))(E - v(1))\,.$$

This shows that we always will have two spectral bands, namely

$$\sigma(H) = \Big[\frac{1}{2}(v(0) + v(1) - \delta),\ \min\{v(0), v(1)\}\Big] \cup \Big[\max\{v(0), v(1)\},\ \frac{1}{2}(v(0) + v(1) + \delta)\Big],$$

where $\delta = 16 + v(0)^2 - 2v(0)v(1) + v(1)^2$ denotes the discriminant of the polynomial $\operatorname{tr}(M(E)) - 2$.

(d) For a potential v with period 3 given by $v(0) = 0$, $v(1) = 1$ and $v(2) = 0$, we get

$$\operatorname{tr}(M(E)) = E^3 - E^2 - 3E + 1.$$

Therefore, the spectrum is given by

$$\sigma(H) = [-\sqrt{3}, -1] \cup [1 - \sqrt{2}, 1] \cup [1 + \sqrt{2}, \sqrt{3}].$$

Corollary 3.6 *Assume Hypothesis 3.1. The spectrum of H consists of at most K bands. More precisely, there are numbers $a_1 \leq b_1 \leq a_2 \leq \cdots \leq a_K \leq b_K$ with*

$$\sigma(H) = \bigcup_{i=1}^{K} [a_i, b_i] .$$

Proof The trace of $M(E)$ gives us a polynomial p_K in the variable E with degree K. A real number E lies in $\sigma(H)$ if and only if $-2 \leq p_K(E) \leq 2$, by Proposition 3.4. The real zeros of the polynomials $p_K \pm 2$ give us the numbers $\{a_i, b_i\}_{i=1,\ldots,K}$. □

3.2 One-Sided Periodic Schrödinger Operators

In this section, we will cut the Schrödinger operator in two different ways to get one-sided operators. If H is a periodic Schrödinger operator on $\ell^p(\mathbb{Z})$, we let H_+ denote its one-sided restriction to the space $\ell^p(\mathbb{Z}_+)$ subject to the Dirichlet boundary condition $x_{-1} = 0$. Analogously, let H_- denote its one-sided restriction to the space $\ell^p(\mathbb{Z}_-)$ subject to the Dirichlet boundary condition $x_1 = 0$. The matrix representation of $H_\pm$ coincides with the respective one-sided compression from Definition 2.6.

The following lemma allows us to restrict ourselves to one of the two compressions, H_+ or H_-.

Lemma 3.7 *Assume Hypothesis 3.1. Let H^{R} denote the Schrödinger operator with the* reversed potential $v^{\mathrm{R}}(n) := v(-n)$, *$n \in \mathbb{Z}$. Then $\sigma(H) = \sigma(H^{\mathrm{R}})$, $\sigma(H_-) = \sigma(H_+^{\mathrm{R}})$, and* $\mathrm{tr}(M_H) = \mathrm{tr}(M_{H^{\mathrm{R}}})$.

Proof Consider the flip operators $\Phi \colon \ell^p(\mathbb{Z}) \to \ell^p(\mathbb{Z})$, $(x_n)_{n\in\mathbb{Z}} \mapsto (x_{-n})_{n\in\mathbb{Z}}$, and $\Phi_- \colon \ell^p(\mathbb{Z}_+) \to \ell^p(\mathbb{Z}_-)$, $(x_n)_{n\in\mathbb{Z}_+} \mapsto (x_{-n})_{n\in\mathbb{Z}_-}$. Clearly, Φ and Φ_- are isomorphisms. The claims for the spectra follow from the identities $H = \Phi H^{\mathrm{R}} \Phi^{-1}$ and

$$H_- = \Phi_- H_+^{\mathrm{R}} \Phi_-^{-1} , \tag{3.5}$$

which is a straightforward calculation.

To prove the trace identity, we use that a transfer matrix $T = T(n, E)$ always has the property:

$$T^{-1} = FTF \quad \text{with} \quad F = \begin{pmatrix} 0 & 1 \\ 1 & 0 \end{pmatrix} .$$

Hence, we also have $FM_HF = T(K-1)^{-1}\cdots T(0)^{-1} = M_{H^{\mathrm{R}}}^{-1}$. Taking the trace yields:

$$\mathrm{tr}(M_H) = \mathrm{tr}(FM_HF) = \mathrm{tr}\left(M_{H^{\mathrm{R}}}^{-1}\right) = \mathrm{tr}\left(M_{H^{\mathrm{R}}}\right),$$

where the last equality holds for every real symplectic matrix. □

Since $H \in \mathrm{Lim}(H_+)$, the spectrum of H_+ contains that of H as a consequence of Proposition 2.10(b). But how much bigger is it? The answer clearly has something to do with the entries of H_+ near the cut-off. To this end, put

$$H_{a..b} := \begin{pmatrix} v(a) & 1 & & & \\ 1 & v(a+1) & \ddots & & \\ & \ddots & \ddots & \ddots & \\ & & \ddots & v(b-1) & 1 \\ & & & 1 & v(b) \end{pmatrix}$$

for $a, b \in \mathbb{Z}_+$ with $a \leq b$. Then indeed, the following proposition allows us to step from the two-sided to the one-sided periodic operator in terms of spectra. It follows directly from [16, Theorem 4.42(i)], where it is even stated for non-self-adjoint operators with three periodic diagonals.

Proposition 3.8 *Assume Hypothesis 3.1 with period $K \geq 2$. Then*

$$\sigma(H_+) = \sigma(H) \cup \left\{E \in \mathbb{R} : M(E)_{2,1} = 0 \text{ and } |M(E)_{1,1}| < 1\right\}. \tag{3.6}$$

For the entries of $M(E)$ we have the formula

$$M(E) = (-1)^{K+1}\begin{pmatrix} -\det(H_{0..K-1} - E) & -\det(H_{1..K-1} - E) \\ \det(H_{0..K-2} - E) & \det(H_{1..K-2} - E) \end{pmatrix}. \tag{3.7}$$

Proof Note that $\sigma(H) = \sigma_{\mathrm{ess}}(H_+)$ by Lemma 3.2(b), Proposition 2.10(a), and Lemma 3.2(a). Since H is self-adjoint on $\ell^2(\mathbb{Z})$, we know that $\sigma(H)$, which is the same for $p \neq 2$, lies on the real axis. This implies that $\mathbb{C} \setminus \sigma_{\mathrm{ess}}(H_+)$ is connected and thus unbounded. With this, identity (3.6) is a direct consequence of [16, Theorem 4.42(i)] as the tridiagonal matrix representation of H_+ has constant upper and lower diagonals. The formula for $M(E)$ is from [16, Section 4.2.3]. □

Note that we have used Proposition 2.10, which is a general result about the essential spectrum of a band operator using its limit operators. However, in the case of periodic Jacobi operators, also other proofs for the essential spectrum $\sigma_{\mathrm{ess}}(H_+)$ are known, see, e.g., [33, Theorem 7.2.1].

Now, we present an interlacing property that holds in general for the eigenvalues of Jacobi matrices. Here, these matrices come from a K-periodic Schrödinger operator H. Then we can consider the following matrices:

$$H_{0..K-2} \qquad \text{and} \qquad H^{\varphi}_{0..K-1} := H_{0..K-1} + \begin{pmatrix} & & \mathrm{e}^{-\mathrm{i}\varphi} \\ & & \\ \mathrm{e}^{\mathrm{i}\varphi} & & \end{pmatrix}$$

for any $\varphi \in [0, 2\pi]$. Both correspond to finite-dimensional restrictions of the operator equation with Dirichlet and periodic boundary conditions, respectively. The eigenvalues of $H^{\varphi}_{0..K-1}$ for all values of φ give us the spectrum of the (two-sided) periodic Schrödinger operator in the following sense

$$\sigma(H) = \bigcup_{\varphi \in [0,2\pi]} \sigma(H^{\varphi}_{0..K-1}) = \bigcup_{j=1}^{K} [E_{2j-1}, E_{2j}],$$

where $E_1, \ldots, E_{2K}$ is the increasing enumeration of the eigenvalues of $H^{0}_{0..K-1}$ and $H^{\pi}_{0..K-1}$ together. Again, one gets the band structure of the periodic Schrödinger operator as shown in Corollary 3.6. For more details about this approach, the so-called *Floquet–Bloch theory* for periodic band operators, see, e.g., [10, Theorem 4.4.9], [25], and [35, Chapter 7]. In addition, one finds the following classical interlacing result, cf. [35, Theorem 4.5].

Proposition 3.9 *Assume Hypothesis 3.1. The eigenvalues of $H_{0..K-2}$ and $H^{\varphi}_{0..K-1}$ are interlacing for every $\varphi \in [0, 2\pi]$, that is*

$$\widetilde{E}_{1,\varphi} \leq E_1 \leq \widetilde{E}_{2,\varphi} \leq E_2 \leq \cdots \leq E_{K-1} \leq \widetilde{E}_{K,\varphi}$$

where $E_1 \leq \cdots \leq E_{K-1}$ and $\widetilde{E}_{1,\varphi} \leq \cdots \leq \widetilde{E}_{K,\varphi}$ denote the eigenvalues of $H_{0..K-2}$ and $H^{\varphi}_{0..K-1}$, respectively.

Proof The proof can be done analogously to [32, Proposition 2.2] by using the *min-max principle*, see, e.g., [17, Theorem 4.2.11] and [31, Theorem 12.1]. □

Applying this result, we can supplement Proposition 3.8 with more information. There we have seen that the spectrum of the restricted operator H_+ can only be larger than the spectrum of H and the difference can only be given by eigenvalues. These eigenvalues of H_+ coincide with those of $H_{0..K-2}$. However, now we conclude using Proposition 3.9 that these eigenvalues necessarily lie inside the gaps of $\sigma(H)$ and that each gap can only add at most one eigenvalue to $\sigma(H_+)$. Since they stem from a cut-off, i.e., imposing a Dirichlet condition, they are often referred to as *Dirichlet eigenvalues*.

Examples 3.10

(a) Consider the 2-periodic Schrödinger operator H from Example 3.5(c). For the spectrum of H_+, we calculate, according to Proposition 3.8,

$$\sigma(H_{0..2-2}) = \sigma\big(v(0)\big) = \{v(0)\}$$

and

$$\det(H_{0..1} - v(0)) = \det\left(\begin{pmatrix} v(0) - v(0) & 1 \\ 1 & v(1) - v(0) \end{pmatrix}\right) = -1,$$

which implies $\sigma(H_+) = \sigma(H)$.

(b) For the 3-periodic Schrödinger operator from Example 3.5(d), we find:

$$\sigma(H_+) = \sigma(H) \cup \Big\{ -\frac{\sqrt{5}-1}{2} \Big\}.$$

As we have seen, Proposition 3.4 with $E = 0$ gives us a sufficient condition for a periodic Schrödinger operator to be invertible. In order to ensure applicability of the FSM, according to Proposition 2.7, we also need to ensure injectivity of one-sided compressions of periodic Schrödinger operators. Note that, for a one-sided Schrödinger operator, the transfer matrices $T(n)$ still follow the same pattern, as the Dirichlet boundary condition is encoded in the vector (x_n) during the start of the recursion from formula (3.2). Also recall that $M := M(0)$ abbreviates the monodromy matrix for energy $E = 0$.

Proposition 3.11 *Assume Hypothesis 3.1. If H is invertible, that is, $|\operatorname{tr}(M)| > 2$, the following are equivalent:*

(i) the compression H_+ is not injective on $\ell^\infty(\mathbb{Z}_+)$,
(ii) the compression H_+ is not invertible on any $\ell^p(\mathbb{Z}_+)$,
(iii) $\binom{1}{0}$ is an eigenvector of M w.r.t. the smaller (in modulus) of the two eigenvalues (the one with $|\lambda_i| < 1$),
(iv) $M_{2,1} = 0$ and $|M_{1,1}| < 1$.

Proof (ii) is independent of $p \in [1, \infty]$ by Proposition 2.3. (i)$\Rightarrow$(ii) is trivial for $p = \infty$, and (ii)$\Rightarrow$(i) follows from Corollary 2.12 for $p = 2$.

(iii)$\Leftrightarrow$(iv) is obvious: the 2×2-matrix M has the eigenvector $\binom{1}{0}$ if and only if $M_{2,1} = 0$. The corresponding eigenvalue is then $M_{1,1}$.

It remains to show the equivalence of (i) and (iii). To this end, assume that (i) holds. This means that $H_+x = 0$ has a bounded solution $x \in \ell^\infty(\mathbb{Z}_+) \setminus \{0\}$. We reflect the homogeneous Dirichlet condition by putting $x_{-1} = 0$, and we let $x_0 \neq 0$

since otherwise the recursion (3.2) leads to $x = 0$. By scaling, we can put $x_0 = 1$. The recursion formula (3.2) for the one-sided Schrödinger operator then gives us

$$\begin{pmatrix} x_K \\ x_{K-1} \end{pmatrix} = M \begin{pmatrix} x_0 \\ x_{-1} \end{pmatrix} = M \begin{pmatrix} 1 \\ 0 \end{pmatrix}$$

and furthermore

$$\begin{pmatrix} x_{nK} \\ x_{nK-1} \end{pmatrix} = M^n \begin{pmatrix} 1 \\ 0 \end{pmatrix} \tag{3.8}$$

for all $n \in \mathbb{Z}_+$.

By assumption, the condition $|\operatorname{tr}(M)| > 2$ holds. As in the proof of Proposition 3.4, this gives for the eigenvalues of M, denoted by λ_1, λ_2, the chain of inequalities $|\lambda_1| > 1 > |\lambda_2|$. Let furthermore (ξ_1, ξ_2) be a corresponding basis of eigenvectors of M. Because x is bounded, also the vector sequence

$$\left(M^n \begin{pmatrix} 1 \\ 0 \end{pmatrix} \right)_{n \in \mathbb{Z}_+}$$

is bounded. Describing $(1, 0)^T$ with respect to the basis of eigenvectors yields

$$M^n \begin{pmatrix} 1 \\ 0 \end{pmatrix} = M^n(\alpha\xi_1 + \beta\xi_2) = \alpha\lambda_1^n\xi_1 + \beta\lambda_2^n\xi_2\,, \qquad n \in \mathbb{Z}_+.$$

The boundedness of this sequence, together with $|\lambda_1| > 1 > |\lambda_2|$, implies that $\alpha = 0$, so that $\binom{1}{0}$ does not have a component into the direction ξ_1, i.e. it is an eigenvector of M for the eigenvalue λ_2. So we arrive at (iii).

Finally, from (iii) we conclude, by the same arguments, the boundedness of the vector sequence (3.8), say $\|\binom{x_{nK}}{x_{nK-1}}\|_\infty < C < \infty$ for all $n \in \mathbb{Z}_+$. Since

$$\begin{pmatrix} x_{nK+j} \\ x_{nK-1+j} \end{pmatrix} = T(j-1) \cdots T(1)T(0) \begin{pmatrix} x_{nK} \\ x_{nK-1} \end{pmatrix}$$

for all $j \in \{1, \ldots, K\}$, we get that

$$|x_i| \le \left\| \begin{pmatrix} x_i \\ x_{i-1} \end{pmatrix} \right\|_\infty \le \max_{j=1,\ldots,K} \|T(j-1) \cdots T(1)T(0)\| C\,, \quad i \in \mathbb{Z}_+\,.$$

Hence, the solution x of the equation $H_+ x = 0$ is bounded, which shows (i). □

Note that $|M_{1,1}| = 1$ is impossible in Proposition 3.11(iv) if $M_{2,1} = 0$ because, by $\det(M) = 1$, then both eigenvalues of M would have to have modulus one, contradicting $|\operatorname{tr}(M)| > 2$. Also note that the equivalence of (ii) and (iv) can also be seen directly by Proposition 3.8. We will mostly use the equivalence of their negations:

$$H_+ \text{ is invertible on all } \ell^p(\mathbb{Z}_+) \quad \Longleftrightarrow \quad M_{2,1} \neq 0 \text{ or } |M_{1,1}| > 1\,. \qquad (3.9)$$

Proposition 3.11 not only yields the criterion (3.9), leading to our efficient test (1.5), it is also the key to the following result about integer-valued potentials.

Proposition 3.12 *Assume Hypothesis 3.1 with $p = \infty$ and $v(n) \in \mathbb{Z}$ for all $n \in \mathbb{Z}$. If $|\operatorname{tr}(M)| > 2$, that means, if H is invertible, then*

(a) H_+ is injective on $\ell^\infty(\mathbb{Z}_+)$ and
(b) H_- is injective on $\ell^\infty(\mathbb{Z}_-)$.

Proof (a) Let H be invertible and assume that our claim, injectivity of H_+ on $\ell^\infty(\mathbb{Z}_+)$, was false. Then, by Proposition 3.11(i)⇔(iv), $M_{2,1} = 0$ and $|M_{1,1}| < 1$. In particular, M is upper triangular and $M_{1,1}$ and $M_{2,2}$ are its eigenvalues. However, $v(n) \in \mathbb{Z}$ for all $n \in \mathbb{Z}$ implies $M \in \mathbb{Z}^{2\times 2}$ and hence, both $M_{1,1}$ and $M_{2,2}$ are integer. Together with $|M_{1,1}| < 1$ this implies that the eigenvalue $M_{1,1}$ is zero, which is impossible since $\det(M) = 1$. (b) Since H is invertible, H^{R} is invertible by Lemma 3.7. Applying (a) to H^{R} in place of H shows that H_+^{R} is injective on $\ell^\infty(\mathbb{Z}_+)$. But by (3.5) from the proof of Lemma 3.7, H_- is injective, too. □

3.3 Periodic Schrödinger Operators with $\{0, \lambda\}$-Valued Potentials

Before we prove our main theorems, we take a closer look at periodic Schrödinger operators with $\{0, 1\}$-valued potentials and their relatives. To balance between the Laplace interaction term and the pointwise multiplication by the potential, one often introduces a so-called *coupling constant* $\lambda > 0$ as a weight for the $\{0, 1\}$-valued potential v. The result is a discrete Schrödinger operator with a $\{0, \lambda\}$-valued potential. In the examples announced in Remarks 1.2 and 1.4, the potential is of this kind.

For $K \in \mathbb{N}$ and $w \in \{0, 1\}^K$, let $v \in \{0, \lambda\}^{\mathbb{Z}}$ be the periodic extension of $\lambda \cdot w$. We consider the discrete Schrödinger operator H with potential v. Note that we do not specify λ yet but keep it as a variable during most of the following computations.

For applicability of the FSM to H, we are, by Proposition 2.7, interested in the invertibility, and hence the spectra, of

(a) H,
(b) all L_+ with $L \in \operatorname{Lim}_-(H)$,
(c) all R_- with $R \in \operatorname{Lim}_+(H)$.

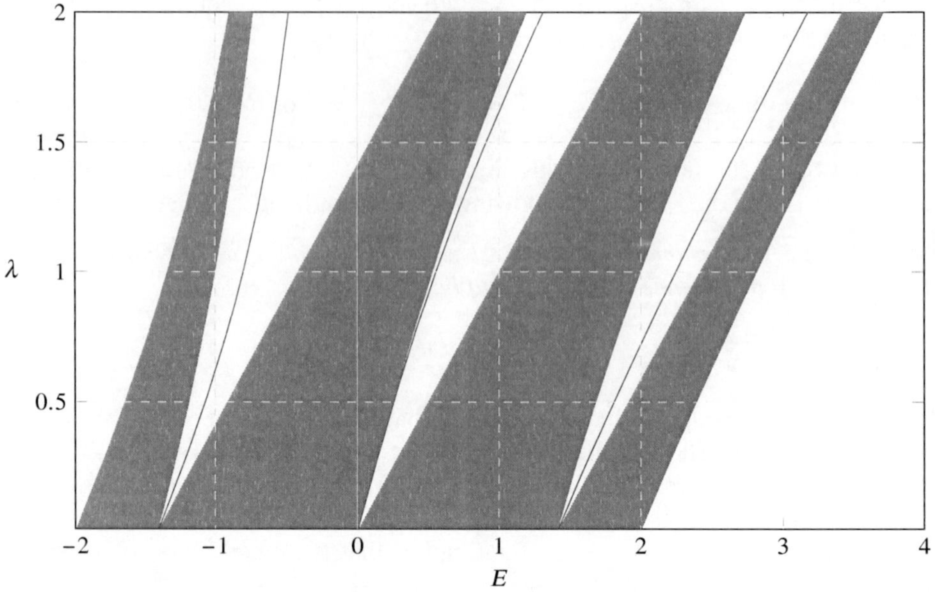

Fig. 1 Union (3.10) of spectra for H with potential $v = \lambda \cdot (1, 1, 0, 1)$, periodically extended, while λ changes along the vertical axis. The spectral bands are shown in grey and the Dirichlet eigenvalues in green. If we look at the vertical line at energy $E = 0$, the plot suggests that, whenever H is invertible (not grey), then all the B_+ and B_- are invertible (not green). In other words: no green line is crossing the vertical line at $E = 0$. Hence, H should be FSM-simple, where the rigorous proof is given in Sect. 4.3

So let us look at the union of spectra

$$\sigma(H) \quad \cup \bigcup_{B \in \mathrm{Lim}(H)} \big(\sigma(B_+) \cup \sigma(B_-)\big) . \tag{3.10}$$

We know from Sect. 3.2 that $\sigma(H)$ consists of at most K closed spectral bands. Further, for $B \in \mathrm{Lim}(H)$, the spectra $\sigma(B_+)$ and $\sigma(B_-)$ are supersets of $\sigma(B) = \sigma(H)$, only larger by at most one Dirichlet eigenvalue per spectral gap. Let us plot the set (3.10), while varying the coupling constant $\lambda > 0$. Figure 1 shows that plot for $w = (1, 1, 0, 1)$.

Figure 2 shows the union (3.10) for $w = (1, 1, 0, 1, 0)$, and Fig. 3 for $w = (1, 1, 0, 1, 0, 1, 0, 1, 1)$.

The latter two choices of w will be revisited in Sect. 4.2.

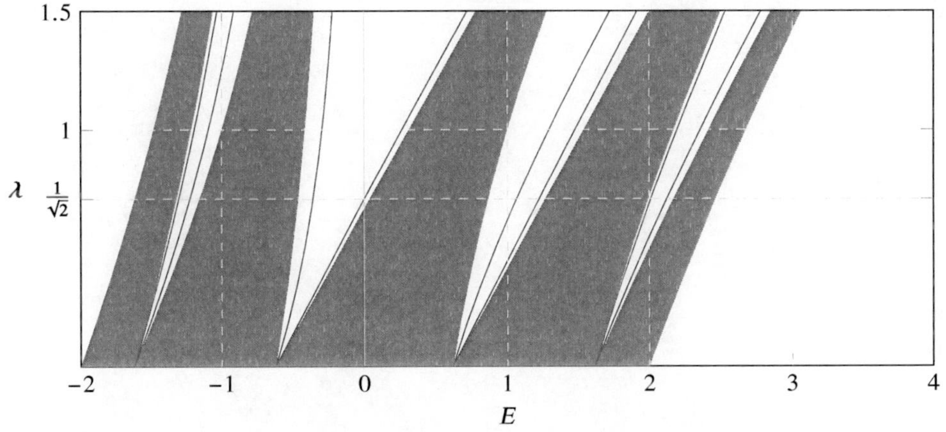

Fig. 2 Union (3.10) of spectra for H with potential $v = \lambda \cdot (1, 1, 0, 1, 0)$, periodically extended, while λ changes along the vertical axis. The spectral bands are shown in grey and the Dirichlet eigenvalues in green. This time we see that one Dirichlet eigenvalue crosses the vertical line at energy $E = 0$ and height $\lambda = \frac{1}{\sqrt{2}}$. So, for this λ, H is invertible (not grey), but the FSM is not applicable (green). In Sect. 4.3, we detect this example algebraically, and in Example 4.4 it is proven that this operator is indeed not FSM-simple

4 Proofs of Our Main Theorems

In this final section, we bring together our results on the invertibility of one- and two-sided periodic Schrödinger operators to conclude, via Proposition 2.7, our Theorems 1.3 and 1.1 about the FSM. In fact, it is advantageous to give the two proofs in this order—as the first and last subsection of Sect. 4. In between, we insert a subsection illustrating the use of Theorem 1.3 with examples and a subsection that deals with an algorithm for the systematic study of $\{0, \lambda\}$-valued potentials. The latter contributes to the proof of Theorem 1.1 which, as already mentioned, forms the last subsection.

The proofs of this section will exploit the following variant of Proposition 2.7 which is made possible by Lemma 3.7 and relies only on one-sided compressions on $\mathbb{Z}_+$ of the limit operators of H and H^{R}:

$$\text{FSM applicable to} H \quad \overset{\text{Prop. 2.7 + Lem. 3.7}}{\Longleftrightarrow} \quad \begin{cases} \text{the following are invertible:} \\ \text{(a) } H \\ \text{(b) all } L_+ \text{ with } L \in \operatorname{Lim}_-(H) \\ \text{(c) all } \widetilde{L}_+ \text{ with } \widetilde{L} \in \operatorname{Lim}_-(H^{\mathrm{R}}) \end{cases} \tag{4.1}$$

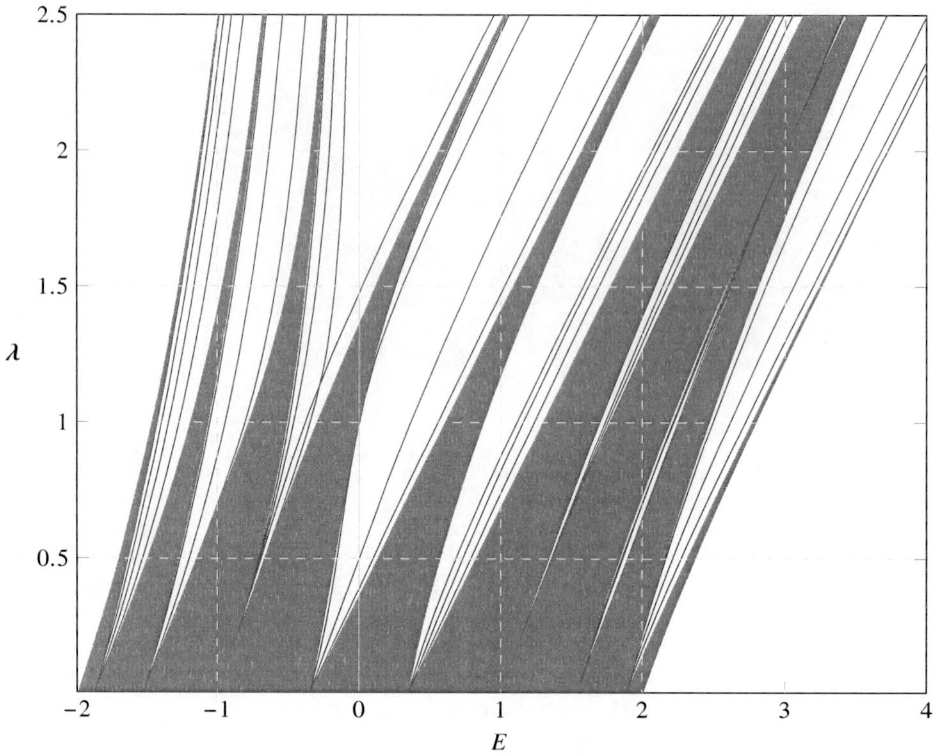

Fig. 3 Union (3.10) of spectra for H with the 9-periodic potential $v = \lambda \cdot (1, 1, 0, 1, 0, 1, 0, 1, 1)$, periodically extended, while λ changes along the vertical axis. The spectral bands are shown in grey and the Dirichlet eigenvalues in green. At $E = 3$ and $\lambda = 2$, two spectral bands merge into one before breaking up again. This time we see that one Dirichlet eigenvalue crosses the vertical line at energy $E = 0$ and height $\lambda = \frac{1}{2}$. So, for $\lambda = \frac{1}{2}$, H is invertible (not grey) but the FSM is not applicable (green). Unlike for periods $K < 9$, this crossing happens at a rational value of λ. In Sect. 4.3, we show how to detect these examples algebraically, and in Example 4.5 it is proven that this operator with $\lambda = \frac{1}{2}$ is indeed not FSM-simple

4.1 Proof of Theorem 1.3

We start with statement (a). Recall that invertibility of H is characterised by the trace formula thanks to Proposition 3.4. Consequently, we assume H to be invertible for the rest of the proof.

In order to complete the proof of (a), we want to use Proposition 3.11 in order to conclude that the invertibility of the one-sided compressions L_+ and $\widetilde{L}_+$ is characterised by the condition

$$M_{2,1} \neq 0 \qquad \text{or} \qquad |M_{1,1}| > 1\,, \tag{4.2}$$

for their respective monodromy matrices.

Indeed, by Lemma 3.2(a), the limit operators L and $\widetilde{L}$ are again periodic Schrödinger operators, and as such they are invertible by Lemma 2.8 as a consequence of the invertibility of H and H^{R}. As a result, Proposition 3.11 applies and all compressions L_+ and $\widetilde{L}_+$ are invertible if and only if their monodromy matrices M_L and $M_{\widetilde{L}}$ are subject to Condition (4.2). Note that due to Lemma 3.2(a), we have a one-to-one correspondence between the monodromy matrices M_L for $L \in \mathrm{Lim}_-(H)$ and the matrices $M^{(j)}$ in the formulation of Theorem 1.3. More precisely, $L = S^{-j}HS^j$ for some $j \in \{0, \ldots, K-1\}$ which yields $M_L = M^{(j)}$. A similar one-to-one correspondence also holds for the matrices $M_{\widetilde{L}}$ and $\widetilde{M}^{(j)}$ with $j \in \{0, \ldots, K-1\}$; this time $\widetilde{M}^{(j)} = M_{\widetilde{L}}$ with limit operator $\widetilde{L} = S^{j-1}H^{\mathrm{R}}S^{-j+1} \in \mathrm{Lim}_-(H^{\mathrm{R}})$.

For statement (b), we employ the second part of Proposition 2.7 which characterises applicability of the FSM to H_+ via invertibility of H_+ and of its limit operators $\widetilde{L}_+$ for $\widetilde{L} \in \mathrm{Lim}_-(H^{\mathrm{R}})$. As in the proof of part (a), the invertibility of H_+ and the limit operators $\widetilde{L}_+$ is characterised by Condition (4.2) through Proposition 3.11.

Finally, we assume that H is FSM-simple and that H_+ is invertible. Then H is invertible by Proposition 3.8 and therefore the FSM is applicable to H, which implies that the FSM is applicable to H_+ as shown above. Hence, H_+ is FSM-simple.

Remark 4.1 Note that the argument in the last part of the proof shows that the implication

$$H \text{ is FSM-simple} \quad \Rightarrow \quad H_+ \text{ is FSM-simple}$$

generalises to all band operators A with the property $A \in \mathrm{Lim}_+(A)$. Indeed, if A is FSM-simple then invertibility of A_+ implies, by $A \in \mathrm{Lim}_+(A) = \mathrm{Lim}(A_+)$ and Lemma 2.8(i)⇒(ii), that A is invertible. Since A is FSM-simple, the FSM is applicable to A and, by Proposition 2.7, it is applicable to A_+.

4.2 Proof of Remarks 1.2 and 1.4

Example 4.2 The 3-periodic Schrödinger operator H with potential given by $v(0) = 2$, $v(1) = \frac{1}{2}$, and $v(2) = \frac{1}{2}$ has monodromy matrices

$$M^{(0)} = \begin{pmatrix} 2 & \frac{3}{4} \\ 0 & \frac{1}{2} \end{pmatrix}, \quad M^{(1)} = \begin{pmatrix} 2 & 0 \\ -\frac{3}{4} & \frac{1}{2} \end{pmatrix}, \quad M^{(2)} = \begin{pmatrix} \frac{1}{2} & 0 \\ 0 & 2 \end{pmatrix}.$$

The trace of $M^{(0)}$ is $2+\frac{1}{2}=\frac{5}{2}>2$, whence H is invertible. The traces of $M^{(1)}$ and $M^{(2)}$ are, not surprisingly, also $\frac{5}{2}$. But now comes the problem:

$$M^{(2)}_{2,1}=0 \qquad \text{and} \qquad |M^{(2)}_{1,1}|=\frac{1}{2}<1$$

so that $M^{(2)}$ fails the test (1.5), whence one particular L_+ with $L \in \mathrm{Lim}_-(H)$ is not invertible, by Proposition 3.11. More precisely, $L_+=(S^{-2}HS^2)_+=(H_l)_+$ with $l=(l_n)=(-3n+2)$.

It is easy to check that $\widetilde{M}^{(2)}=M^{(2)}$ so that also $\widetilde{M}^{(2)}$ fails (1.5). This means that $R_-=H_-=(H_r)_-$ with $r=(r_n)=(3n+1)$ is not invertible.

By Theorem 1.3(a), the FSM is not applicable to H—although H is invertible. Hence, H is not FSM-simple, as already announced in Remark 1.2(b).

One further checks that the matrices $\widetilde{M}^{(0)}=M^{(1)}$ and $\widetilde{M}^{(1)}=M^{(0)}$ pass the test (1.5), which allows one to conclude that the applicability of the FSM only fails because of the two "bad" compressions of the limit operators H_l and H_r identified above. However, with the detailed knowledge about the underlying "bad" sequences $(l_n)=(-3n+2)$ and $(r_n)=(3n+1)$ corresponding to these limit operators, one can adapt the FSM, see, e.g., [22, 24, 28]. In particular, one can show that the *adapted FSM* for H with $A_n=(a_{ij})_{i,j=l'_n}^{r'_n}$ is applicable if the cut-off sequences (l'_n) and (r'_n) asymptotically avoid the sequences $(l_n)=(-3n+2)$ and $(r_n)=(3n+1)$, respectively.

Remark 4.3 If v is the periodic extension of a palindrome, as in Example 4.2 above (use the palindrome $(\frac{1}{2},2,\frac{1}{2})$) and Example 4.5 below, then

$$\{M^{(j)} : j=0,\dots,K-1\} = \{\widetilde{M}^{(j)} : j=0,\dots,K-1\},$$

so that, by Theorem 1.3(a) and (b), the FSM is applicable to H if and only if it is applicable to H_+.

Example 4.4 The 5-periodic Schrödinger operator H with potential given by repeating $\frac{1}{\sqrt{2}}(1,1,0,1,0)$ has five monodromy matrices $M^{(0)}$, $M^{(1)}$, $M^{(2)}$, $M^{(3)}$, $M^{(4)}$ and five reversed order monodromy matrices $\widetilde{M}^{(0)}$, $\widetilde{M}^{(1)}$, $\widetilde{M}^{(2)}$, $\widetilde{M}^{(3)}$, $\widetilde{M}^{(4)}$.

All ten matrices have the trace $\frac{3}{\sqrt{2}}>2$, showing that H is invertible, but the monodromy matrix

$$M^{(0)} = \begin{pmatrix} -\frac{1}{\sqrt{2}} & -1 \\ 0 & -\sqrt{2} \end{pmatrix} = \widetilde{M}^{(3)}$$

fulfils $M^{(0)}_{2,1}=0$ and $|M^{(0)}_{1,1}|<1$. Hence the FSM is not applicable, where one L_+ and one R_- are not invertible. The other matrices $M^{(j)}$ and $\widetilde{M}^{(j)}$ pass the test (1.5), though. So, as announced in Remark 1.2(a), this operator H is not FSM-simple.

Note that a similar 7-periodic example was discussed in the Bachelor thesis [36], which initially led to this study of the periodic case in a more structured way.

Example 4.5 Consider the 9-periodic Schrödinger operator H with potential given by repeating $\frac{1}{2}(1, 1, 0, 1, 0, 1, 0, 1, 1)$. All monodromy matrices (reflected or not) have trace equal to $-\frac{5}{2} \notin [-2, 2]$, whence H is invertible. Like in Example 4.2, the symmetry of the finite word above leads to symmetry of v and, if v is accordingly shifted, to $M^{(j)} = \widetilde{M}^{(j)}$ for all j.

The $(2, 1)$-entry of $M^{(j)}$ is zero for $j = 0$ and for $j = 1$:

$$M^{(0)} = \begin{pmatrix} -\frac{1}{2} & 0 \\ 0 & -2 \end{pmatrix} = \widetilde{M}^{(0)}, \qquad M^{(1)} = \begin{pmatrix} -2 & -\frac{3}{4} \\ 0 & -\frac{1}{2} \end{pmatrix} = \widetilde{M}^{(1)}.$$

Furthermore, the $(1, 1)$-entry of $M^{(0)}$ is less than one in modulus. Consequently, the FSM does not apply to H, so that H is not FSM-simple.

Example 4.6 Consider the 9-periodic Schrödinger operator H with potential given by repeating $\frac{1}{\sqrt{2}}(1, 1, 1, 0, 1, 1, 0, 1, 0)$. We readily check that

$$M^{(1)} = \begin{pmatrix} -\frac{1}{\sqrt{2}} & 2 \\ 0 & -\sqrt{2} \end{pmatrix}$$

and therefore, as in Example 4.4, the FSM is not applicable to H. However, in contrast to Example 4.4, there is no $j \in \{0, \dots, K-1\}$ such that $\widetilde{M}^{(j)} = M^{(1)}$. In fact, one can check that the conditions of Theorem 1.3(b) are satisfied, i.e. that the FSM is applicable to H_+. For an extensive list of all matrices $M^{(j)}$ and $\widetilde{M}^{(j)}$ see [12].

In this computation, we assume that the finite vector $\frac{1}{\sqrt{2}}(1, 1, 1, 0, 1, 1, 0, 1, 0)$ forms entries $v(0), \dots, v(8)$ of the potential. If we place it at $v(-1), \dots, v(7)$ then the corresponding $M^{(0)}$, instead of $M^{(1)}$, fails the test (1.5), so that H_+ is not invertible, whence the FSM is not applicable to it.

4.3 Systematic Studies of $\{0, \lambda\}$-Valued Potentials

In this section, we present an algorithm to find non-FSM-simple examples like Examples 4.4 and 4.5. Because our algorithm does, in a certain sense, a systematic search, it can even be used to prove a positive result (for certain periods K) by *not* finding such examples.

Recall from Sect. 3.3 that for $K \in \mathbb{N}$ and $w \in \{0, 1\}^K$, the periodic extension of $\lambda \cdot w$ is denoted by $v \in \{0, \lambda\}^{\mathbb{Z}}$ and the corresponding periodic Schrödinger operator by H.

Table 1 Procedure for $K = 3$

w	M	zeros of $M_{2,1}$	$\operatorname{tr}(M)$	at zeros
$(0, 0, 0)$	$\begin{pmatrix} 0 & 1 \\ -1 & 0 \end{pmatrix}$	$\varnothing$	0	
$(1, 0, 0)$	$\begin{pmatrix} \lambda & 1 \\ -1 & 0 \end{pmatrix}$	$\varnothing$	λ	
$(0, 1, 0)$	$\begin{pmatrix} 0 & 1 \\ -1 & \lambda \end{pmatrix}$	$\varnothing$	λ	
$(1, 1, 0)$	$\begin{pmatrix} \lambda & 1 \\ \lambda^2-1 & \lambda \end{pmatrix}$	± 1	2λ	± 2
$(0, 0, 1)$	$\begin{pmatrix} \lambda & 1 \\ -1 & 0 \end{pmatrix}$	$\varnothing$	λ	
$(1, 0, 1)$	$\begin{pmatrix} 2\lambda & 1 \\ -1 & 0 \end{pmatrix}$	$\varnothing$	2λ	
$(0, 1, 1)$	$\begin{pmatrix} \lambda & -\lambda^2+1 \\ -1 & \lambda \end{pmatrix}$	$\varnothing$	2λ	
$(1, 1, 1)$	$\begin{pmatrix} -\lambda^3+2\lambda & -\lambda^2+1 \\ \lambda^2-1 & \lambda \end{pmatrix}$	± 1	$-\lambda^3 + 3\lambda$	± 2

Without loss of generality, assume $K \geq 2$ and compute the monodromy matrix $M = M(0)$ in (3.4). Now the four entries of M are given by polynomials in λ and, because of (3.9), we are particularly interested in values λ where $M_{2,1} = 0$. Up to period $K = 9$, we can compute these zeros λ exactly by radicals for the following reason: the polynomial $M_{2,1}(\lambda)$ has degree at most K. One can see, by induction over K, that $M_{2,1}(\lambda)$ is always entirely even or odd. Hence, polynomials of degree 6 or 8 can be reduced by substitution $\mu = \lambda^2$ to degree 3 or 4. For degree 5, 7, and 9, one can factor out one λ and apply the substitution $\mu = \lambda^2$ to the rest. So one always ends up with a polynomial of degree at most 4 and can solve in radicals by the standard formulas.

Let us demonstrate the procedure for period $K = 3$ in Table 1.

We conclude that, for $K = 3$, all $w \in \{0, 1\}^3$, and all $\lambda \in \mathbb{R}$, the implication

$$M_{2,1} = 0 \quad \overset{\text{Table 1}}{\Longrightarrow} \quad |\operatorname{tr}(M)| = 2 \quad \overset{3.4}{\Longrightarrow} \quad H \text{ is not invertible} \tag{4.3}$$

holds. The contraposition of (4.3) then shows

$$H \text{ invertible} \quad \overset{(4.3)}{\Longrightarrow} \quad M_{2,1} \neq 0 \quad \overset{(3.9)}{\Longrightarrow} \quad H_+ \text{ is invertible.} \tag{4.4}$$

By Lemma 3.2(a), all $L \in \operatorname{Lim}_-(H)$ and all $\widetilde{L} \in \operatorname{Lim}_-(H^{\mathrm{R}})$ are again 3-periodic Schrödinger operators with $\{0, \lambda\}$-valued potential that are invertible if H is invertible by Lemma 2.8. We apply the reasoning (4.4) with L and $\widetilde{L}$ in place of H and derive that all corresponding compressions L_+ and $\widetilde{L}_+$ are invertible if H is invertible.

Summarising, we conclude by Proposition 2.7 that the FSM is applicable to H if H is invertible, i.e., H is FSM-simple. This settles the case $K = 3$ in Theorem 1.1(b).

For $K = 4$, the corresponding table is easily computed and shows the same implications. As an illustration, Fig. 1 shows the 4-periodic example given by $w = (1, 1, 0, 1)$. In this case, we get

$$M = \begin{pmatrix} -2\lambda^2 + 1 & -2\lambda \\ \lambda & 1 \end{pmatrix},$$

so that $M_{2,1} = 0$ if and only if $\lambda = 0$. In this case, however $\mathrm{tr}(M) = 2$. For $w = (0, 0, 0, 0)$, one gets $M = \left(\begin{smallmatrix} 1 & 0 \\ 0 & 1 \end{smallmatrix}\right)$, so that $M_{2,1} = 0$ for all $\lambda \in \mathbb{R}$, but also here, $\mathrm{tr}(M) = 2$. After checking the other 14 cases of $w \in \{0, 1\}^4$ in the same manner, we get the same result as in (4.3).

The number of cases for period K is obviously 2^K. Because of this exponential growth, we have written code in *SageMath*, see [30]. Note that actually all computations are purely symbolic; the zeros are computed exactly up to $K = 9$ and all tables and the code are well-documented in [12]. An interim summary at this point could be:

If $K \in \{1, 2, 3, 4\}$ and $\lambda \in \mathbb{R}$, then all K-periodic discrete Schrödinger operators with a $\{0, \lambda\}$-valued potential are FSM-simple.

Note that, for $K = 5$, one finds that H defined by $\lambda = \frac{1}{\sqrt{2}}$ and $w = (1, 1, 0, 1, 0)$ is not FSM-simple, as shown in Example 4.4 and Fig. 2. Also for $w = (1, 1, 0, 1, 1)$ and $w = (1, 1, 1, 1, 1)$, the entry $M_{2,1}$ has irrational zeros with $|\mathrm{tr}(M)| > 2$. However, we find that the only non-FSM-simple examples with $K = 5$ have irrational λ. This behaviour continues up to period $K = 8$, proving the statement:

If $K \in \{5, 6, 7, 8\}$ and $\lambda \in \mathbb{Q}$, then all K-periodic discrete Schrödinger operators H with a $\{0, \lambda\}$-valued potential are FSM-simple.

By comparison, we find Examples 4.5 and 4.6 at period $K = 9$.

4.4 Proof of Theorem 1.1

We restrict ourselves to the two-sided infinite case here and show that H on $\ell^p(\mathbb{Z})$ is FSM-simple in the respective situations. By Theorem 1.3, it follows that also the one-sided compression H_+ is FSM-simple.

For (a): let the potential be integer-valued, and suppose that H is invertible. We have to show that the FSM is applicable to H. We prove applicability of the FSM via Proposition 2.7. As H is invertible, it remains to show that all L_+ with $L \in \mathrm{Lim}_-(H)$ and all R_- with $R \in \mathrm{Lim}_+(H)$ are invertible. We do this in three steps, as follows:

(A) All $L \in \mathrm{Lim}_-(H)$ and all $R \in \mathrm{Lim}_+(H)$ are again periodic Schrödinger operators with integer-valued potentials by Lemma 3.2(a), and all are invertible as a consequence of Lemma 2.8 since H is invertible.

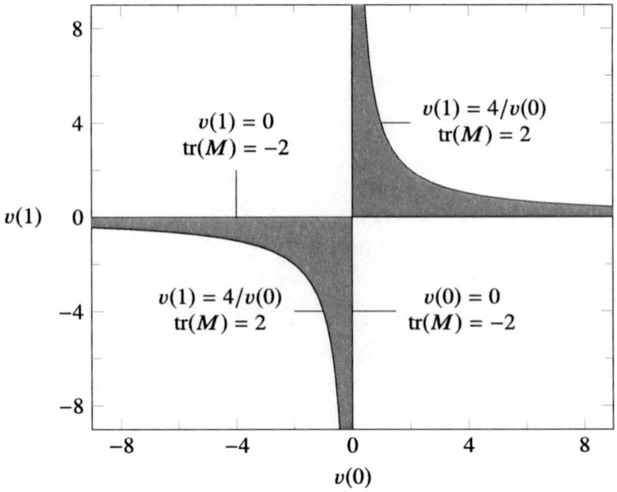

Fig. 4 Example 3.5(c) for $E = 0$ gives $\mathrm{tr}(M) = -2 + v(0)v(1)$. For all points $(v(0), v(1))$ lying outside of the closed grey region, we have $|\,\mathrm{tr}(M)| > 2$. This means $v(0)v(1) \notin [0, 4]$ and, hence, the FSM for the 2-periodic Schrödinger operator H with potential $(v(0), v(1))$ is applicable

(B) Applying Proposition 3.12 to all such L and R in place of H gives that each L_+ is injective on $\ell^\infty(\mathbb{Z}_+)$ and each R_- is injective on $\ell^\infty(\mathbb{Z}_-)$.

(C) It remains to apply Corollary 2.12 to every such L and R, which are, if considered as acting on $\ell^2(\mathbb{Z})$, self-adjoint invertible band operators, to see that all L_+ with $L \in \mathrm{Lim}_-(H)$ and all R_- with $R \in \mathrm{Lim}_+(H)$ are invertible.

This completes the proof of part (a).

For (b): see Sect. 4.3.

For (c): let H be a 2-periodic Schrödinger operator with real potential v, not necessarily integer-valued. If $L \in \mathrm{Lim}_-(H)$ and $\widetilde{L} \in \mathrm{Lim}_-(H^{\mathrm{R}})$, then also L and $\widetilde{L}$ are 2-periodic as a consequence of Lemma 3.2(a). By Example 3.10(a), with L and $\widetilde{L}$ in place of H, we conclude

$$\sigma(L_+) = \sigma(L) = \sigma(H) \qquad \text{and} \qquad \sigma(\widetilde{L}_+) = \sigma(\widetilde{L}) = \sigma(H^{\mathrm{R}}) = \sigma(H).$$

So if H is invertible, all L_+ and $\widetilde{L}_+$ are invertible, too, and the FSM is applicable to H, by Proposition 2.7. □

For a 2-periodic potential, Fig. 4 shows a plot of the critical region outside of which the trace condition is fulfilled and the FSM is applicable to H.

Acknowledgments The authors would like to thank the anonymous referee for his or her interest and helpful comments on our manuscript.

References

1. G. Baxter, A norm inequality for a 'finite-section' Wiener-Hopf equation. Illinois J. Math. **7**, 97–103 (1963)

2. S.N. Chandler-Wilde, M. Lindner, Sufficiency of Favard's condition for a class of band-dominated operators on the axis. J. Funct. Anal. **254**, 1146–1159 (2008)
3. S.N. Chandler-Wilde, M. Lindner, *Limit Operators, Collective Compactness, and the Spectral Theory of Infinite Matrices* (American Mathematical Society, 2011)
4. S.N. Chandler-Wilde, M. Lindner, Coburn's lemma and the finite section method for random Jacobi operators. J. Funct. Anal. **270**, 802–841 (2016)
5. D. Damanik, Gordon-type arguments in the spectral theory of one-dimensional quasi-crystals, in *Directions in Mathematical Quasicrystals*, ed. by M. Baake, R.V. Moody. CRM Monograph Series, vol. 13 (American Mathematical Society, 2000), pp. 277–304
6. D. Damanik, Strictly ergodic subshifts and associated operators, in *Spectral Theory and Mathematical Physics: A Festschrift in Honor of Barry Simon's 60th Birthday*, ed.by F. Gesztesy, P. Deift, C. Galvez, P. Perry, W. Schlag. Proc. Sympos. Pure Math., vol. 76 (American Mathematical Society, 2007), pp. 505–538
7. D. Damanik, A. Gorodetski, W. Yessen, The Fibonacci Hamiltonian. Invent. Math. **206**, 629–692 (2016)
8. D. Damanik, Schrödinger operators with dynamically defined potentials. Ergodic Theory Dyn. Syst. **37**, 1681–1764 (2017)
9. D. Damanik, L. Fang, H. Jun, Schrödinger operators generated by locally constant functions on the Fibonacci subshift. Ann. Henri Poincaré **22**, 1459–1498 (2021)
10. E.B. Davies, *Linear Operators and their Spectra* (Cambridge University Press, 2007)
11. M. Embree, J. Fillman, Spectra of discrete two-dimensional periodic Schrödinger operators with small potentials. J. Spectr. Theory **9**, 1063–1087 (2019)
12. F. Gabel, D. Gallaun, J. Großmann, M. Lindner, R. Ukena, *Analysis Code for Finite Sections of Periodic Schrödinger Operators* (TUHH Universitätsbibliothek, 2021). Available at https://doi.org/10.15480/336.3828
13. F. Gabel, D. Gallaun, J. Großmann, M. Lindner, R. Ukena, *Finite Section Method for Aperiodic Schrödinger Operators*, preprint, 2023. Available at https://doi.org/10.48550/arXiv.2104.00711
14. I. Gohberg, I.A. Feldman, *Convolution Equations and Projection Methods for their Solution* (American Mathematical Society, 1974)
15. R. Hagen, S. Roch, B. Silbermann, *C*-Algebras and Numerical Analysis* (CRC Press, 2000)
16. R. Hagger, Fredholm theory with applications to random operators, Ph.D. thesis, Technische Universität Hamburg, 2016. Available at https://doi.org/10.15480/882.1272
17. R. Horn, C. Johnson, *Matrix Analysis* (Cambridge University Press, 1985)
18. V.G. Kurbatov, *Functional Differential Operators and Equations* (Springer, 1999)
19. Y. Last, B. Simon, Eigenfunctions, transfer matrices, and absolutely continuous spectrum of one-dimensional Schrödinger operators. Invent. Math. **135**, 329–367 (1999)
20. M. Lindner, *Infinite Matrices and their Finite Sections: An Introduction to the Limit Operator Method* (Birkhäuser, 2006)
21. M. Lindner, Fredholm theory and stable approximation of band operators and their generalisations, Habilitation thesis, Technische Universität Chemnitz, 2009. Available at https://nbn-resolving.org/urn:nbn:de:bsz:ch1-200901182
22. M. Lindner, The finite section method and stable subsequences. Appl. Numer. Math. **60**, 501–512 (2010)
23. M. Lindner, M. Seidel, An affirmative answer to a core issue on limit operators. J. Funct. Anal. **267**, 901–917 (2014)
24. M. Lindner, H. Söding, Finite sections of the Fibonacci Hamiltonian, in *The Diversity and Beauty of Applied Operator Theory*, ed. by A. Böttcher, D. Potts, P. Stollmann, D. Wenzel. Oper. Theory Adv. Appl., vol. 268 (Birkhäuser, 2018), pp. 381–396
25. C. Puelz, M. Embree, J. Fillman, Spectral approximation for quasiperiodic Jacobi operators. Integral Equations Oper. Theory **82**, 533–554 (2015)
26. V.S. Rabinovich, S. Roch, B. Silbermann, Fredholm theory and finite section method for band-dominated operators. Integral Equations Oper. Theory **30**, 452–495 (1998)

27. V.S. Rabinovich, S. Roch, B. Silbermann, *Limit Operators and their Applications in Operator Theory* (Birkhäuser, 2004)
28. V.S. Rabinovich, S. Roch, B. Silbermann, On finite sections of band-dominated operators, in *Operator Algebras, Operator Theory and Applications*, ed. by M.A. Bastos, A.B. Lebre, F.-O. Speck, I. Gohberg. Oper. Theory Adv. Appl., vol. 181 (Birkhäuser, 2008), pp. 303–344
29. M. Reed, B. Simon, *Methods of Modern Mathematical Physics IV: Analysis of Operators* (Academic Press, 1978)
30. SageMath, *The Sage Mathematics Software System (Version 9.1)* (Zenodo, 2020). Available at https://doi.org/10.5281/zenodo.4066866
31. K. Schmüdgen, *Unbounded Self-adjoint Operators on Hilbert Space* (Springer, 2012)
32. B. Simon, *Sturm Oscillation and Comparison Theorems* (Birkhäuser, 2005)
33. B. Simon, *Szegő's Theorem and its Descendants* (Princeton University Press, 2011)
34. A. Sütő, Schrödinger difference equation with deterministic ergodic potentials, in *Beyond Quasicrystals*, ed. by F. Axel, D. Gratias. Centre de Physique des Houches, vol. 3 (Springer, 1995), pp. 481–549
35. G. Teschl, *Jacobi Operators and Completely Integrable Nonlinear Lattices* (American Mathematical Society, 2000)
36. L. Weber, One-dimensional quasicrystals, finite sections and invertibility (Eindimensionale Quasikristalle, endliche Abschnitte und Invertierbarkeit), Bachelor thesis, Technische Universität Hamburg, 2018
37. W. Yessen, Spectral analysis of tridiagonal Fibonacci Hamiltonians. J. Spectr. Theory **3**, 101–128 (2013)

The Jacobi Operator and Its Donoghue m-Functions

Fritz Gesztesy, Mateusz Piorkowski, and Jonathan Stanfill

Abstract In this paper we construct Donoghue m-functions for the Jacobi differential operator in $L^2\big((-1,1); (1-x)^\alpha(1+x)^\beta dx\big)$, associated to the differential expression

$$\tau_{\alpha,\beta} = -(1-x)^{-\alpha}(1+x)^{-\beta}(d/dx)\big((1-x)^{\alpha+1}(1+x)^{\beta+1}\big)(d/dx),$$
$$x \in (-1,1),\ \ \alpha, \beta \in \mathbb{R},$$

whenever at least one endpoint, $x = \pm 1$, is in the limit circle case. In doing so, we provide a full treatment of the Jacobi operator's m-functions corresponding to coupled boundary conditions whenever both endpoints are in the limit circle case, a topic not covered in the literature.

Keywords Singular Sturm–Liouville operators · Jacobi equation · Boundary values · Boundary conditions · Donoghue m-functions

---

F. Gesztesy (✉)
Department of Mathematics, Baylor University, Waco, TX, USA
e-mail: Fritz_Gesztesy@baylor.edu
http://www.baylor.edu/math/index.php?id=935340

J. Stanfill
Department of Mathematics, The Ohio State University, Columbus, OH, USA
e-mail: stanfill.13@osu.edu
https://u.osu.edu/stanfill-13/

M. Piorkowski
Department of Mathematics, KU Leuven, Leuven, Belgium
e-mail: Mateusz.Piorkowski@kuleuven.be
https://www.kuleuven.be/wieiswie/en/person/00153535

Y. Choi et al. (eds.), *Operators, Semigroups, Algebras and Function Theory*,
Operator Theory: Advances and Applications 292,
https://doi.org/10.1007/978-3-031-38020-4_7

1 Introduction

This paper should be regarded as a sequel to the recent [33] in which the Donoghue m-function was derived for singular Sturm–Liouville operators. To illustrate the theory, we now apply it to a representative example, the Jacobi differential operator associated with $L^2\big((-1,1); (1-x)^{\alpha}(1+x)^{\beta}dx\big)$-realizations of the differential expression,

$$\begin{aligned}
&\tau_{\alpha,\beta} = -(1-x)^{-\alpha}(1+x)^{-\beta}(d/dx)\big((1-x)^{\alpha+1}(1+x)^{\beta+1}\big)(d/dx),\\
&x \in (-1,1),\ \alpha,\beta \in \mathbb{R},
\end{aligned} \tag{1.1}$$

whenever at least one endpoint, $x = \pm 1$, is in the limit circle case (see, e.g. [1, 4, 8, Ch. 22], [20, Sect. 23], [38, Ch. 4], [44, Sects. VII.6.1, XIV.2], [62, Ch. 18], [72, Ch. IV]). In particular, this provides a full treatment of m-functions corresponding to coupled boundary conditions whenever both endpoints are in the limit circle case, a new result.

To set the stage we briefly discuss abstract Donoghue m-functions following [29, 31, 33], and [28]. Given a self-adjoint extension A of a densely defined, closed, symmetric operator $\dot{A}$ in $\mathcal{H}$ (a complex, separable Hilbert space) with equal deficiency indices and the deficiency subspace $\mathcal{N}_i$ of $\dot{A}$ in $\mathcal{H}$, with

$$\mathcal{N}_i = \ker\big((\dot{A})^* - iI_{\mathcal{H}}\big), \quad \dim(\mathcal{N}_i) = k \in \mathbb{N} \cup \{\infty\}, \tag{1.2}$$

the Donoghue m-operator $M_{A,\mathcal{N}_i}^{Do}(\,\cdot\,) \in \mathcal{B}(\mathcal{N}_i)$ associated with the pair $(A, \mathcal{N}_i)$ is given by

$$\begin{aligned}
M_{A,\mathcal{N}_i}^{Do}(z) &= P_{\mathcal{N}_i}(zA + I_{\mathcal{H}})(A - zI_{\mathcal{H}})^{-1}P_{\mathcal{N}_i}\big|_{\mathcal{N}_i}\\
&= zI_{\mathcal{N}_i} + (z^2+1)P_{\mathcal{N}_i}(A - zI_{\mathcal{H}})^{-1}P_{\mathcal{N}_i}\big|_{\mathcal{N}_i}, \quad z \in \mathbb{C}\backslash\mathbb{R},
\end{aligned} \tag{1.3}$$

with $I_{\mathcal{N}_i}$ the identity operator in $\mathcal{N}_i$, and $P_{\mathcal{N}_i}$ the orthogonal projection in $\mathcal{H}$ onto $\mathcal{N}_i$. The special case $k = 1$, was discussed in detail by Donoghue [17]; for the case $k \in \mathbb{N}$ we refer to [26].

More generally, given a self-adjoint extension A of $\dot{A}$ in $\mathcal{H}$ and a closed, linear subspace $\mathcal{N}$ of $\mathcal{N}_i$, the Donoghue m-operator $M_{A,\mathcal{N}}^{Do}(\,\cdot\,) \in \mathcal{B}(\mathcal{N})$ associated with the pair $(A, \mathcal{N})$ is defined by

$$\begin{aligned}
M_{A,\mathcal{N}}^{Do}(z) &= P_{\mathcal{N}}(zA + I_{\mathcal{H}})(A - zI_{\mathcal{H}})^{-1}P_{\mathcal{N}}\big|_{\mathcal{N}}\\
&= zI_{\mathcal{N}} + (z^2+1)P_{\mathcal{N}}(A - zI_{\mathcal{H}})^{-1}P_{\mathcal{N}}\big|_{\mathcal{N}}, \quad z \in \mathbb{C}\backslash\mathbb{R},
\end{aligned} \tag{1.4}$$

with $I_{\mathcal{N}}$ the identity operator in $\mathcal{N}$ and $P_{\mathcal{N}}$ the orthogonal projection in $\mathcal{H}$ onto $\mathcal{N}$.

Since $M^{Do}_{A,\mathcal{N}}(z)$ is analytic for $z \in \mathbb{C}\backslash\mathbb{R}$ and satisfies (see [31, Theorem 5.3])

$$[\mathrm{Im}(z)]^{-1}\mathrm{Im}\big(M^{Do}_{A,\mathcal{N}}(z)\big) \\ \geq 2\Big[\big(|z|^2+1\big) + \big[\big(|z|^2-1\big)^2 + 4(\mathrm{Re}(z))^2\big]^{1/2}\Big]^{-1} I_{\mathcal{N}}, \quad z \in \mathbb{C}\backslash\mathbb{R}, \tag{1.5}$$

$M^{Do}_{A,\mathcal{N}}(\cdot)$ is a $\mathcal{B}(\mathcal{N})$-valued Nevanlinna–Herglotz function. Thus, $M^{Do}_{A,\mathcal{N}}(\cdot)$ admits the representation

$$M^{Do}_{A,\mathcal{N}}(z) = \int_{\mathbb{R}} d\Omega^{Do}_{A,\mathcal{N}}(\lambda)\bigg[\frac{1}{\lambda - z} - \frac{\lambda}{\lambda^2+1}\bigg], \quad z \in \mathbb{C}\backslash\mathbb{R}, \tag{1.6}$$

where the $\mathcal{B}(\mathcal{N})$-valued measure $\Omega^{Do}_{A,\mathcal{N}}(\cdot)$ satisfies

$$\Omega^{Do}_{A,\mathcal{N}}(\lambda) = (\lambda^2+1)(P_{\mathcal{N}} E_A(\lambda) P_{\mathcal{N}}|_{\mathcal{N}}), \tag{1.7}$$

$$\int_{\mathbb{R}} d\Omega^{Do}_{A,\mathcal{N}}(\lambda)\,(1+\lambda^2)^{-1} = I_{\mathcal{N}}, \tag{1.8}$$

$$\int_{\mathbb{R}} d(\xi, \Omega^{Do}_{A,\mathcal{N}}(\lambda)\xi)_{\mathcal{N}} = \infty \text{ for all } \xi \in \mathcal{N}\backslash\{0\}, \tag{1.9}$$

with $E_A(\cdot)$ the family of strongly right-continuous spectral projections of A in $\mathcal{H}$ (see [29] for details). Operators of the type $M^{Do}_{A,\mathcal{N}}(\cdot)$ and some of its variants have attracted considerable attention in the literature. They appear to go back to Krein [45] (see also [46]), Saakjan [71], and independently, Donoghue [17]. The interested reader can find a wealth of additional information in the context of (1.3)–(1.9) in [2, 3, 5, 6, 10–12, 15, 16, 25, 26, 28–33, 37, 47, 49, 51–54, 56–58, 63, 65, 70], and the references therein.

Without going into further details (see [31, Corollary 5.8] for details) we note that the prime reason for the interest in $M^{Do}_{A,\mathcal{N}_i}(\cdot)$ lies in the fundamental fact that the entire spectral information of A contained in its family of spectral projections $E_A(\cdot)$, is already encoded in the $\mathcal{B}(\mathcal{N}_i)$-valued measure $\Omega^{Do}_{A,\mathcal{N}_i}(\cdot)$ (including multiplicity properties of the spectrum of A) if and only if $\dot{A}$ is completely non-self-adjoint in $\mathcal{H}$ (that is, if and only if $\dot{A}$ has no invariant subspace on which it is self-adjoint, see [31, Lemma 5.4]).

We also note that a particularly attractive feature of the Donoghue m-operator, that distinguishes it from the Weyl–Titchmarsh–Kodaira m-operator, consists of the explicit appearance of the resolvent $(A - zI_{\mathcal{H}})^{-1}$, $z \in \mathbb{C}\backslash\mathbb{R}$, in its definition (1.3) (resp., (1.4)).

We conclude these introductory remarks with a brief sketch of the relation between $M^{Do}_{A,\mathcal{N}}(\cdot)$ and an associated γ-field, familiar from the abstract theory of self-adjoint extensions of closed symmetric operators in connection with the notion of boundary triplets. To keep the discussion as short as possible, we restrict ourselves in the following to a relatively prime pair (A_1, A_2) of self-adjoint extensions of a closed symmetric operator $\dot{A}$ with equal (finite or infinite) deficiency indices (such that $\operatorname{dom}(A_1) \cap \operatorname{dom}(A_2) = \operatorname{dom}(\dot{A})$). We start by recalling Krein's resolvent formula as described in [28],

$$\begin{aligned}&(A_2 - zI_{\mathcal{H}})^{-1} = (A_1 - zI_{\mathcal{H}})^{-1} \\ &\quad + \gamma_{A_1,\mathcal{N}_i}(z)\big[\tan(\alpha_{1,2}) - M^{Do}_{A_1,\mathcal{N}_i}(z)\big]^{-1}\gamma_{A_1,\mathcal{N}_i}(\overline{z})^*, \quad z \in \rho(A_1) \cap \rho(A_2),\end{aligned} \tag{1.10}$$

where the γ-field $\gamma_{A_1,\mathcal{N}_i}(\cdot)$ is given by

$$\begin{aligned}&\gamma_{A_1,\mathcal{N}_i}(z) = (A_1 - iI_{\mathcal{H}})(A_1 - zI_{\mathcal{H}})^{-1}P_{\mathcal{N}_i}, \quad z \in \rho(A_1), \\ &\gamma_{A_1,\mathcal{N}_i}(\overline{z})^* = P_{\mathcal{N}_i}(A_1 + iI_{\mathcal{H}})(A_1 - zI_{\mathcal{H}})^{-1}, \quad z \in \rho(A_1),\end{aligned} \tag{1.11}$$

and the operator $\alpha_{1,2}$ in $\mathcal{N}_i$ is introduced via

$$e^{-2i\alpha_{1,2}} = -C_{A_2}C_{A_1}^{-1}|_{\mathcal{N}_i}, \tag{1.12}$$

where C_B denotes the (unitary) Cayley transform of the self-adjoint operator B in $\mathcal{H}$, that is,

$$C_B = (B + iI_{\mathcal{H}})(B - iI_{\mathcal{H}})^{-1}, \tag{1.13}$$

and one notes that $\mathcal{N}_i$ is an invariant subspace for $C_{A_2}C_{A_1}^{-1}$.

One then verifies the following formulas, familiar from the theory of boundary triplets (see, e.g., [7, Sect. 2.3]),

$$\gamma_{A_1,\mathcal{N}_i}(z) = (A_1 - \zeta I_{\mathcal{H}})(A_1 - zI_{\mathcal{H}})^{-1}\gamma_{A_1,\mathcal{N}_i}(\zeta), \quad z, \zeta \in \rho(A_1), \tag{1.14}$$

$$M^{Do}_{A_1,\mathcal{N}_i}(z) - M^{Do}_{A_1,\mathcal{N}_i}(\zeta)^* = (z - \overline{\zeta})\gamma_{A_1,\mathcal{N}_i}(\zeta)^*\gamma_{A_1,\mathcal{N}_i}(z), \quad z, \zeta \in \rho(A_1), \tag{1.15}$$

$$M^{Do}_{A_1,\mathcal{N}_i}(z)^* = M^{Do}_{A_1,\mathcal{N}_i}(\overline{z}), \quad z \in \rho(A_1), \tag{1.16}$$

$$\begin{aligned}&(z - \zeta_1)(\zeta_1 - \zeta_2)(\zeta_2 - z)\,\gamma_{A_1,\mathcal{N}_i}(\overline{\zeta}_1)^*(A_1 - zI_{\mathcal{H}})^{-1}\gamma_{A_1,\mathcal{N}_i}(\zeta_2) \\ &\quad = (\zeta_1 - \zeta_2)M^{Do}_{A_1,\mathcal{N}_i}(z) + (z - \zeta_2)M^{Do}_{A_1,\mathcal{N}_i}(\zeta_1) + (\zeta_1 - z)M^{Do}_{A_1,\mathcal{N}_i}(\zeta_2), \\ &\qquad z, \zeta_1, \zeta_2 \in \rho(A_1).\end{aligned} \tag{1.17}$$

In the remainder of this paper, we will exclusively focus on the particular case $\mathcal{N} = \mathcal{N}_i = \ker\big((\dot{A})^* - iI_{\mathcal{H}}\big)$, with $\dot{A}$ being a singular Sturm–Liouville operator.

Turning to the content of each section, we discuss the necessary background in connection to singular Sturm–Liouville operators in Sect. 2. In Sects. 3 and 4 we recall the Donoghue m-functions in the two limit circle and one limit circle endpoint cases, respectively, following [33]. The Jacobi operator and its Donoghue

m-functions are the topic of Sect. 5, with Appendices A–C providing a detailed treatment of solutions of the Jacobi differential equation and the associated hypergeometric differential equations.

Finally, some comments on some of the basic notation used throughout this paper. If T is a linear operator mapping (a subspace of) a Hilbert space into another, then $\operatorname{dom}(T)$ and $\ker(T)$ denote the domain and kernel (i.e., null space) of T. The spectrum and resolvent set of a closed linear operator in a Hilbert space will be denoted by $\sigma(\,\cdot\,)$ and $\rho(\,\cdot\,)$, respectively. Moreover, we denote the scalar product and norm in $L^2((a,b);rdx)$ by $(\,\cdot\,,\,\cdot\,)_{L^2((a,b);rdx)}$ (linear in the second argument) and $\|\,\cdot\,\|_{L^2((a,b);rdx)}$.

2 Some Background

In this section we briefly recall the basics of singular Sturm–Liouville operators. The material is standard and can be found, for instance, in [7, Ch. 6], [14, Chs. 8, 9], [18, 19, Sects. 13.6, 13.9, 13.10], [27, Ch. 4], [39, Ch. III], [59, 60, Ch. V], [64, Ch. 6], [73, Ch. 9], [74, Sect. 8.3], [75, Ch. 13], [77, Chs. 4, 6–8].

Throughout this section we make the following assumptions:

Hypothesis 2.1 Let $(a,b) \subseteq \mathbb{R}$ and suppose that p, q, r are (Lebesgue) measurable functions on (a,b) such that the following items (i)–(iii) hold:

(i) $r > 0$ a.e. on (a,b), $r \in L^1_{loc}((a,b);dx)$.
(ii) $p > 0$ a.e. on (a,b), $1/p \in L^1_{loc}((a,b);dx)$.
(iii) q is real-valued a.e. on (a,b), $q \in L^1_{loc}((a,b);dx)$.

Given Hypothesis 2.1, we study Sturm–Liouville operators associated with the general, three-coefficient differential expression τ of the form

$$\tau = \frac{1}{r(x)}\left[-\frac{d}{dx}p(x)\frac{d}{dx} + q(x)\right] \text{ for a.e. } x \in (a,b) \subseteq \mathbb{R}. \tag{2.1}$$

If $f \in AC_{loc}((a,b))$, then the quasi-derivative of f is defined to be $f^{[1]} := pf'$. Moreover, the Wronskian of two functions $f, g \in AC_{loc}((a,b))$ is defined by

$$W(f,g)(x) = f(x)g^{[1]}(x) - f^{[1]}(x)g(x) \text{ for a.e. } x \in (a,b). \tag{2.2}$$

Assuming Hypothesis 2.1, the *maximal operator* T_{max} in $L^2((a,b);rdx)$ associated with τ is defined by

$$\begin{aligned} &T_{max} f = \tau f, \\ &f \in \operatorname{dom}(T_{max}) = \big\{g \in L^2((a,b);rdx) \,\big|\, g, g^{[1]} \in AC_{loc}((a,b)); \\ &\qquad\qquad \tau g \in L^2((a,b);rdx)\big\}. \end{aligned} \tag{2.3}$$

The *preminimal operator* $\dot{T}_{min}$ in $L^2((a, b); rdx)$ associated with τ is defined by

$$\begin{aligned}
&\dot{T}_{min} f = \tau f,\\
&f \in \operatorname{dom}\big(\dot{T}_{min}\big) = \big\{g \in L^2((a, b); rdx) \,\big|\, g, g^{[1]} \in AC_{loc}((a, b));\\
&\qquad\qquad \operatorname{supp}(g) \subset (a, b) \text{ is compact}; \ \tau g \in L^2((a, b); rdx)\big\}.
\end{aligned} \tag{2.4}$$

One can prove that $\dot{T}_{min}$ is closable, and one then defines the *minimal operator* T_{min} by $T_{min} = \overline{\dot{T}_{min}}$.

Still assuming Hypothesis 2.1, one can prove the following basic fact,

$$\big(\dot{T}_{min}\big)^* = T_{max}, \tag{2.5}$$

and hence T_{max} is closed. Moreover, $\dot{T}_{min}$ is essentially self-adjoint if and only if T_{max} is symmetric, and then $\overline{\dot{T}_{min}} = T_{min} = T_{max}$.

The celebrated Weyl alternative can be stated as follows:

Theorem 2.2 (Weyl's Alternative) *Assume Hypothesis 2.1. Then the following alternative holds: Either*

(i) *for every $z \in \mathbb{C}$, all solutions u of $(\tau - z)u = 0$ are in $L^2((a, b); rdx)$ near b (resp., near a),*

or,

(ii) *for every $z \in \mathbb{C}$, there exists at least one solution u of $(\tau - z)u = 0$ which is not in $L^2((a, b); rdx)$ near b (resp., near a). In this case, for each $z \in \mathbb{C}\backslash\mathbb{R}$, there exists precisely one solution u_b (resp., u_a) of $(\tau - z)u = 0$ (up to constant multiples) which lies in $L^2((a, b); rdx)$ near b (resp., near a).*

This yields the limit circle/limit point classification of τ at an interval endpoint as follows.

Definition 2.3 Assume Hypothesis 2.1.
In case (i) in Theorem 2.2, τ is said to be in the *limit circle case* at b (resp., a). (Frequently, τ is then called *quasi-regular* at b (resp., a).)
In case (ii) in Theorem 2.2, τ is said to be in the *limit point case* at b (resp., a).
If τ is in the limit circle case at a and b then τ is also called *quasi-regular* on (a, b).

The next result links self-adjointness of T_{min} (resp., T_{max}) and the limit point property of τ at both endpoints. Here, and throughout, we shall employ the notation

$$\mathcal{N}_z = \ker\big(T_{max} - zI_{L^2((a,b);rdx)}\big), \quad z \in \mathbb{C}. \tag{2.6}$$

Theorem 2.4 *Assume Hypothesis 2.1, then the following items (i) and (ii) hold:*
(i) *If τ is in the limit point case at a (resp., b), then*

$$W(f, g)(a) = 0 \; (\textit{resp., } W(f, g)(b) = 0) \textit{ for all } f, g \in \operatorname{dom}(T_{max}). \tag{2.7}$$

(ii) Let $T_{min} = \overline{\dot{T}_{min}}$. *Then*

$$n_\pm(T_{min}) = \dim\left(\mathcal{N}_{\pm i}\right)$$

$$= \begin{cases} 2 & \text{if } \tau \text{ is in the limit circle case at } a \text{ and } b, \\ 1 & \text{if } \tau \text{ is in the limit circle case at } a \\ & \text{and in the limit point case at } b, \text{ or vice versa}, \\ 0 & \text{if } \tau \text{ is in the limit point case at } a \text{ and } b. \end{cases} \tag{2.8}$$

In particular, T_{min} *(resp.,* T_{max} *) is self-adjoint if and only if* τ *is in the limit point case at* a *and* b*, in which case* $T_{min} = T_{max}$.

All self-adjoint extensions of T_{min} are then described as follows:

Theorem 2.5 *Assume Hypothesis 2.1 and that* τ *is in the limit circle case at* a *and* b *(i.e.,* τ *is quasi-regular on* (a, b) *). In addition, assume that* $v_j \in \mathrm{dom}(T_{max})$, $j = 1, 2$, *satisfy*

$$W(\overline{v_1}, v_2)(a) = W(\overline{v_1}, v_2)(b) = 1, \quad W(\overline{v_j}, v_j)(a) = W(\overline{v_j}, v_j)(b) = 0, \quad j = 1, 2. \tag{2.9}$$

(E.g., real-valued solutions v_j, $j = 1, 2$, *of* $(\tau - \lambda)u = 0$ *with* $\lambda \in \mathbb{R}$, *such that* $W(v_1, v_2) = 1$.*) For* $g \in \mathrm{dom}(T_{max})$ *we introduce the generalized boundary values*

$$\begin{aligned} \widetilde{g}_1(a) &= -W(v_2, g)(a), & \widetilde{g}_1(b) &= -W(v_2, g)(b), \\ \widetilde{g}_2(a) &= W(v_1, g)(a), & \widetilde{g}_2(b) &= W(v_1, g)(b). \end{aligned} \tag{2.10}$$

Then the following items (i)*–*(iii) *hold:*

(i) *All self-adjoint extensions* $T_{\gamma,\delta}$ *of* T_{min} *with separated boundary conditions are of the form*

$$T_{\gamma,\delta} f = \tau f, \quad \gamma, \delta \in [0, \pi), \tag{2.11}$$

$$f \in \mathrm{dom}(T_{\gamma,\delta}) = \left\{ g \in \mathrm{dom}(T_{max}) \,\middle|\, \begin{array}{l} \cos(\gamma)\widetilde{g}_1(a) + \sin(\gamma)\widetilde{g}_2(a) = 0, \\ \cos(\delta)\widetilde{g}_1(b) + \sin(\delta)\widetilde{g}_2(b) = 0 \end{array} \right\}.$$

(ii) *All self-adjoint extensions* $T_{\varphi,R}$ *of* T_{min} *with coupled boundary conditions are of the type*

$$T_{\varphi,R} f = \tau f,$$

$$f \in \mathrm{dom}(T_{\varphi,R}) = \left\{ g \in \mathrm{dom}(T_{max}) \,\middle|\, \begin{pmatrix} \widetilde{g}_1(b) \\ \widetilde{g}_2(b) \end{pmatrix} = e^{i\varphi} R \begin{pmatrix} \widetilde{g}_1(a) \\ \widetilde{g}_2(a) \end{pmatrix} \right\}, \tag{2.12}$$

where $\varphi \in [0, \pi)$, *and* R *is a real* 2×2 *matrix with* $\det(R) = 1$ (*i.e.,* $R \in SL(2, \mathbb{R})$).

(iii) *Every self-adjoint extension of* T_{min} *is either of type* (i) (*i.e., separated*) *or of type* (ii) (*i.e., coupled*).

One can now detail the characterization of $\mathrm{dom}(T_{min})$ by

$$T_{min} f = \tau f, \tag{2.13}$$
$$f \in \mathrm{dom}(T_{min}) = \big\{g \in \mathrm{dom}(T_{max}) \,\big|\, \widetilde{g}_1(a) = \widetilde{g}_2(a) = \widetilde{g}_1(b) = \widetilde{g}_2(b) = 0\big\}.$$

Finally, we turn to the characterization of generalized boundary values in the case where T_{min} is bounded from below following [32] and [60].

We briefly recall the basics of oscillation theory with particular emphasis on principal and nonprincipal solutions, a notion originally due to Leighton and Morse [50] (see also Rellich [66, 67] and Hartman and Wintner [36, Appendix]). Our outline below follows [13], [18, Sects. 13.6, 13.9, 13.10], [27, Ch. 7], [35, 60, Ch. XI], [77, Chs. 4, 6–8].

Definition 2.6 Assume Hypothesis 2.1.

(i) Fix $c \in (a, b)$ and $\lambda \in \mathbb{R}$. Then $\tau - \lambda$ is called *nonoscillatory* at a (resp., b), if every real-valued solution $u(\lambda, \cdot)$ of $\tau u = \lambda u$ has finitely many zeros in (a, c) (resp., (c, b)). Otherwise, $\tau - \lambda$ is called *oscillatory* at a (resp., b).

(ii) Let $\lambda_0 \in \mathbb{R}$. Then T_{min} is called bounded from below by λ_0, and one writes $T_{min} \geq \lambda_0 I_{L^2((a,b);rdx)}$, if

$$\big(u, [T_{min} - \lambda_0 I_{L^2((a,b);rdx)}]u\big)_{L^2((a,b);rdx)} \geq 0, \quad u \in \mathrm{dom}(T_{min}). \tag{2.14}$$

The following is a key result.

Theorem 2.7 *Assume Hypothesis 2.1. Then the following items* (i)–(iii) *are equivalent:*

(i) T_{min} (*and hence any symmetric extension of* T_{min}) *is bounded from below.*

(ii) *There exists a* $\nu_0 \in \mathbb{R}$ *such that for all* $\lambda < \nu_0$, $\tau - \lambda$ *is nonoscillatory at* a *and* b.

(iii) *For fixed* $c, d \in (a, b)$, $c \leq d$, *there exists a* $\nu_0 \in \mathbb{R}$ *such that for all* $\lambda < \nu_0$, $\tau u = \lambda u$ *has* (*real-valued*) *nonvanishing solutions* $u_a(\lambda, \cdot) \neq 0$, $\widehat{u}_a(\lambda, \cdot) \neq 0$ *in the neighborhood* $(a, c]$ *of* a, *and* (*real-valued*) *nonvanishing solutions* $u_b(\lambda, \cdot) \neq 0$, $\widehat{u}_b(\lambda, \cdot) \neq 0$ *in the neighborhood* $[d, b)$ *of* b, *such that*

$$W(\widehat{u}_a(\lambda, \cdot), u_a(\lambda, \cdot)) = 1, \quad u_a(\lambda, x) = o(\widehat{u}_a(\lambda, x)) \text{ as } x \downarrow a, \tag{2.15}$$
$$W(\widehat{u}_b(\lambda, \cdot), u_b(\lambda, \cdot)) = 1, \quad u_b(\lambda, x) = o(\widehat{u}_b(\lambda, x)) \text{ as } x \uparrow b, \tag{2.16}$$

$$\int_a^c dx\, p(x)^{-1} u_a(\lambda, x)^{-2} = \int_d^b dx\, p(x)^{-1} u_b(\lambda, x)^{-2} = \infty, \tag{2.17}$$

$$\int_a^c dx\, p(x)^{-1} \widehat{u}_a(\lambda, x)^{-2} < \infty, \quad \int_d^b dx\, p(x)^{-1} \widehat{u}_b(\lambda, x)^{-2} < \infty. \tag{2.18}$$

Definition 2.8 Assume Hypothesis 2.1, suppose that T_{min} is bounded from below, and let $\lambda \in \mathbb{R}$. Then $u_a(\lambda, \cdot)$ (resp., $u_b(\lambda, \cdot)$) in Theorem 2.7 (iii) is called a *principal* (or *minimal*) solution of $\tau u = \lambda u$ at a (resp., b). A real-valued solution $\breve{u}_a(\lambda, \cdot)$ (resp., $\breve{u}_b(\lambda, \cdot)$) of $\tau u = \lambda u$ linearly independent of $u_a(\lambda, \cdot)$ (resp., $u_b(\lambda, \cdot)$) is called *nonprincipal* at a (resp., b). In particular, $\widehat{u}_a(\lambda, \cdot)$ (resp., $\widehat{u}_b(\lambda, \cdot)$) in (2.15)–(2.18) are nonprincipal solutions at a (resp., b).

Next, we revisit in Theorem 2.5 how the generalized boundary values are utilized in the description of all self-adjoint extensions of T_{min} in the case where T_{min} is bounded from below.

Theorem 2.9 ([32], Theorem 4.5) *Assume Hypothesis 2.1 and that τ is in the limit circle case at a and b (i.e., τ is quasi-regular on (a, b)). In addition, assume that $T_{min} \geq \lambda_0 I_{L^2((a,b);rdx)}$ for some $\lambda_0 \in \mathbb{R}$, and denote by $u_a(\lambda_0, \cdot)$ and $\widehat{u}_a(\lambda_0, \cdot)$ (resp., $u_b(\lambda_0, \cdot)$ and $\widehat{u}_b(\lambda_0, \cdot)$) principal and nonprincipal solutions of $\tau u = \lambda_0 u$ at a (resp., b), satisfying*

$$W(\widehat{u}_a(\lambda_0, \cdot), u_a(\lambda_0, \cdot)) = W(\widehat{u}_b(\lambda_0, \cdot), u_b(\lambda_0, \cdot)) = 1. \tag{2.19}$$

Introducing $v_j \in \mathrm{dom}(T_{max})$, $j = 1, 2$, via

$$v_1(x) = \begin{cases} \widehat{u}_a(\lambda_0, x), & \text{for } x \text{ near } a, \\ \widehat{u}_b(\lambda_0, x), & \text{for } x \text{ near } b, \end{cases} \quad v_2(x) = \begin{cases} u_a(\lambda_0, x), & \text{for } x \text{ near } a, \\ u_b(\lambda_0, x), & \text{for } x \text{ near } b, \end{cases} \tag{2.20}$$

one obtains for all $g \in \mathrm{dom}(T_{max})$,

$$\begin{aligned} \widetilde{g}(a) &= \widetilde{g}_1(a) = -W(u_a(\lambda_0, \cdot), g)(a) = \lim_{x \downarrow a} \frac{g(x)}{\widehat{u}_a(\lambda_0, x)}, \\ \widetilde{g}(b) &= \widetilde{g}_1(b) = -W(u_b(\lambda_0, \cdot), g)(b) = \lim_{x \uparrow b} \frac{g(x)}{\widehat{u}_b(\lambda_0, x)}, \end{aligned} \tag{2.21}$$

$$\begin{aligned} \widetilde{g}'(a) &= \widetilde{g}_2(a) = W(\widehat{u}_a(\lambda_0, \cdot), g)(a) = \lim_{x \downarrow a} \frac{g(x) - \widetilde{g}(a)\widehat{u}_a(\lambda_0, x)}{u_a(\lambda_0, x)}, \\ \widetilde{g}'(b) &= \widetilde{g}_2(b) = W(\widehat{u}_b(\lambda_0, \cdot), g)(b) = \lim_{x \uparrow b} \frac{g(x) - \widetilde{g}(b)\widehat{u}_b(\lambda_0, x)}{u_b(\lambda_0, x)}. \end{aligned} \tag{2.22}$$

In particular, the limits on the right-hand sides in (2.21), (2.22) *exist.*

The Friedrichs extension T_F of T_{min} now permits a particularly simple characterization in terms of the generalized boundary values $\widetilde{g}(a)$, $\widetilde{g}(b)$ as derived by Kalf [41, Remark 3], Niessen and Zettl [60, Theorem 4.2], and Rosenberger [69, Theorem 3] (see also [25, 40, 42, 55, 67, 68, 76]):

Theorem 2.10 *Assume Hypothesis 2.1 and that τ is in the limit circle case at a and b (i.e., τ is quasi-regular on (a, b)). In addition, assume that $T_{min} \geq \lambda_0 I_{L^2((a,b);rdx)}$ for some $\lambda_0 \in \mathbb{R}$. Then the Friedrichs extension $T_F = T_{0,0}$ of T_{min} is characterized by*

$$T_F f = \tau f, \quad f \in \operatorname{dom}(T_F) = \big\{g \in \operatorname{dom}(T_{max}) \,\big|\, \widetilde{g}(a) = \widetilde{g}(b) = 0\big\}. \tag{2.23}$$

3 Donoghue m-functions: Two Limit Circle Endpoints

The Donoghue m-functions in the case where τ is in the limit circle case at a and b is the primary topic of this section following [33, Sect. 6].

Hypothesis 3.1 In addition to Hypothesis 2.1 assume that τ is in the limit circle case at a and b. Moreover, for $z \in \rho(T_{0,0})$, let $\{u_j(z, \cdot)\}_{j=1,2}$ denote solutions to $\tau u = zu$ which satisfy the boundary conditions

$$\begin{aligned} &\widetilde{u}_1(z, a) = 0, \quad \widetilde{u}_1(z, b) = 1, \\ &\widetilde{u}_2(z, a) = 1, \quad \widetilde{u}_2(z, b) = 0. \end{aligned} \tag{3.1}$$

Assume Hypotheses 3.1. By Theorem 2.5 or 2.9, the following statements (i)–(iii) hold.

(i) If $\gamma, \delta \in [0, \pi)$, then the operator $T_{\gamma,\delta}$ defined by

$$T_{\gamma,\delta} f = T_{max} f, \tag{3.2}$$

$$f \in \operatorname{dom}(T_{\gamma,\delta}) = \left\{g \in \operatorname{dom}(T_{max}) \,\middle|\, \begin{array}{l} \cos(\gamma)\widetilde{g}(a) + \sin(\gamma)\widetilde{g}'(a) = 0, \\ \cos(\delta)\widetilde{g}(b) + \sin(\delta)\widetilde{g}'(b) = 0 \end{array}\right\},$$

is a self-adjoint extension of T_{min}.

(ii) If $\varphi \in [0, \pi)$ and $R \in \mathrm{SL}(2, \mathbb{R})$, then the operator $T_{\varphi,R}$ defined by

$$T_{\varphi,R} f = T_{max} f, \tag{3.3}$$

$$f \in \operatorname{dom}(T_{\varphi,R}) = \left\{g \in \operatorname{dom}(T_{max}) \,\middle|\, \begin{pmatrix} \widetilde{g}(b) \\ \widetilde{g}'(b) \end{pmatrix} = e^{i\varphi} R \begin{pmatrix} \widetilde{g}(a) \\ \widetilde{g}'(a) \end{pmatrix} \right\},$$

is a self-adjoint extension of T_{min}.

(iii) If T is a self-adjoint extension of T_{min}, then either $T = T_{\gamma,\delta}$ for some $\gamma, \delta \in [0, \pi)$, or $T = T_{\varphi,R}$ for some $\varphi \in [0, \pi)$ and some $R \in \mathrm{SL}(2, \mathbb{R})$.

Notational Convention *To describe all possible self-adjoint boundary conditions associated with self-adjoint extensions of T_{min} effectively, we will frequently employ the notation $T_{A,B}$, $M_{A,B}^{Do}(\cdot)$, etc., where A, B represents γ, δ in the case of separated boundary conditions and φ, R in the context of coupled boundary conditions.*

Choosing $\gamma = \delta = 0$ in (3.2) yields the self-adjoint extension with Dirichlet-type boundary conditions at a and b, equivalently, the Friedrichs extension T_F of T_{min}:

$$\operatorname{dom}(T_{0,0}) = \operatorname{dom}(T_F) = \{g \in \operatorname{dom}(T_{max}) \,|\, \widetilde{g}(a) = \widetilde{g}(b) = 0\}. \tag{3.4}$$

Since the coefficients of the Sturm–Liouville differential expression are real, the following conjugation property holds:

$$\overline{u_j(z, \cdot)} = u_j(\overline{z}, \cdot), \quad z \in \rho(T_{0,0}),\ j \in \{1, 2\}. \tag{3.5}$$

Applying (3.1), one computes

$$\begin{aligned} &W(u_1(z, \cdot), u_2(z, \cdot))(a) = -\widetilde{u}_1'(z, a), \\ &W(u_1(z, \cdot), u_2(z, \cdot))(b) = \widetilde{u}_2'(z, b), \quad z \in \rho(T_{0,0}). \end{aligned} \tag{3.6}$$

In particular, since the Wronskian of two solutions is constant,

$$\widetilde{u}_2'(z, b) = -\widetilde{u}_1'(z, a), \quad z \in \rho(T_{0,0}). \tag{3.7}$$

We begin by recalling the orthonormal basis for $\mathcal{N}_{\pm i}$ given by $\{v_j(\pm i, \cdot)\}_{j=1,2}$,

$$v_1(\pm i, \cdot) = c_1(\pm i) u_1(\pm i, \cdot), \tag{3.8}$$

$$\begin{aligned} v_2(\pm i, \cdot) &= c_2(\pm i)\Bigg[u_2(\pm i, \cdot) - \frac{(u_1(\pm i, \cdot), u_2(\pm i, \cdot))_{L^2((a,b);rdx)}}{\|u_1(\pm i, \cdot)\|^2_{L^2((a,b);rdx)}} u_1(\pm i, \cdot)\Bigg] \\ &= c_2(\pm i)\Bigg[u_2(\pm i, \cdot) - \frac{\operatorname{Im}\big(\widetilde{u}_2'(i, b)\big)}{\operatorname{Im}\big(\widetilde{u}_1'(i, b)\big)} u_1(\pm i, \cdot)\Bigg], \end{aligned} \tag{3.9}$$

with

$$c_1(\pm i) = \|u_1(\pm i, \cdot)\|^{-1}_{L^2((a,b);rdx)} = \big[\mp \operatorname{Im}\big(\widetilde{u}_1'(\pm i, b)\big)\big]^{-1/2}, \tag{3.10}$$

$$\begin{aligned} c_2(\pm i) &= \left\| u_2(\pm i, \cdot) - \frac{\operatorname{Im}\big(\widetilde{u}_2'(i, b)\big)}{\operatorname{Im}\big(\widetilde{u}_1'(i, b)\big)} u_1(\pm i, \cdot) \right\|^{-1}_{L^2((a,b);rdx)} \\ &= \left[\pm \operatorname{Im}\big(\widetilde{u}_2'(\pm i, a)\big) \pm \frac{\big[\operatorname{Im}\big(\widetilde{u}_2'(\pm i, b)\big)\big]^2}{\operatorname{Im}\big(\widetilde{u}_1'(\pm i, b)\big)} \right]^{-1/2}. \end{aligned} \tag{3.11}$$

The Donoghue m-function $M^{Do}_{T_{A,B},\mathcal{N}_i}(\,\cdot\,)$ with $T_{A,B}$ any self-adjoint extension of T_{min} is provided next (cf. Theorems 6.1–6.3 in [33]).

Theorem 3.2 *Assume Hypothesis 3.1 and let* $\{v_j(i,\,\cdot\,)\}_{j=1,2}$ *be the orthonormal basis for* $\mathcal{N}_i$ *defined in* (3.8)–(3.11)*. The Donoghue m-function* $M^{Do}_{T_{0,0},\mathcal{N}_i}(\,\cdot\,)$: $\mathbb{C}\backslash\mathbb{R} \to \mathcal{B}(\mathcal{N}_i)$ *for* $T_{0,0}$ *satisfies*

$$M^{Do}_{T_{0,0},\mathcal{N}_i}(\pm i) = \pm i I_{\mathcal{N}_i}, \tag{3.12}$$

$$\begin{aligned} M^{Do}_{T_{0,0},\mathcal{N}_i}(z) &= -\sum_{j,k=1}^{2} [i\delta_{j,k} + W_{j,k}(z)](v_k(i,\,\cdot\,),\,\cdot\,)_{L^2((a,b);rdx)} v_j(i,\,\cdot\,)\big|_{\mathcal{N}_i}, \\ &= -iI_{\mathcal{N}_i} - \sum_{j,k=1}^{2} W_{j,k}(z)\big(v_k(i,\,\cdot\,),\,\cdot\,\big)_{L^2((a,b);rdx)} v_j(i,\,\cdot\,)\big|_{\mathcal{N}_i}, \\ &\qquad z \in \mathbb{C}\backslash\mathbb{R},\ z \neq \pm i, \end{aligned}$$

where the matrix $\big(W_{j,k}(\,\cdot\,)\big)_{j,k=1}^2$, $z \in \mathbb{C}\backslash\mathbb{R}$, $z \neq \pm i$, *is given by*

$$W_{1,1}(z) = [c_1(i)]^2\big[\widetilde{u}_1'(z,b) - \widetilde{u}_1'(-i,b)\big], \tag{3.13}$$

$$\begin{aligned} W_{1,2}(z) = c_1(i)c_2(i)\bigg\{&\frac{\operatorname{Im}\big(\widetilde{u}_2'(i,b)\big)}{\operatorname{Im}\big(\widetilde{u}_1'(i,b)\big)}[\widetilde{u}_1'(-i,b) - \widetilde{u}_1'(z,b)] \\ &+ \widetilde{u}_2'(z,b) + \widetilde{u}_1'(-i,a)\bigg\}, \end{aligned} \tag{3.14}$$

$$\begin{aligned} W_{2,1}(z) = -c_1(i)c_2(i)\bigg\{&\frac{\operatorname{Im}\big(\widetilde{u}_2'(i,b)\big)}{\operatorname{Im}\big(\widetilde{u}_1'(i,b)\big)}[\widetilde{u}_1'(z,b) - \widetilde{u}_1'(-i,b)] \\ &+ \widetilde{u}_2'(-i,b) + \widetilde{u}_1'(z,a)\bigg\}, \end{aligned} \tag{3.15}$$

$$\begin{aligned} W_{2,2}(z) = [c_2(i)]^2\bigg\{&\bigg[\widetilde{u}_2'(-i,b) - \widetilde{u}_2'(z,b) \\ &+ \frac{\operatorname{Im}\big(\widetilde{u}_2'(i,b)\big)}{\operatorname{Im}\big(\widetilde{u}_1'(i,b)\big)}[\widetilde{u}_1'(z,b) - \widetilde{u}_1'(-i,b)]\bigg]\frac{\operatorname{Im}\big(\widetilde{u}_2'(i,b)\big)}{\operatorname{Im}\big(\widetilde{u}_1'(i,b)\big)} \\ &+ \widetilde{u}_2'(-i,a) - \widetilde{u}_2'(z,a) + \frac{\operatorname{Im}\big(\widetilde{u}_2'(i,b)\big)}{\operatorname{Im}\big(\widetilde{u}_1'(i,b)\big)}[\widetilde{u}_1'(z,a) - \widetilde{u}_1'(-i,a)]\bigg\}. \end{aligned} \tag{3.16}$$

Furthermore, the following items (i)–(v) *hold.*

(*i*) *If* $\gamma, \delta \in (0, \pi)$, *then the Donoghue m-function* $M^{Do}_{T_{\gamma,\delta}, \mathcal{N}_i}(\,\cdot\,) : \mathbb{C}\backslash\mathbb{R} \to \mathcal{B}(\mathcal{N}_i)$ *for* $T_{\gamma,\delta}$ *satisfies*

$$\begin{aligned}
&M^{Do}_{T_{\gamma,\delta}, \mathcal{N}_i}(\pm i) = \pm i I_{\mathcal{N}_i}, \\
&M^{Do}_{T_{\gamma,\delta}, \mathcal{N}_i}(z) = M^{Do}_{T_{0,0}, \mathcal{N}_i}(z) \\
&\quad + (i - z) \sum_{j,k,\ell=1}^{2} \big[K_{\gamma,\delta}(z)^{-1}\big]_{j,k} W^{Kr}_{\ell,k}(z) (u_j(\overline{z}, \,\cdot\,), \,\cdot\,)_{L^2((a,b);r dx)} v_\ell(i, \,\cdot\,)\big|_{\mathcal{N}_i}, \\
&\qquad z \in \mathbb{C}\backslash\mathbb{R}, \ z \neq \pm i,
\end{aligned} \tag{3.17}$$

where the invertible matrix $K_{\gamma,\delta}(\,\cdot\,)$ *and* $\big(W^{Kr}_{\ell,k}(\,\cdot\,)\big)_{\ell,k=1}^{2}$ *are given by*

$$K_{\gamma,\delta}(z) = \begin{pmatrix} \cot(\delta) + \widetilde{u}_1'(z, b) & -\widetilde{u}_1'(z, a) \\ \widetilde{u}_2'(z, b) & -\cot(\gamma) - \widetilde{u}_2'(z, a) \end{pmatrix}, \tag{3.18}$$

$$W^{Kr}_{1,1}(z) = c_1(i)\big[\widetilde{u}_1'(z, b) - \widetilde{u}_1'(-i, b)\big], \tag{3.19}$$

$$W^{Kr}_{1,2}(z) = c_1(i)\big[\widetilde{u}_2'(z, b) + \widetilde{u}_1'(-i, a)\big], \tag{3.20}$$

$$\begin{aligned}
W^{Kr}_{2,1}(z) &= \widetilde{v}_2(-i, b)\widetilde{u}_1'(z, b) - \widetilde{v}_2'(-i, b) - \widetilde{v}_2(-i, a)\widetilde{u}_1'(z, a) \\
&= -c_2(i)\bigg\{\frac{\operatorname{Im}\big(\widetilde{u}_2'(i, b)\big)}{\operatorname{Im}\big(\widetilde{u}_1'(i, b)\big)}\big[\widetilde{u}_1'(z, b) - \widetilde{u}_1'(-i, b)\big] \\
&\qquad + \widetilde{u}_2'(-i, b) + \widetilde{u}_1'(z, a)\bigg\},
\end{aligned} \tag{3.21}$$

$$\begin{aligned}
W^{Kr}_{2,2}(z) &= \widetilde{v}_2(-i, b)\widetilde{u}_2'(z, b) - \widetilde{v}_2(-i, a)\widetilde{u}_2'(z, a) + \widetilde{v}_2'(-i, a) \\
&= c_2(i)\bigg\{\widetilde{u}_2'(-i, a) - \widetilde{u}_2'(z, a) - \frac{\operatorname{Im}\big(\widetilde{u}_2'(i, b)\big)}{\operatorname{Im}\big(\widetilde{u}_1'(i, b)\big)}\big[\widetilde{u}_2'(z, b) + \widetilde{u}_1'(-i, a)\big]\bigg\}.
\end{aligned} \tag{3.22}$$

(*ii*) *If* $\varphi \in [0, \pi)$ *and* $R \in \mathrm{SL}(2, \mathbb{R})$ *with* $R_{1,2} \neq 0$, *then the Donoghue m-function* $M^{Do}_{T_{\varphi,R}, \mathcal{N}_i}(\,\cdot\,) : \mathbb{C}\backslash\mathbb{R} \to \mathcal{B}(\mathcal{N}_i)$ *for* $T_{\varphi,R}$ *satisfies*

$$\begin{aligned}
&M^{Do}_{T_{\varphi,R}, \mathcal{N}_i}(\pm i) = \pm i I_{\mathcal{N}_i}, \\
&M^{Do}_{T_{\varphi,R}, \mathcal{N}_i}(z) = M^{Do}_{T_{0,0}, \mathcal{N}_i}(z) \\
&\quad + (i - z) \sum_{j,k,\ell=1}^{2} \big[K_{\varphi,R}(z)^{-1}\big]_{j,k} W^{Kr}_{\ell,k}(z) (u_j(\overline{z}, \,\cdot\,), \,\cdot\,)_{L^2((a,b);r dx)} v_\ell(i, \,\cdot\,)\big|_{\mathcal{N}_i}, \\
&\qquad z \in \mathbb{C}\backslash\mathbb{R}, \ z \neq \pm i,
\end{aligned} \tag{3.23}$$

where $\big(W^{Kr}_{\ell,k}(\,\cdot\,)\big)_{\ell,k=1}^{2}$ *is once again given in* (3.19)–(3.22) *and the invertible matrix* $K_{\varphi,R}(\,\cdot\,)$ *is given by*

$$K_{\varphi,R}(z) = \begin{pmatrix} -\dfrac{R_{2,2}}{R_{1,2}} + \widetilde{u}_1'(z,b) & \dfrac{e^{-i\varphi}}{R_{1,2}} - \widetilde{u}_1'(z,a) \\ \dfrac{e^{i\varphi}}{R_{1,2}} + \widetilde{u}_2'(z,b) & -\dfrac{R_{1,1}}{R_{1,2}} - \widetilde{u}_2'(z,a) \end{pmatrix}. \tag{3.24}$$

(*iii*) *If* $\gamma \in (0,\pi)$, *then the Donoghue m-function* $M^{Do}_{T_{\gamma,0},\mathcal{N}_i}(\,\cdot\,) : \mathbb{C}\backslash\mathbb{R} \to \mathcal{B}(\mathcal{N}_i)$ *for* $T_{\gamma,0}$ *satisfies*

$$\begin{aligned} &M^{Do}_{T_{\gamma,0},\mathcal{N}_i}(\pm i) = \pm i I_{\mathcal{N}_i}, \\ &M^{Do}_{T_{\gamma,0},\mathcal{N}_i}(z) = M^{Do}_{T_{0,0},\mathcal{N}_i}(z) \\ &\quad + \frac{z-i}{\cot(\gamma) + \widetilde{u}_2'(z,a)} (u_2(\overline{z},\,\cdot\,),\,\cdot\,)_{L^2((a,b);rdx)} \sum_{\ell=1}^{2} W^{Kr}_{\ell,2}(z) v_\ell(i,\,\cdot\,)\big|_{\mathcal{N}_i}, \\ &\qquad z \in \mathbb{C}\backslash\mathbb{R},\ z \neq \pm i, \end{aligned} \tag{3.25}$$

where $\cot(\gamma) + \widetilde{u}_2'(z,a) \neq 0$ *and the scalars* $\big\{W^{Kr}_{\ell,2}(\,\cdot\,)\big\}_{\ell=1,2}$ *are given by* (3.20) *and* (3.22).

(*iv*) *If* $\delta \in (0,\pi)$, *then the Donoghue m-function* $M^{Do}_{T_{0,\delta},\mathcal{N}_i}(\,\cdot\,) : \mathbb{C}\backslash\mathbb{R} \to \mathcal{B}(\mathcal{N}_i)$ *for* $T_{0,\delta}$ *satisfies*

$$\begin{aligned} &M^{Do}_{T_{0,\delta},\mathcal{N}_i}(\pm i) = \pm i I_{\mathcal{N}_i}, \\ &M^{Do}_{T_{0,\delta},\mathcal{N}_i}(z) = M^{Do}_{T_{0,0},\mathcal{N}_i}(z) \\ &\quad - \frac{z-i}{\cot(\delta) + \widetilde{u}_1'(z,b)} (u_1(\overline{z},\,\cdot\,),\,\cdot\,)_{L^2((a,b);rdx)} \sum_{\ell=1}^{2} W^{Kr}_{\ell,1}(z) v_\ell(i,\,\cdot\,)\big|_{\mathcal{N}_i}, \\ &\qquad z \in \mathbb{C}\backslash\mathbb{R},\ z \neq \pm i, \end{aligned} \tag{3.26}$$

where $\cot(\delta) + \widetilde{u}_1'(z,b) \neq 0$ *and the scalars* $\big\{W^{Kr}_{\ell,1}(\,\cdot\,)\big\}_{\ell=1,2}$ *are given by* (3.19) *and* (3.21).

(v) *If* $\varphi \in [0, \pi)$ *and* $R \in \mathrm{SL}(2, \mathbb{R})$ *with* $R_{1,2} = 0$*, then the Donoghue* m*-function* $M^{Do}_{T_{\varphi,R}, \mathcal{N}_i}(\,\cdot\,) : \mathbb{C}\backslash\mathbb{R} \to \mathcal{B}(\mathcal{N}_i)$ *for* $T_{\varphi,R}$ *satisfies*

$$M^{Do}_{T_{\varphi,R}, \mathcal{N}_i}(\pm i) = \pm i I_{\mathcal{N}_i},$$

$$M^{Do}_{T_{\varphi,R}, \mathcal{N}_i}(z) = M^{Do}_{T_{0,0}, \mathcal{N}_i}(z) - \frac{z-i}{k_{\varphi,R}(z)}(u_{\varphi,R}(\overline{z}, \,\cdot\,), \,\cdot\,)_{L^2((a,b);rdx)} \tag{3.27}$$

$$\times \sum_{\ell=1}^{2} \big[e^{-i\varphi} R_{2,2} W^{Kr}_{\ell,2}(z) + W^{Kr}_{\ell,1}(z)\big] v_\ell(i, \,\cdot\,)\big|_{\mathcal{N}_i},$$

$$z \in \mathbb{C}\backslash\mathbb{R},\ z \neq \pm i,$$

where the matrix $\big(W^{Kr}_{\ell,k}(\,\cdot\,)\big)^2_{\ell,k=1}$ *is once again given in* (3.19)–(3.22) *and the nonzero scalar* $k_{\varphi,R}(\,\cdot\,)$ *is given by*

$$k_{\varphi,R}(z) = -R_{2,1}R_{2,2} - e^{i\varphi} R_{2,2}\widetilde{u}'_{\varphi,R}(z, a) + \widetilde{u}'_{\varphi,R}(z, b), \tag{3.28}$$

where

$$u_{\varphi,R}(\zeta, \,\cdot\,) = e^{-i\varphi} R_{2,2} u_2(\zeta, \,\cdot\,) + u_1(\zeta, \,\cdot\,), \quad \zeta \in \rho(T_{0,0}). \tag{3.29}$$

Remark 3.3 For the Krein extension, T_{0,R_K}, under the additional assumption that $T_{min} \geq \varepsilon I_{L^2((a,b);rdx)}$ for some $\varepsilon > 0$, applying [24, Theorem 3.5 (ii)], one computes for the matrix K_{0,R_K},

$$K_{0,R_K}(z) = \begin{pmatrix} \widetilde{u}'_1(z, b) - \widetilde{u}'_1(0, b) & \widetilde{u}'_1(0, a) - \widetilde{u}'_1(z, a) \\ \widetilde{u}'_2(z, b) - \widetilde{u}'_2(0, b) & \widetilde{u}'_2(0, a) - \widetilde{u}'_2(z, a) \end{pmatrix},$$

$$z \in \rho(T_{0,0}) \cap \rho(T_{0,R_K}); \tag{3.30}$$

in this case one has $0 \in \sigma(T_{0,R_K})$. ◇

4 Donoghue m-functions: One Limit Circle Endpoint

In this section we recall the Donoghue m-functions in the case where τ is in the limit circle case at precisely one endpoint (which we choose to be a without loss of generality) following [33, Sect. 5].

Hypothesis 4.1 In addition to Hypothesis 2.1 assume that τ is in the limit circle case at a and in the limit point case at b. Moreover, for $z \in \rho(T_0)$, let $\psi(z, \,\cdot\,)$

denote the unique solution to $(\tau - z)y = 0$ that satisfies $\psi(z, \cdot) \in L^2((a, b); rdx)$ and $\widetilde{\psi}(z, a) = 1$.

Assume Hypothesis 4.1. By Theorem 2.5 or 2.9, the following statements (i) and (ii) hold.

(i) If $\gamma \in [0, \pi)$, then the operator T_γ defined by

$$\begin{aligned} &T_\gamma f = T_{max} f, \\ &f \in \operatorname{dom}(T_\gamma) = \{g \in \operatorname{dom}(T_{max}) \mid \cos(\gamma)\widetilde{g}(a) + \sin(\gamma)\widetilde{g}'(a) = 0\}, \end{aligned} \tag{4.1}$$

is a self-adjoint extension of T_{min}.

(ii) If T is a self-adjoint extension of T_{min}, then $T = T_\gamma$ for some $\gamma \in [0, \pi)$.

Statements analogous to (i) and (ii) hold if τ is in the limit point case at a and in the limit circle case at b; for brevity we omit the details.

Choosing $\gamma = 0$ in (4.1) yields the self-adjoint extension T_0 with a Dirichlet-type boundary condition at a:

$$\operatorname{dom}(T_0) = \{g \in \operatorname{dom}(T_{max}) \mid \widetilde{g}(a) = 0\}. \tag{4.2}$$

Since the coefficients p, q, and r are real-valued, the solution $\psi(z, \cdot)$ has the following conjugation property:

$$\overline{\psi(z, \cdot)} = \psi(\overline{z}, \cdot), \quad z \in \rho(T_0). \tag{4.3}$$

We now turn to the Donoghue m-function $M_{T_\gamma, \mathcal{N}_i}^{Do}(\cdot)$ with T_γ any self-adjoint extension of T_{min} (cf. Theorems 5.1 and 5.2 in [33]).

Theorem 4.2 *Assume Hypothesis 4.1 and let* $\gamma \in [0, \pi)$. *The Donoghue m-function* $M_{T_\gamma, \mathcal{N}_i}^{Do}(\cdot) : \mathbb{C}\backslash\mathbb{R} \to \mathcal{B}(\mathcal{N}_i)$ *for* T_γ *satisfies*

$$\begin{aligned} &M_{T_\gamma, \mathcal{N}_i}^{Do}(\pm i) = \pm i I_{\mathcal{N}_i}, \quad \gamma \in [0, \pi), \\ &M_{T_0, \mathcal{N}_i}^{Do}(z) = \bigg[-i + \frac{\widetilde{\psi}'(z, a) - \widetilde{\psi}'(-i, a)}{\operatorname{Im}\big(\widetilde{\psi}'(i, a)\big)}\bigg] I_{\mathcal{N}_i}, \quad z \in \mathbb{C}\backslash\mathbb{R},\ z \neq \pm i, \\ &M_{T_\gamma, \mathcal{N}_i}^{Do}(z) = M_{T_0, \mathcal{N}_i}^{Do}(z) \\ &\quad + (i - z)\frac{\widetilde{\psi}'(z, a) - \widetilde{\psi}'(-i, a)}{\cot(\gamma) + \widetilde{\psi}'(z, a)} (\psi(\overline{z}, \cdot), \cdot)_{L^2((a,b);rdx)} \psi(i, \cdot)\bigg|_{\mathcal{N}_i}, \\ &\qquad \gamma \in (0, \pi),\ z \in \mathbb{C}\backslash\mathbb{R},\ z \neq \pm i. \end{aligned} \tag{4.4}$$

5 The Jacobi Operator and Its Donoghue m-Functions

We now turn to the principal topic of this paper, the Jacobi differential expression

$$\tau_{\alpha,\beta} = -(1-x)^{-\alpha}(1+x)^{-\beta}(d/dx)\big((1-x)^{\alpha+1}(1+x)^{\beta+1}\big)(d/dx),$$
$$x \in (-1,1),\ \alpha,\beta \in \mathbb{R}, \tag{5.1}$$

that is, in connection with Sect. 2 one now has

$$\begin{aligned} &a=-1, \quad b=1, \\ &p(x)=p_{\alpha,\beta}(x)=(1-x)^{\alpha+1}(1+x)^{\beta+1}, \quad q(x)=q_{\alpha,\beta}(x)=0, \\ &r(x)=r_{\alpha,\beta}(x)=(1-x)^{\alpha}(1+x)^{\beta}, \quad x\in(-1,1), \quad \alpha,\beta\in\mathbb{R} \end{aligned} \tag{5.2}$$

(see, e.g. [1, 4, 8, Ch. 22], [20, Sect. 23], [38, Ch. 4], [44, Sects. VII.6.1, XIV.2], [62, Ch. 18], [72, Ch. IV]).

L^2-realizations of $\tau_{\alpha,\beta}$ are thus most naturally associated with the Hilbert space $L^2((-1,1); r_{\alpha,\beta}dx)$. However, occasionally the weight function is absorbed into the Hilbert space leading to an equivalent differential expression in the Hilbert space $L^2((-1,1); dx)$ (cf. [18, pp. 1510–1520], [20, Sect. 24], [34]). For more recent developments see, for instance, [21–23, 43, 48].

To decide the limit point/limit circle classification of $\tau_{\alpha,\beta}$ at the interval endpoints ± 1, it suffices to note that if y_1 is a given solution of $\tau y = 0$, then a 2nd linearly independent solution y_2 of $\tau y = 0$ is obtained via the standard formula

$$y_2(x) = y_1(x)\int_c^x dx'\, p(x')^{-1} y_1(x')^{-2}, \quad c, x \in (a,b). \tag{5.3}$$

Returning to the concrete Jacobi case at hand, one can choose

$$\begin{aligned} &y_1(x)=1, \quad x\in(-1,1), \\ &y_2(x)=\int_0^x dx'\,(1-x')^{-1-\alpha}(1+x')^{-1-\beta}, \quad x\in(-1,1), \\ &= \begin{cases} 2^{-1-\alpha}\beta^{-1}(1+x)^{-\beta}[1+O(1+x)]+O(1), \\ \qquad\qquad \alpha\in\mathbb{R},\ \beta\in\mathbb{R}\backslash\{0\},\ \text{as } x\downarrow -1, \\ -2^{-1-\alpha}\ln(1+x)+O(1), \quad \alpha\in\mathbb{R},\ \beta=0,\ \text{as } x\downarrow -1, \\ 2^{-1-\beta}\alpha^{-1}(1-x)^{-\alpha}[1+O(1-x)]+O(1), \\ \qquad\qquad \alpha\in\mathbb{R}\backslash\{0\},\ \beta\in\mathbb{R},\ \text{as } x\uparrow +1, \\ -2^{-1-\beta}\ln(1-x)+O(1), \quad \alpha=0,\ \beta\in\mathbb{R},\ \text{as } x\uparrow +1. \end{cases} \end{aligned} \tag{5.4}$$

Thus, one has the classification,

$$\tau_{\alpha,\beta} \text{ is } \begin{cases} \text{regular at } -1 \text{ if and only if } \alpha \in \mathbb{R}, \beta \in (-1,0), \\ \text{in the limit circle case and singular at } -1 \text{ if and only if} \\ \quad \alpha \in \mathbb{R},\ \beta \in [0,1), \\ \text{in the limit point case at } -1 \text{ if and only if } \alpha \in \mathbb{R},\ \beta \in \mathbb{R}\backslash(-1,1), \\ \text{regular at } +1 \text{ if and only if } \alpha \in (-1,0),\ \beta \in \mathbb{R}, \\ \text{in the limit circle case and singular at } +1 \text{ if and only if} \\ \quad \alpha \in [0,1),\ \beta \in \mathbb{R}, \\ \text{in the limit point case at } +1 \text{ if and only if } \alpha \in \mathbb{R}\backslash(-1,1), \beta \in \mathbb{R}. \end{cases} \tag{5.5}$$

The maximal and preminimal operators, $T_{max,\alpha,\beta}$ and $\dot{T}_{min,\alpha,\beta}$, associated to $\tau_{\alpha,\beta}$ in $L^2((-1,1); r_{\alpha,\beta}dx)$ are then given by

$$\begin{aligned} &T_{max,\alpha,\beta} f = \tau_{\alpha,\beta} f, \\ &f \in \operatorname{dom}(T_{max,\alpha,\beta}) = \big\{g \in L^2((-1,1); r_{\alpha,\beta}dx) \,\big|\, g, g^{[1]} \in AC_{loc}((-1,1)); \\ &\qquad\qquad\qquad\qquad\qquad\qquad \tau_{\alpha,\beta} g \in L^2((-1,1); r_{\alpha,\beta}dx)\big\}, \end{aligned} \tag{5.6}$$

and

$$\begin{aligned} &\dot{T}_{min,\alpha,\beta} f = \tau_{\alpha,\beta} f, \\ &f \in \operatorname{dom}\big(\dot{T}_{min,\alpha,\beta}\big) = \big\{g \in L^2((-1,1); r_{\alpha,\beta}dx) \,\big|\, g, g^{[1]} \in AC_{loc}((-1,1)); \\ &\qquad\qquad \operatorname{supp}(g) \subset (-1,1) \text{ is compact; } \tau_{\alpha,\beta} g \in L^2((-1,1); r_{\alpha,\beta}dx)\big\}. \end{aligned} \tag{5.7}$$

The fact (5.4) naturally leads to principal and nonprincipal solutions $u_{\pm 1,\alpha,\beta}(0,x)$ and $\widehat{u}_{\pm 1,\alpha,\beta}(0,x)$ of $\tau_{\alpha,\beta} y = 0$ near ± 1 as follows:

$$\begin{aligned} &u_{-1,\alpha,\beta}(0,x) = \begin{cases} -2^{-\alpha-1}\beta^{-1}(1+x)^{-\beta}[1+O(1+x)], & \beta \in (-\infty,0), \\ 1, & \beta \in [0,\infty), \end{cases} \\ &\widehat{u}_{-1,\alpha,\beta}(0,x) = \begin{cases} 1, & \beta \in (-\infty,0), \\ -2^{-\alpha-1}\ln((1+x)/2), & \beta = 0, \\ 2^{-\alpha-1}\beta^{-1}(1+x)^{-\beta}[1+O(1+x)], & \beta \in (0,\infty), \end{cases} \\ &\qquad\qquad\qquad\qquad\qquad\qquad\qquad\qquad\qquad\qquad \alpha \in \mathbb{R}, \end{aligned} \tag{5.8}$$

and

$$u_{+1,\alpha,\beta}(0,x) = \begin{cases} 2^{-\beta-1}\alpha^{-1}(1-x)^{-\alpha}[1+O(1-x)], & \alpha \in (-\infty, 0), \\ 1, & \alpha \in [0,\infty), \end{cases}$$

$$\widehat{u}_{+1,\alpha,\beta}(0,x) = \begin{cases} 1, & \alpha \in (-\infty, 0), \\ 2^{-\beta-1}\ln((1-x)/2), & \alpha = 0, \\ -2^{-\beta-1}\alpha^{-1}(1-x)^{-\alpha}[1+O(1-x)], & \alpha \in (0,\infty), \end{cases}$$

$$\beta \in \mathbb{R}. \qquad (5.9)$$

Combining the fact (5.5) with Theorem 2.2, $\dot{T}_{min,\alpha,\beta}$ is essentially self-adjoint in $L^2((-1,1); r_{\alpha,\beta}dx)$ if and only if $\alpha, \beta \in \mathbb{R}\backslash(-1,1)$. Thus, boundary values for $T_{max,\alpha,\beta}$ at -1 exist if and only if $\alpha \in \mathbb{R}$, $\beta \in (-1,1)$, and similarly, boundary values for $T_{max,\alpha,\beta}$ at $+1$ exist if and only if $\alpha \in (-1,1)$, $\beta \in \mathbb{R}$.

Employing the principal and nonprincipal solutions (5.8), (5.9) at ± 1, according to (2.22), (2.23), generalized boundary values for $g \in \mathrm{dom}(T_{max,\alpha,\beta})$ are of the form

$$\widetilde{g}(-1) = \begin{cases} g(-1), & \beta \in (-1,0), \\ -2^{\alpha+1}\lim_{x\downarrow -1} g(x)/\ln((1+x)/2), & \beta = 0, \\ \beta 2^{\alpha+1}\lim_{x\downarrow -1}(1+x)^{\beta} g(x), & \beta \in (0,1), \end{cases}$$

$$\widetilde{g}'(-1) = \begin{cases} g^{[1]}(-1), & \beta \in (-1,0), \\ \lim_{x\downarrow -1}\big[g(x) + \widetilde{g}(-1)2^{-\alpha-1}\ln((1+x)/2)\big], & \beta = 0, \\ \lim_{x\downarrow -1}\big[g(x) - \widetilde{g}(-1)2^{-\alpha-1}\beta^{-1}(1+x)^{-\beta}\big], & \beta \in (0,1), \end{cases}$$

$$\alpha \in \mathbb{R}, \qquad (5.10)$$

$$\widetilde{g}(1) = \begin{cases} g(1), & \alpha \in (-1,0), \\ 2^{\beta+1}\lim_{x\uparrow 1} g(x)/\ln((1-x)/2), & \alpha = 0, \\ -\alpha 2^{\beta+1}\lim_{x\uparrow 1}(1-x)^{\alpha} g(x), & \alpha \in (0,1), \end{cases}$$

$$\widetilde{g}'(1) = \begin{cases} g^{[1]}(1), & \alpha \in (-1,0), \\ \lim_{x\uparrow 1}\big[g(x) - \widetilde{g}(1)2^{-\beta-1}\ln((1-x)/2)\big], & \alpha = 0, \\ \lim_{x\uparrow 1}\big[g(x) + \widetilde{g}(1)2^{-\beta-1}\alpha^{-1}(1-x)^{-\alpha}\big], & \alpha \in (0,1), \end{cases} \quad \beta \in \mathbb{R}. \qquad (5.11)$$

As a result, the minimal operator $T_{min,\alpha,\beta}$ associated to $\tau_{\alpha,\beta}$, that is, $T_{min,\alpha,\beta} = \overline{\dot{T}_{min,\alpha,\beta}}$, is thus given by

$$\begin{aligned}
&T_{min,\alpha,\beta} f = \tau_{\alpha,\beta} f, \qquad\qquad (5.12)\\
&f \in \mathrm{dom}(T_{min,\alpha,\beta}) = \big\{g \in L^2((-1,1); r_{\alpha,\beta}dx) \,\big|\, g, g^{[1]} \in AC_{loc}((-1,1));\\
&\qquad \widetilde{g}(-1) = \widetilde{g}'(-1) = \widetilde{g}(1) = \widetilde{g}'(1) = 0;\ \tau_{\alpha,\beta} g \in L^2((-1,1); r_{\alpha,\beta}dx)\big\}.
\end{aligned}$$

For a detailed treatment of solutions of the Jacobi differential equation and the associated hypergeometric differential equations we refer to Appendices A–C.

Remark 5.1 We now mention a few special cases of interest. The Legendre equation ($\alpha = \beta = 0$) has frequently been discussed in the literature, see, for instance, [32] and the extensive list of references cited therein. The Gegenbauer, or ultraspherical, equation (see, e.g., [1, Ch. 22], [62, Ch. 18], [72, Ch. IV]) can be realized by choosing the parameters $\alpha = \beta = \mu - 1/2$, noting at the endpoints $x = \pm 1$, τ_μ is regular for $\mu \in (-1/2, 1/2)$, in the limit circle case and singular for $\mu \in [1/2, 3/2)$, and in the limit point case for $\mu \in \mathbb{R}\backslash(-1/2, 3/2)$. The Chebyshev equations of the first and second kinds are two more important special cases, with the first kind realized by choosing $\mu = 0$ in the Gegenbauer equation, or $\alpha = \beta = -1/2$ in the Jacobi equation (see, e.g., [1, Ch. 22], [62, Ch. 18], [72, Ch. IV]), whereas the second kind is realized by choosing $\mu = 1$ in the Gegenbauer equation, or $\alpha = \beta = 1/2$ in the Jacobi equation (see, e.g., [1, Ch. 22], [62, Ch. 18], [72, Ch. IV]). ◇

We now determine the solutions $\phi_{0,\alpha,\beta}(z,\,\cdot\,)$ and $\theta_{0,\alpha,\beta}(z,\,\cdot\,)$ of $\tau_{\alpha,\beta} u = zu$, $z \in \mathbb{C}$, that are subject to the conditions

$$\begin{aligned}
&\widetilde{\phi}_{0,\alpha,\beta}(z,-1) = 0, \quad \widetilde{\phi}'_{0,\alpha,\beta}(z,-1) = 1,\\
&\widetilde{\theta}_{0,\alpha,\beta}(z,-1) = 1, \quad \widetilde{\theta}'_{0,\alpha,\beta}(z,-1) = 0. \qquad (5.13)
\end{aligned}$$

In particular, one obtains from (C.1),

$$\begin{aligned}
&\phi_{0,\alpha,\beta}(z,x) = \begin{cases} -2^{-\alpha-1}\beta^{-1} y_{2,\alpha,\beta,-1}(z,x), & \beta \in (-1,0),\\ y_{1,\alpha,\beta,-1}(z,x), & \beta \in [0,1), \end{cases}\\
&\theta_{0,\alpha,\beta}(z,x) = \begin{cases} y_{1,\alpha,\beta,-1}(z,x), & \beta \in (-1,0),\\ -2^{-\alpha-1} y_{2,\alpha,0,-1}(z,x), & \beta = 0,\\ 2^{-\alpha-1}\beta^{-1} y_{2,\alpha,\beta,-1}(z,x), & \beta \in (0,1), \end{cases}\\
&\qquad \alpha \in \mathbb{R},\ z \in \mathbb{C},\ x \in (-1,1). \qquad (5.14)
\end{aligned}$$

5.1 The Regular and Limit Circle Case $\alpha, \beta \in (-1, 1)$

In this section we compute the Donoghue m-function when the Jacobi problem considered is either in the regular or limit circle case at ± 1.

Using (5.13), the solutions in (3.1) for this example are given by

$$\begin{aligned} u_{1,\alpha,\beta}(z,x) &= \phi_{0,\alpha,\beta}(z,x)/\widetilde{\phi}_{0,\alpha,\beta}(z,1) \\ &= \begin{cases} y_{2,\alpha,\beta,-1}(z,x)/\widetilde{y}_{2,\alpha,\beta,-1}(z,1), & \beta \in (-1,0), \\ y_{1,\alpha,\beta,-1}(z,x)/\widetilde{y}_{1,\alpha,\beta,-1}(z,1), & \beta \in [0,1), \end{cases} \end{aligned} \tag{5.15}$$

$$u_{2,\alpha,\beta}(z,x) = \theta_{0,\alpha,\beta}(z,x) - [\widetilde{\theta}_{0,\alpha,\beta}(z,1)/\widetilde{\phi}_{0,\alpha,\beta}(z,1)]\phi_{0,\alpha,\beta}(z,x)$$

$$= \begin{cases} y_{1,\alpha,\beta,-1}(z,x) - [\widetilde{y}_{1,\alpha,\beta,-1}(z,1)/\widetilde{y}_{2,\alpha,\beta,-1}(z,1)]y_{2,\alpha,\beta,-1}(z,x), \\ \qquad\qquad\qquad \beta \in (-1,0), \\ -2^{-\alpha-1}\{y_{2,\alpha,0,-1}(z,x) - [\widetilde{y}_{2,\alpha,0,-1}(z,1)/\widetilde{y}_{1,\alpha,0,-1}(z,1)]y_{1,\alpha,0,-1}(z,x)\}, \\ \qquad\qquad\qquad \beta = 0, \\ 2^{-\alpha-1}\beta^{-1}\{y_{2,\alpha,\beta,-1}(z,x) \\ -[\widetilde{y}_{2,\alpha,\beta,-1}(z,1)/\widetilde{y}_{1,\alpha,\beta,-1}(z,1)]y_{1,\alpha,\beta,-1}(z,x)\}, \quad \beta \in (0,1), \end{cases}$$

$$\alpha \in (-1,1), \ z \in \mathbb{C}, \ x \in (-1,1),$$

where the generalized boundary values are given in (C.2)–(C.4). Hence substituting (5.15) into (3.8)–(3.11) and applying Theorem 3.2 yields the (Nevanlinna–Herglotz) Donoghue m-function $M^{Do}_{T_{A,B,\alpha,\beta},\mathcal{N}_i}(\cdot)$ for any self-adjoint extension $T_{A,B,\alpha,\beta}$ of T_{min} with $\alpha, \beta \in (-1, 1)$.

As an example of coupled boundary conditions, we consider the Krein–von Neumann extension following Example 4.3 found in [24]. For $\alpha, \beta \in (-1, 1)$, the following five cases are associated with a strictly positive minimal operator $T_{min,\alpha,\beta}$ and we now provide the corresponding choices of $R_{K,\alpha,\beta}$ for the Krein–von Neumann extension $T_{0,R_K,\alpha,\beta}$ of $T_{min,\alpha,\beta}$:

$$T_{0,R_K,\alpha,\beta} f = \tau_{\alpha,\beta} f, \tag{5.16}$$

$$f \in \operatorname{dom}(T_{0,R_K,\alpha,\beta}) = \left\{ g \in \operatorname{dom}(T_{max,\alpha,\beta}) \,\middle|\, \begin{pmatrix} \widetilde{g}(1) \\ \widetilde{g}'(1) \end{pmatrix} = R_{K,\alpha,\beta} \begin{pmatrix} \widetilde{g}(-1) \\ \widetilde{g}'(-1) \end{pmatrix} \right\},$$

$$R_{K,\alpha,\beta} = \begin{cases} \begin{pmatrix} 1 & 2^{-\alpha-\beta-1}\dfrac{\Gamma(-\alpha)\Gamma(-\beta)}{\Gamma(-\alpha-\beta)} \\ 0 & 1 \end{pmatrix}, & \alpha, \beta \in (-1,0), \\ \begin{pmatrix} -2^{-\alpha-\beta-1}\dfrac{\Gamma(-\alpha)\Gamma(-\beta)}{\Gamma(-\alpha-\beta)} & 1 \\ -1 & 0 \end{pmatrix}, & \alpha \in (-1,0),\ \beta \in (0,1), \\ \begin{pmatrix} 0 & -1 \\ 1 & 2^{-\alpha-\beta-1}\dfrac{\Gamma(-\alpha)\Gamma(-\beta)}{\Gamma(-\alpha-\beta)} \end{pmatrix}, & \alpha \in (0,1),\ \beta \in (-1,0), \\ \begin{pmatrix} 0 & -1 \\ 1 & -2^{-\beta-1}[\gamma_E + \psi(-\beta)] \end{pmatrix}, & \alpha = 0,\ \beta \in (-1,0) \\ \begin{pmatrix} 2^{-\alpha-1}[\gamma_E + \psi(-\alpha)] & 1 \\ -1 & 0 \end{pmatrix}, & \alpha \in (-1,0),\ \beta = 0, \end{cases} \tag{5.17}$$

where we interpret $1/\Gamma(0) = 0$, $\psi(\cdot) = \Gamma'(\cdot)/\Gamma(\cdot)$ denotes the Digamma function, and $\gamma_E = -\psi(1) = 0.57721\ldots$ represents Euler's constant. Obviously, $\det(R_{K,\alpha,\beta}) = 1$ in all five cases. Furthermore, as $R_{1,2} \neq 0$ for each case, Theorem 3.2 (ii) applies and one obtains the Donoghue m-function $M^{Do}_{T_{0,R_{K,\alpha,\beta}},\mathcal{N}_i}(\cdot)$ for the Krein–von Neumann extension $T_{0,R_{K,\alpha,\beta}}$ by utilizing (5.15) and (5.17) as well as the explicit form of $K_{0,R_K}(\cdot)$ in (3.30). We note once again that $M^{Do}_{T_{0,R_{K,\alpha,\beta}},\mathcal{N}_i}(\cdot)$ is a Nevanlinna–Herglotz function.

In the remaining four cases not covered by (5.17), given by all combinations of $\alpha = 0$, $\beta = 0$, $\alpha \in (0,1)$, and $\beta \in (0,1)$, one observes that [24, Theorem 3.5] is not applicable as the underlying minimal operator, $T_{min,\alpha,\beta}$, is nonnegative but not strictly positive. In particular, the Jacobi polynomials satisfy Friedrichs boundary conditions for $\alpha, \beta \in [0,1)$, hence $0 \in \sigma(T_{F,\alpha,\beta})$, $\alpha, \beta \in [0,1)$ and $T_{min,\alpha,\beta} \geq 0$ is nonnegative, but not strictly positive when $\alpha, \beta \in [0,1)$.

5.2 Precisely One Interval Endpoint in the Limit Point Case

In this section we determine the Donoghue m-function in all situations where precisely one interval endpoint is in the limit point case. We will focus on the case when $\alpha \in (-\infty,-1]$ or $\alpha \in [1,\infty)$, so that the right endpoint $x = 1$ represents the limit point case. The converse situation can be obtained by reflection with respect to the origin (i.e., considering the transform $(-1,1) \ni x \mapsto -x \in (-1,1)$).

We recall from [32, Sect. 6] that the Weyl–Titchmarsh–Kodaira solution and m-function corresponding to the Friedrichs (resp., Dirichlet) boundary condition at

$x = -1$ is determined via the requirement

$$\psi_{0,\alpha,\beta}(z,\,\cdot\,) = \theta_{0,\alpha,\beta}(z,\,\cdot\,) + m_{0,\alpha,\beta}(z)\phi_{0,\alpha,\beta}(z,\,\cdot\,) \in L^2((c,1); r_{\alpha,\beta}dx),$$
$$z \in \mathbb{C}\backslash\sigma(T_{F,\alpha,\beta}),\ \alpha \in (-\infty,-1] \cup [1,\infty),\ \beta \in (-1,1),\ c \in (-1,1). \tag{5.18}$$

In particular, since $\widetilde{\psi}'_{0,\alpha,\beta}(z,-1) = m_{0,\alpha,\beta}(z)$ one finds from Theorem 4.2,

$$M^{Do}_{T_{0,\alpha,\beta},\mathcal{N}_i}(z) = \bigg[-i + \frac{m_{0,\alpha,\beta}(z) - m_{0,\alpha,\beta}(-i)}{\operatorname{Im}(m_{0,\alpha,\beta}(i))}\bigg] I_{\mathcal{N}_i},$$
$$M^{Do}_{T_{\gamma,\alpha,\beta},\mathcal{N}_i}(z) = M^{Do}_{T_{0,\alpha,\beta},\mathcal{N}_i}(z) + (i-z)\frac{m_{0,\alpha,\beta}(z) - m_{0,\alpha,\beta}(-i)}{\cot(\gamma) + m_{0,\alpha,\beta}(z)} \tag{5.19}$$
$$\times (\psi_{0,\alpha,\beta}(\overline{z},\,\cdot\,),\,\cdot\,)_{L^2((a,b);rdx)}\psi_{0,\alpha,\beta}(i,\,\cdot\,)\big|_{\mathcal{N}_i},\quad \gamma \in (0,\pi),$$
$$\alpha \in (-\infty,-1] \cup [1,\infty),\ \beta \in (-1,1),\ z \in \mathbb{C}\backslash\mathbb{R},$$

where $\psi_{0,\alpha,\beta}(z,\,\cdot\,)$ and $m_{0,\alpha,\beta}(z,\,\cdot\,)$ are given by the following:

(I) The Case $\alpha \in [1,\infty)$ and $\beta \in (-1,0)$:

$$\psi_{0,\alpha,\beta}(z,x) = y_{1,\alpha,\beta,-1}(z,x) - 2^{-\alpha-1}\beta^{-1}y_{2,\alpha,\beta,-1}(z,x)m_{0,\alpha,\beta}(z),$$
$$m_{0,\alpha\,\beta}(z) = 2^{1+\alpha+\beta}\beta\frac{\Gamma(1+\beta)}{\Gamma(1-\beta)} \tag{5.20}$$
$$\times \frac{\Gamma([1+\alpha-\beta+\sigma_{\alpha,\beta}(z)]/2)\Gamma([1+\alpha-\beta-\sigma_{\alpha,\beta}(z)]/2)}{\Gamma([1+\alpha+\beta+\sigma_{\alpha,\beta}(z)]/2)\Gamma([1+\alpha+\beta-\sigma_{\alpha,\beta}(z)]/2)},$$
$$z \in \rho(T_{F,\alpha,\beta}),\ \alpha \in [1,\infty),\ \beta \in (-1,0),$$
$$\sigma(T_{F,\alpha,\beta}) = \{(n-\beta)(n+1+\alpha)\}_{n\in\mathbb{N}_0},\quad \alpha \in [1,\infty),\ \beta \in (-1,0),$$

with

$$\sigma_{\alpha,\beta}(z) = \big[(1+\alpha+\beta)^2 + 4z\big]^{1/2}. \tag{5.21}$$

(II) The Case $\alpha \in [1,\infty)$ and $\beta = 0$:

$$\psi_{0,\alpha,0}(z,x) = -2^{-\alpha-1}y_{2,\alpha,0,-1}(z,x) + y_{1,\alpha,0,-1}(z,x)m_{0,\alpha,0}(z),$$
$$m_{0,\alpha,0}(z) = -2^{-\alpha-1}\{2\gamma_E + \psi([1+\alpha+\sigma_{\alpha,0}(z)]/2) \tag{5.22}$$
$$+ \psi([1+\alpha-\sigma_{\alpha,0}(z)]/2)\},\quad z \in \rho(T_{F,\alpha,0}),\ \alpha \in [1,\infty),\ \beta = 0,$$
$$\sigma(T_{F,\alpha,0}) = \{n(n+1+\alpha)\}_{n\in\mathbb{N}_0},\quad \alpha \in [1,\infty),\ \beta = 0.$$

(III) The Case $\alpha \in [1,\infty)$ and $\beta \in (0,1)$:

$$\begin{aligned}
&\psi_{0,\alpha,\beta}(z,x) = 2^{-\alpha-1}\beta^{-1}y_{2,\alpha,\beta,-1}(z,x) + y_{1,\alpha,\beta,-1}(z,x)m_{0,\alpha,\beta}(z),\\
&m_{0,\alpha\,\beta}(z) = \beta^{-1}2^{-1-\alpha-\beta}\frac{-\Gamma(1-\beta)}{\Gamma(1+\beta)}\\
&\qquad\times\frac{\Gamma([1+\alpha+\beta+\sigma_{\alpha,\beta}(z)]/2)\Gamma([1+\alpha+\beta-\sigma_{\alpha,\beta}(z)]/2)}{\Gamma([1+\alpha-\beta+\sigma_{\alpha,\beta}(z)]/2)\Gamma((1+\alpha-\beta-\sigma_{\alpha,\beta}(z))/2)},\\
&\qquad z\in\rho(T_{F,\alpha,\beta}),\ \alpha\in[1,\infty),\ \beta\in(0,1),\\
&\sigma(T_{F,\alpha,\beta}) = \{n(n+1+\alpha+\beta)\}_{n\in\mathbb{N}_0},\quad \alpha\in[1,\infty),\ \beta\in(0,1).
\end{aligned}\tag{5.23}$$

(IV) The Case $\alpha \in (-\infty,-1]$ and $\beta \in (-1,0)$:

$$\begin{aligned}
&\psi_{0,\alpha,\beta}(z,x) = y_{1,\alpha,\beta,-1}(z,x) - 2^{-\alpha-1}\beta^{-1}y_{2,\alpha,\beta,-1}(z,x)m_{0,\alpha,\beta}(z),\\
&m_{0,\alpha\,\beta}(z) = 2^{1+\alpha+\beta}\beta\frac{\Gamma(1+\beta)}{\Gamma(1-\beta)}\\
&\qquad\times\frac{\Gamma([1-\alpha-\beta+\sigma_{\alpha,\beta}(z)]/2)\Gamma([1-\alpha-\beta-\sigma_{\alpha,\beta}(z)]/2)}{\Gamma([1+\beta-\alpha+\sigma_{\alpha,\beta}(z)]/2)\Gamma([1+\beta-\alpha-\sigma_{\alpha,\beta}(z)]/2)},\\
&\qquad z\in\rho(T_{F,\alpha,\beta}),\ \alpha\in(-\infty,-1],\ \beta\in(-1,0),\\
&\sigma(T_{F,\alpha,\beta}) = \{(n-\alpha-\beta)(n+1)\}_{n\in\mathbb{N}_0},\quad \alpha\in(-\infty,-1],\ \beta\in(-1,0).
\end{aligned}\tag{5.24}$$

(V) The Case $\alpha \in (-\infty,-1]$ and $\beta = 0$:

$$\begin{aligned}
&\psi_{0,\alpha,0}(z,x) = -2^{-\alpha-1}y_{2,\alpha,0,-1}(z,x) + y_{1,\alpha,0,-1}(z,x)m_{0,\alpha,0}(z),\\
&m_{0,\alpha,0}(z) = -2^{-\alpha-1}\{2\gamma_E + \psi([1-\alpha+\sigma_{\alpha,0}(z)]/2)\\
&\quad+\psi([1-\alpha-\sigma_{\alpha,0}(z)]/2)\},\quad z\in\rho(T_{F,\alpha,0}),\ \alpha\in(-\infty,-1],\ \beta=0,\\
&\sigma(T_{F,\alpha,0}) = \{(n-\alpha)(n+1)\}_{n\in\mathbb{N}_0},\quad \alpha\in(-\infty,-1],\ \beta=0.
\end{aligned}\tag{5.25}$$

(VI) The Case $\alpha \in (-\infty,-1]$ and $\beta \in (0,1)$:

$$\begin{aligned}
&\psi_{0,\alpha,\beta}(z,x) = 2^{-\alpha-1}\beta^{-1}y_{2,\alpha,\beta,-1}(z,x) + y_{1,\alpha,\beta,-1}(z,x)m_{0,\alpha,\beta}(z),\\
&m_{0,\alpha\,\beta}(z) = -\beta^{-1}2^{-1-\alpha-\beta}\frac{\Gamma(1-\beta)}{\Gamma(1+\beta)}\\
&\qquad\times\frac{\Gamma([1+\beta-\alpha+\sigma_{\alpha,\beta}(z)]/2)\Gamma([1+\beta-\alpha-\sigma_{\alpha,\beta}(z)]/2)}{\Gamma([1-\alpha-\beta+\sigma_{\alpha,\beta}(z)]/2)\Gamma([1-\alpha-\beta-\sigma_{\alpha,\beta}(z)]/2)},\\
&\qquad z\in\rho(T_{F,\alpha,\beta}),\ \alpha\in(-\infty,-1],\ \beta\in(0,1),\\
&\sigma(T_{F,\alpha,\beta}) = \{(n-\alpha)(n+1+\beta)\}_{n\in\mathbb{N}_0},\quad \alpha\in(-\infty,-1],\ \beta\in(0,1).
\end{aligned}\tag{5.26}$$

Appendix A: The Hypergeometric and Jacobi Differential Equations

In this appendix we provide the connection between the hypergeometric differential equation (cf. [1, Sect. 15.5])

$$\xi(1-\xi)\ddot{w}(\xi) + [c - (a+b+1)\xi]\dot{w}(\xi) - abw(\xi) = 0, \quad \xi \in (0,1), \tag{A.1}$$

(where $^{\boldsymbol{\cdot}} = d/d\xi$) and the Jacobi differential equation

$$\tau_{\alpha,\beta} y(z,x) = -(1-x^2)y''(z,x) + [\alpha - \beta + (\alpha+\beta+2)x]y'(z,x) = zy(z,x), \\ \alpha, \beta \in \mathbb{R},\ x \in (-1,1), \tag{A.2}$$

(where $' = d/dx$). Making the substitution $\xi = (1+x)/2$ in (A.2) yields

$$\xi(1-\xi)\ddot{y}(z,\xi) + [\beta + 1 - (\alpha+\beta+2)\xi]\dot{y}(z,\xi) + zy(z,\xi) = 0, \\ \alpha, \beta \in \mathbb{R},\ \xi \in (0,1), \tag{A.3}$$

which is equal to (A.1) once one identifies,

$$a = [1+\alpha+\beta+\sigma_{\alpha,\beta}(z)]/2, \quad b = [1+\alpha+\beta-\sigma_{\alpha,\beta}(z)]/2, \quad c = 1+\beta, \\ \sigma_{\alpha,\beta}(z) = \big[(1+\alpha+\beta)^2 + 4z\big]^{1/2}. \tag{A.4}$$

At the endpoint $x = -1$ of the Jacobi equation the substitution used to arrive at (A.3) yields $\xi = 0$, hence we next consider solutions of (A.1) near $\xi = 0$ (cf. [1, Eqs. 15.5.3, 15.5.4]) (analogous solutions near $\xi = 1$ are found in (A.13))

$$w_{1,0}(\xi) = F(a,b;c;\xi) = \sum_{n\in\mathbb{N}_0} \frac{(a)_n (b)_n}{(c)_n} \frac{\xi^n}{n!}, \quad a,b \in \mathbb{C},\ c \in \mathbb{C}\backslash(-\mathbb{N}_0), \\ w_{2,0}(\xi) = \xi^{1-c} F(a-c+1, b-c+1; 2-c; \xi), \quad a,b \in \mathbb{C},\ (c-1) \in \mathbb{C}\backslash\mathbb{N}, \\ \xi \in (0,1). \tag{A.5}$$

Here $F(\,\cdot\,,\,\cdot\,;\,\cdot\,;\,\cdot\,)$ (frequently written as ${}_2F_1(\,\cdot\,,\,\cdot\,;\,\cdot\,;\,\cdot\,)$) denotes the hypergeometric function (see, e.g., [1, Ch. 15]), $\psi(\,\cdot\,) = \Gamma'(\,\cdot\,)/\Gamma(\,\cdot\,)$ the Digamma function, $\gamma_E = -\psi(1) = 0.57721\ldots$ represents Euler's constant, and

$$(\zeta)_0 = 1, \quad (\zeta)_n = \Gamma(\zeta+n)/\Gamma(\zeta),\ n \in \mathbb{N}, \quad \zeta \in \mathbb{C}\backslash(-\mathbb{N}_0), \tag{A.6}$$

abbreviates Pochhammer's symbol (see, e.g., [1, Ch. 6]).

In addition,

$$w_{1,0} \text{ and } w_{2,0} \text{ are linearly independent if } c \in \mathbb{C}\backslash\mathbb{Z}, \tag{A.7}$$

which can be seen by noticing the different behaviors of $w_{1,0}(\xi)$, $w_{2,0}(\xi)$ around $\xi = 0$. One notes that only the case $c = 1 + \beta \in (0, 2)$ is needed. Thus, for $c = 1$ we will use instead

$$\begin{aligned} w_{1,0}(\xi) &= F(a, b; 1; \xi), \quad a, b \in \mathbb{C}, \\ w_{2,0}^{\ln}(\xi) &= F(a, b; 1; \xi)\ln(\xi) + \sum_{n\in\mathbb{N}} \frac{(a)_n(b)_n}{(n!)^2}\xi^n \\ &\times [\psi(a+n) - \psi(a) + \psi(b+n) - \psi(b) - 2\psi(n+1) - 2\gamma_E], \; a, b \in \mathbb{C}\backslash(-\mathbb{N}_0), \\ &\qquad \xi \in (0, 1), \end{aligned} \tag{A.8}$$

where the superscipt "ln" indicates the presence of a logarithmic term (familiar from Frobenius theory).

Using (A.4) in formulas (A.5) and (A.8), one obtains for the solutions of the Jacobi differential equation $\tau_{\alpha,\beta}y(z,\,\cdot\,) = zy(z,\,\cdot\,)$ (cf. (A.2)) near $x = -1$,

$$y_{1,\alpha,\beta,-1}(z, x) = F(a_{\alpha,\beta,\sigma_{\alpha,\beta}(z)}, a_{\alpha,\beta,-\sigma_{\alpha,\beta}(z)}; 1 + \beta; (1 + x)/2), \quad \beta \in \mathbb{R}\backslash(-\mathbb{N}), \tag{A.9}$$

$$y_{2,\alpha,\beta,-1}(z, x) = (1 + x)^{-\beta} F(a_{\alpha,-\beta,\sigma_{\alpha,\beta}(z)}, a_{\alpha,-\beta,-\sigma_{\alpha,\beta}(z)}; 1 - \beta; (1 + x)/2), \quad \beta \in \mathbb{R}\backslash\mathbb{N}_0, \tag{A.10}$$

$$\begin{aligned} y_{2,\alpha,0,-1}(z, x) &= F(a_{\alpha,0,\sigma_{\alpha,0}(z)}, a_{\alpha,0,-\sigma_{\alpha,0}(z)}; 1; (1 + x)/2)\ln((1 + x)/2) \\ &+ \sum_{n\in\mathbb{N}} \frac{(a_{\alpha,0,\sigma_{\alpha,0}(z)})_n(a_{\alpha,0,-\sigma_{\alpha,0}(z)})_n}{2^n(n!)^2}(1 + x)^n \\ &\times [\psi(a_{\alpha,0,\sigma_{\alpha,0}(z)} + n) - \psi(a_{\alpha,0,\sigma_{\alpha,0}(z)}) + \psi(a_{\alpha,0,-\sigma_{\alpha,0}(z)} + n) \\ &- \psi(a_{\alpha,0,-\sigma_{\alpha,0}(z)}) - 2\psi(n + 1) - 2\gamma_E], \quad \beta = 0, \\ &\alpha \in \mathbb{R},\; z \in \mathbb{C},\; x \in (-1, 1), \end{aligned} \tag{A.11}$$

where we abbreviated

$$a_{\mu,\nu,\pm\sigma} = [1 + \mu + \nu \pm \sigma]/2, \quad \mu, \nu, \sigma \in \mathbb{C}. \tag{A.12}$$

Again one observes that for $z \in \mathbb{C}$, $y_{1,\alpha,\beta,-1}(z,\,\cdot\,)$ and $y_{2,\alpha,\beta,-1}(z,\,\cdot\,)$ are linearly independent for $\alpha \in \mathbb{R}$, $\beta \in \mathbb{R}\backslash\mathbb{Z}$. Similarly, for $z \in \mathbb{C}$, $y_{1,\alpha,0,-1}(z,\,\cdot\,)$ and $y_{2,\alpha,0,-1}(z,\,\cdot\,)$ are linearly independent for $\alpha \in \mathbb{R}$.

In precisely the same manner additional solutions of (A.1) are given by

$$\begin{aligned} &w_{1,1}(\xi) = F(a, b; a+b-c+1; 1-\xi), \quad a, b \in \mathbb{C},\ c-a-b \in \mathbb{C}\backslash\mathbb{N}, \\ &w_{2,1}(\xi) = (1-\xi)^{c-a-b} F(c-a, c-b; c-a-b+1; 1-\xi), \\ &\qquad\qquad a, b \in \mathbb{C},\ a+b-c \in \mathbb{C}\backslash\mathbb{N}, \end{aligned} \tag{A.13}$$

and for $a+b-c=0$,

$$\begin{aligned} &w_{1,1}(\xi) = F(a, b; 1; 1-\xi), \quad a, b \in \mathbb{C}, \\ &w_{2,1}^{\ln}(\xi) = F(a, b; 1; 1-\xi)\ln(1-\xi) + \sum_{n\in\mathbb{N}} \frac{(a)_n (b)_n}{(n!)^2}(1-\xi)^n \\ &\qquad \times [\psi(a+n) - \psi(a) + \psi(b+n) - \psi(b) - 2\psi(n+1) - 2\gamma_E], \\ &\qquad\qquad a, b \in \mathbb{C},\ \xi \in (0, 1), \end{aligned} \tag{A.14}$$

which are obtained from (A.5) and (A.8) by the change of variables

$$(a,\ b,\ c,\ \xi) \to (a,\ b,\ a+b-c+1,\ 1-\xi). \tag{A.15}$$

Together with the identification $x = (1+\xi)/2$ and (A.4) one obtains the following solutions of $\tau_{\alpha,\beta} y(z,\ \cdot) = zy(z,\ \cdot)$ near $x = +1$,

$$\begin{aligned} y_{1,\alpha,\beta,+1}(z, x) = F(a_{\alpha,\beta,\sigma_{\alpha,\beta}(z)}, a_{\alpha,\beta,-\sigma_{\alpha,\beta}(z)}; 1+\alpha; (1-x)/2), \\ \alpha \in \mathbb{R}\backslash(-\mathbb{N}), \end{aligned} \tag{A.16}$$

$$\begin{aligned} y_{2,\alpha,\beta,+1}(z, x) = (1-x)^{-\alpha} F(a_{-\alpha,\beta,\sigma_{\alpha,\beta}(z)}, a_{-\alpha,\beta,-\sigma_{\alpha,\beta}(z)}; 1-\alpha; (1-x)/2) \\ \alpha \in \mathbb{R}\backslash\mathbb{N}, \end{aligned} \tag{A.17}$$

$$\begin{aligned} y_{2,0,\beta,+1}(z, x) &= F(a_{0,\beta,\sigma_{0,\beta}(z)}, a_{0,\beta,-\sigma_{0,\beta}(z)}; 1; (1-x)/2)\ln((1-x)/2) \\ &+ \sum_{n\in\mathbb{N}} \frac{(a_{0,\beta,\sigma_{0,\beta}(z)})_n (a_{0,\beta,-\sigma_{0,\beta}(z)})_n}{2^n (n!)^2}(1-x)^n \\ &\times [\psi(a_{0,\beta,\sigma_{0,\beta}(z)} + n) - \psi(a_{0,\beta,\sigma_{0,\beta}(z)}) + \psi(a_{0,\beta,-\sigma_{0,\beta}(z)} + n) \\ &- \psi(a_{0,\beta,-\sigma_{0,\beta}(z)}) - 2\psi(n+1) - 2\gamma_E], \quad \alpha = 0, \\ &\beta \in \mathbb{R},\ z \in \mathbb{C},\ x \in (-1, 1). \end{aligned} \tag{A.18}$$

Again, for $z \in \mathbb{C}$, $y_{1,\alpha,\beta,+1}(z,\ \cdot)$ and $y_{2,\alpha,\beta,+1}(z,\ \cdot)$ are linearly independent for $\alpha \in \mathbb{R}\backslash\mathbb{Z}$, $\beta \in \mathbb{R}$. Similarly, for $z \in \mathbb{C}$, $y_{1,0,\beta,+1}(z,\ \cdot)$ and $y_{2,0,\beta,+1}(z,\ \cdot)$ are linearly independent for $\beta \in \mathbb{R}$.

In the limit point case at $x = 1$, where $\alpha \in (-\infty, -1] \cup [1, \infty)$, one only needs the principal solutions, which are $y_{1,\alpha,\beta,+1}(z, \cdot)$ for $\alpha \geq 1$ and $y_{2,\alpha,\beta,+1}(z, \cdot)$ for $\alpha \leq -1$. Thus, one concludes from (A.16) and (A.17) that these cases are already covered, and one does not have to define an additional solution for $\alpha \in \mathbb{Z}\backslash\{0\}$.

Since $(a_{\alpha,\beta,\sigma_{\alpha,\beta}(z)})_n(a_{\alpha,\beta,-\sigma_{\alpha,\beta}(z)})_n$, $n \in \mathbb{N}_0$, depends polynomially on $z \in \mathbb{C}$, one infers that

$$\text{for fixed } x \in (0,1),\ y_{j,\alpha,\beta,\pm 1}(z,x),\ j = 1,2, \text{ are entire with respect to } z \in \mathbb{C}. \tag{A.19}$$

Moreover $y_{j,\alpha,\beta,\pm 1}(z, x)$ satisfy the relations (cf. (A.28))

$$y_{1,\alpha,\beta,-1}(z,x) = (1+x)^{-\beta} y_{2,\alpha,-\beta,-1}(z + (1+\alpha)\beta, x), \quad \alpha \in \mathbb{R},\ \beta \in \mathbb{R}\backslash\{0\}, \tag{A.20}$$

$$y_{2,\alpha,\beta,-1}(z,x) = (1+x)^{-\beta} y_{1,\alpha,-\beta,-1}(z + (1+\alpha)\beta, x), \quad \alpha \in \mathbb{R},\ \beta \in \mathbb{R}\backslash\{0\}, \tag{A.21}$$

$$y_{1,\alpha,\beta,+1}(z,x) = (1-x)^{-\alpha} y_{2,-\alpha,\beta,+1}(z + (1+\beta)\alpha, x), \quad \alpha \in \mathbb{R}\backslash\{0\},\ \beta \in \mathbb{R}, \tag{A.22}$$

$$y_{2,\alpha,\beta,+1}(z,x) = (1-x)^{-\alpha} y_{1,-\alpha,\beta,+1}(z + (1+\beta)\alpha, x), \quad \alpha \in \mathbb{R}\backslash\{0\},\ \beta \in \mathbb{R}, \tag{A.23}$$

where we used the fact

$$\sigma_{\alpha,\beta}(z) = \begin{cases} \sigma_{\alpha,-\beta}(z + (1+\alpha)\beta), \\ \sigma_{-\alpha,\beta}(z + (1+\beta)\alpha), \\ \sigma_{-\alpha,-\beta}(z + \alpha + \beta). \end{cases} \tag{A.24}$$

Remark A.1 We conclude this appendix by briefly discussing Jacobi polynomials and quasi-rational eigenfunctions. The nth Jacobi polynomial is defined as (see [61, Eq. 18.5.7])

$$P_n^{\alpha,\beta}(x) := \frac{(\alpha+1)_n}{n!} F(-n, n+\alpha+\beta+1; \alpha+1;\ (1-x)/2),$$
$$n \in \mathbb{N}_0,\ -\alpha \notin \mathbb{N},\ (-n-\alpha-\beta-1) \notin \mathbb{N}, \tag{A.25}$$

and can be defined by continuity for all parameters $\alpha, \beta \in \mathbb{R}$. Note that $P_n^{\alpha,\beta}(x)$ is a polynomial of degree at most n, and has strictly smaller degree if and only if $-n-\alpha-\beta \in \{1,\dots,n\}$ (cf. [72, p. 64]). It satisfies the equation

$$\tau_{\alpha,\beta} P_n^{\alpha,\beta}(x) = \lambda_n^{\alpha,\beta} P_n^{\alpha,\beta}(x), \tag{A.26}$$

with

$$\lambda_n^{\alpha,\beta} := n(n+1+\alpha+\beta). \tag{A.27}$$

In particular, one can verify that the Jacobi polynomials are solutions of the Jacobi differential equation (A.2) with Neumann boundary conditions at $x = +1$ (resp. $x = -1$) if $\alpha \in (-1,0)$ (resp. $\beta \in (-1,0)$) and Friedrichs boundary conditions if $\alpha \geq 0$ (resp. $\beta \geq 0$).

More generally, all quasi-rational solutions, meaning the logarithmic derivative being rational, can be derived from the Jacobi polynomials together with

$$\begin{aligned}
&(1+x)^{-\beta} \circ \tau_{\alpha,-\beta} \circ (1+x)^{\beta} = \tau_{\alpha,\beta} + (1+\alpha)\beta, \\
&(1-x)^{-\alpha} \circ \tau_{-\alpha,\beta} \circ (1-x)^{\alpha} = \tau_{\alpha,\beta} + (1+\beta)\alpha, \\
&(1-x)^{-\alpha}(1+x)^{-\beta} \circ \tau_{-\alpha,-\beta} \circ (1-x)^{\alpha}(1+x)^{\beta} = \tau_{\alpha,\beta} + \alpha + \beta,
\end{aligned} \tag{A.28}$$

where $(1+x)^{\pm\beta}$ and $(1-x)^{\pm\alpha}$ are regarded as formal multiplication operators. This is summarized in Table A.1, which is taken from [9].

Here $(1-x)^{-\alpha} P_n^{-\alpha,\beta}(x)$ satisfy at $x = +1$ the Friedrichs boundary condition for $\alpha \leq 0$ and Neumann for $\alpha \in (0,1)$, while at $x = -1$ they satisfy the Friedrichs for $\beta \geq 0$ and Neumann for $\beta \in (-1,0)$. For $(1+x)^{-\beta} P_n^{\alpha,-\beta}(x)$ the roles of α and β interchange compared to the last case, meaning Friedrichs at $x = +1$ for $\alpha \geq 0$, Neumann for $\alpha \in (-1,0)$, and at $x = -1$, Friedrichs for $\beta \leq 0$, Neumann for $\beta \in (0,1)$. Finally $(1-x)^{-\alpha}(1+x)^{-\beta} P_n^{-\alpha,-\beta}(x)$ satisfy at $x = +1$ (resp. $x = -1$) the Friedrichs boundary condition for $\alpha \leq 0$ (resp. $\beta \leq 0$) and Neumann for $\alpha \in (0,1)$ (resp. $\beta \in (0,1)$). ◇

Table A.1 Formal quasi-rational eigensolutions of $\tau_{\alpha,\beta}$

Eigenfunctions	Eigenvalues
$P_n^{\alpha,\beta}(x)$	$n(n+1+\alpha+\beta)$
$(1-x)^{-\alpha} P_n^{-\alpha,\beta}(x)$	$n(n+1-\alpha+\beta) - \alpha(1+\beta)$
$(1+x)^{-\beta} P_n^{\alpha,-\beta}(x)$	$n(n+1+\alpha-\beta) - \beta(1+\alpha)$
$(1-x)^{-\alpha}(1+x)^{-\beta} P_n^{-\alpha,-\beta}(x)$	$n(n+1-\alpha-\beta) - (\alpha+\beta)$

Appendix B: Connection Formulas

In this appendix we provide the connection formulas utilized to find the solution behaviors in Sect. 5.2. We express them using $w_{1,0}(\xi)$ and $w_{2,0}(\xi)$ $\big(w_{2,0}^{\ln}(\xi)\big)$ and their analogs $w_{1,1}(\xi)$ and $w_{2,1}(\xi)$ $\big(w_{2,1}^{\ln}(\xi)\big)$ at the endpoint $\xi = 1$.

We recall the relations (A.4) connecting the parameters a, b, c and α, β.

(I) *The case* $\alpha \in \mathbb{R}\backslash\mathbb{Z}$, $\beta \in (-1,1)\backslash\{0\}$, *that is,* $c \in (0,2)\backslash\{1\}$, $a+b-c \in \mathbb{R}\backslash\mathbb{Z}$:
The two connection formulas are given by (cf. [62, Eq. 15.10.21–22])

$$w_{1,0}(\xi) = \frac{\Gamma(c)\Gamma(c-a-b)}{\Gamma(c-a)\Gamma(c-b)} w_{1,1}(\xi) + \frac{\Gamma(c)\Gamma(a+b-c)}{\Gamma(a)\Gamma(b)} w_{2,1}(\xi), \tag{B.1}$$

$$\begin{aligned} w_{2,0}(\xi) = {}& \frac{\Gamma(2-c)\Gamma(c-a-b)}{\Gamma(1-a)\Gamma(1-b)} w_{1,1}(\xi) \\ &+ \frac{\Gamma(2-c)\Gamma(a+b-c)}{\Gamma(a-c+1)\Gamma(b-c+1)} w_{2,1}(\xi). \end{aligned} \tag{B.2}$$

One notes that poles occur on the right-hand side of (B.1), (B.2) whenever $(a+b-c) \in \mathbb{Z}$. Using (A.15) one can also express $w_{1,1}(\xi)$ or $w_{2,1}(\xi)$ as a linear combination of $w_{1,0}(\xi)$ and $w_{2,0}(\xi)$:

$$\begin{aligned} w_{1,1}(\xi) = {}& \frac{\Gamma(a+b-c+1)\Gamma(1-c)}{\Gamma(a-c+1)\Gamma(b-c+1)} w_{1,0}(\xi) \\ &+ \frac{\Gamma(a+b-c+1)\Gamma(c-1)}{\Gamma(a)\Gamma(b)} w_{2,0}(\xi), \end{aligned} \tag{B.3}$$

$$\begin{aligned} w_{2,1}(\xi) = {}& \frac{\Gamma(1+c-a-b)\Gamma(1-c)}{\Gamma(1-a)\Gamma(1-b)} w_{1,0}(\xi) \\ &+ \frac{\Gamma(1+c-a-b)\Gamma(c-1)}{\Gamma(c-a)\Gamma(c-b)} w_{2,0}(\xi), \end{aligned} \tag{B.4}$$

though, in the following we shall only write down one pair of connection formulas for brevity.

(II) *The case* $\alpha = 0$, $\beta \in \mathbb{R}\backslash\mathbb{Z}$, *that is,* $c \in \mathbb{R}\backslash\mathbb{Z}$, $a+b=c$:
The solution $w_{1,0}(\xi) = F(a,b;a+b;\xi)$ can be expanded at $\xi = 1$ (cf. [1, Eq. 15.3.10]):

$$\begin{aligned} F(a,b;a+b;\xi) = {}& \frac{\Gamma(a+b)}{\Gamma(a)\Gamma(b)} \\ &\times \sum_{n\in\mathbb{N}_0} \frac{(a)_n(b)_n}{(n!)^2} [2\psi(n+1) - \psi(a+n) - \psi(b+n) \\ &\qquad - \ln(1-\xi)](1-\xi)^n. \end{aligned} \tag{B.5}$$

Meanwhile, two linearly independent solutions at $\xi = 1$ are taken from (A.14). The connection formula for $w_{1,1}(\xi)$ is given by (B.3) with $a + b = c$. To obtain a second connection formula one compares the expansion of $w_{2,1}^{\ln}(\xi)$ at $\xi = 1$ with the expansion of $F(a, b; a + b; \xi)$ at $\xi = 1$, using (B.5), and then obtains

$$\begin{aligned} w_{2,1}^{\ln}(\xi) = &- [\psi(1-a) + \psi(1-b) + 2\gamma_E] \frac{\Gamma(1-a-b)}{\Gamma(1-a)\Gamma(1-b)} w_{1,0}(\xi) \\ &- [\psi(a) + \psi(b) + 2\gamma_E] \frac{\Gamma(a+b-1)}{\Gamma(a)\Gamma(b)} w_{2,0}(\xi). \end{aligned} \tag{B.6}$$

(III) *The case $\alpha \in \mathbb{R}\backslash\mathbb{Z}$, $\beta = 0$, that is, $c = 1$, $a + b \in \mathbb{R}\backslash\mathbb{Z}$:*
This case is analogous to the previous case, with the roles of α and β interchanged. Concretely, this means that the connection formulas (B.5) and (B.6) must be changed through the renaming (A.15) with $c \to a+b-c+1 = a+b$, as $c = 1$. As c does not appear in (B.5) and (B.6) (it was eliminated via $c = a+b$), one can adopt the aforementioned formulas directly, only changing the second index in the w's[1]

$$w_{1,0}(\xi) = \frac{\Gamma(1-a-b)}{\Gamma(1-a)\Gamma(1-b)} w_{1,1}(\xi) + \frac{\Gamma(a+b-1)}{\Gamma(a)\Gamma(b)} w_{2,1}(\xi), \tag{B.7}$$

$$\begin{aligned} w_{2,0}^{\ln}(\xi) = &-[\psi(1-a) + \psi(1-b) + 2\gamma_E] \frac{\Gamma(1-a-b)}{\Gamma(1-a)\Gamma(1-b)} w_{1,1}(\xi) \\ &- [\psi(a) + \psi(b) + 2\gamma_E] \frac{\Gamma(a+b-1)}{\Gamma(a)\Gamma(b)} w_{2,1}(\xi). \end{aligned} \tag{B.8}$$

(IV) *The case $\alpha = \beta = 0$, that is, $a + b = c = 1$:*
For $\alpha = 0$ and $\beta = 0$ the Jacobi differential expression (5.1) becomes the Legendre differential expression. Since this case was treated in detail in [32], we shall only present the connection formulas for completeness.

The special solutions $w_{1,i}(\xi)$ and $w_{2,i}^{\ln}(\xi)$ for $i = 1, 2$ are given by (A.8) and (A.14), respectively. Note that the following relations hold

$$w_{1,1}(\xi) = w_{1,0}(1-\xi), \qquad w_{2,1}^{\ln}(\xi) = w_{2,0}^{\ln}(1-\xi). \tag{B.9}$$

Using [1, Eq. 15.3.10] together with $w_{1,0}(\xi) = F(a, b; a+b; \xi)$ and Euler's famous reflection formula, $\Gamma(z)\Gamma(1-z) = \pi \csc(\pi z)$ (cf. [1, Eq. 6.1.17]), one obtains

$$w_{1,0}(\xi) = -\pi^{-1} \sin(\pi a)\big([\psi(a) + \psi(b) + 2\gamma_E] w_{1,1}(\xi) + w_{2,1}^{\ln}(\xi)\big). \tag{B.10}$$

[1] Formula (B.7) could have been obtained directly from (B.1) by setting $c = 1$.

The two relations (B.9) immediately imply

$$w_{1,1}(\xi) = -\pi^{-1}\sin(\pi a)\big([\psi(a)+\psi(b)+2\gamma_E]w_{1,0}(\xi) + w_{2,0}^{\ln}(\xi)\big), \tag{B.11}$$

$$\begin{aligned} w_{2,1}^{\ln}(\xi) &= \pi^{-1}\sin(\pi a)\big[\big([\psi(a)+\psi(b)+2\gamma_E]^2 - \pi^2[\sin(\pi a)]^{-2}\big)w_{1,0}(\xi) \\ &\quad + [\psi(a)+\psi(b)+2\gamma_E]w_{2,0}^{\ln}(\xi)\big]. \end{aligned} \tag{B.12}$$

Appendix C: Behavior of $y_{j,\alpha,\beta,\mp 1}(z,x)$, $j=1,2$, near $x=\pm 1$

In this appendix we focus on the generalized boundary values for the solutions $y_{j,\alpha,\beta,-1}(z,x)$, $j=1,2$ at $x=\mp 1$. One obtains for $z\in\mathbb{C}$,

$$\widetilde{y}_{1,\alpha,\beta,-1}(z,-1) = \begin{cases} 1, & \beta\in(-1,0), \\ 0, & \beta=0, \\ 0, & \beta\in(0,1), \end{cases}$$

$$\widetilde{y}'_{1,\alpha,\beta,-1}(z,-1) = \begin{cases} 0, & \beta\in(-1,0), \\ 1, & \beta=0, \\ 1, & \beta\in(0,1), \end{cases}$$

$$\widetilde{y}_{2,\alpha,\beta,-1}(z,-1) = \begin{cases} 0, & \beta\in(-1,0), \\ -2^{\alpha+1}, & \beta=0, \\ \beta 2^{\alpha+1}, & \beta\in(0,1), \end{cases}$$

$$\widetilde{y}'_{2,\alpha,\beta,-1}(z,-1) = \begin{cases} -\beta 2^{\alpha+1}, & \beta\in(-1,0), \\ 0, & \beta=0, \\ 0, & \beta\in(0,1), \end{cases} \qquad \alpha\in\mathbb{R}, \tag{C.1}$$

and employing connection formulas for the endpoint $x=+1$,

$$\widetilde{y}_{1,\alpha,\beta,-1}(z,1) = \begin{cases} \dfrac{\Gamma(1+\beta)\Gamma(-\alpha)}{\Gamma(a_{-\alpha,\beta,\sigma_{\alpha,\beta}(z)})\Gamma(a_{-\alpha,\beta,-\sigma_{\alpha,\beta}(z)})}, & \alpha\in(-1,0), \\ \dfrac{-2^{1+\alpha+\beta}\Gamma(1+\alpha)\Gamma(1+\beta)}{\Gamma(a_{\alpha,\beta,\sigma_{\alpha,\beta}(z)})\Gamma(a_{\alpha,\beta,-\sigma_{\alpha,\beta}(z)})}, & \alpha\in[0,1), \end{cases}$$

$$\widetilde{y}'_{1,\alpha,\beta,-1}(z,1) = \begin{cases} \dfrac{2^{1+\alpha+\beta}\Gamma(1+\alpha)\Gamma(1+\beta)}{\Gamma(a_{\alpha,\beta,\sigma_{\alpha,\beta}(z)})\Gamma(a_{\alpha,\beta,-\sigma_{\alpha,\beta}(z)})}, & \alpha \in (-1,0), \\ \dfrac{-\Gamma(1+\beta)}{\Gamma(a_{0,\beta,\sigma_{0,\beta}(z)})\Gamma(a_{0,\beta,-\sigma_{0,\beta}(z)})}[2\gamma_E & \\ \quad +\psi(a_{0,\beta,\sigma_{0,\beta}(z)}) + \psi(a_{0,\beta,-\sigma_{0,\beta}(z)})], & \alpha = 0, \\ \dfrac{\Gamma(1+\beta)\Gamma(-\alpha)}{\Gamma(a_{-\alpha,\beta,\sigma_{\alpha,\beta}(z)})\Gamma(a_{-\alpha,\beta,-\sigma_{\alpha,\beta}(z)})}, & \alpha \in (0,1), \end{cases}$$

$$\beta \in (-1,1), \tag{C.2}$$

$$\widetilde{y}_{2,\alpha,\beta,-1}(z,1) = \begin{cases} \dfrac{2^{-\beta}\Gamma(1-\beta)\Gamma(-\alpha)}{\Gamma(a_{-\alpha,-\beta,\sigma_{\alpha,\beta}(z)})\Gamma(a_{-\alpha,-\beta,-\sigma_{\alpha,\beta}(z)})}, & \alpha \in (-1,0), \\ \dfrac{-2^{\alpha+1}\Gamma(1+\alpha)\Gamma(1-\beta)}{\Gamma(a_{\alpha,-\beta,\sigma_{\alpha,\beta}(z)})\Gamma(a_{\alpha,-\beta,-\sigma_{\alpha,\beta}(z)})}, & \alpha \in [0,1), \end{cases}$$

$$\widetilde{y}'_{2,\alpha,\beta,-1}(z,1) = \begin{cases} \dfrac{2^{\alpha+1}\Gamma(1+\alpha)\Gamma(1-\beta)}{\Gamma(a_{\alpha,-\beta,\sigma_{\alpha,\beta}(z)})\Gamma(a_{\alpha,-\beta,-\sigma_{\alpha,\beta}(z)})}, & \alpha \in (-1,0), \\ \dfrac{-2^{-\beta}\Gamma(1-\beta)}{\Gamma(a_{0,-\beta,\sigma_{0,\beta}(z)})\Gamma(a_{0,-\beta,-\sigma_{0,\beta}(z)})}[2\gamma_E & \\ \quad +\psi(a_{0,-\beta,\sigma_{0,\beta}(z)}) + \psi(a_{0,-\beta,-\sigma_{0,\beta}(z)})], & \alpha = 0, \\ \dfrac{2^{-\beta}\Gamma(1-\beta)\Gamma(-\alpha)}{\Gamma(a_{-\alpha,-\beta,\sigma_{\alpha,\beta}(z)})\Gamma(a_{-\alpha,-\beta,-\sigma_{\alpha,\beta}(z)})}, & \alpha \in (0,1), \end{cases}$$

$$\beta \in (-1,1)\backslash\{0\}, \tag{C.3}$$

$$\widetilde{y}_{2,\alpha,0,-1}(z,1) = \begin{cases} \dfrac{-[2\gamma_E + \psi(a_{-\alpha,0,\sigma_{\alpha,0}(z)}) + \psi(a_{-\alpha,0,-\sigma_{\alpha,0}(z)})]\Gamma(-\alpha)}{\Gamma(a_{-\alpha,0,\sigma_{\alpha,0}(z)})\Gamma(a_{-\alpha,0,-\sigma_{\alpha,0}(z)})}, & \\ & \alpha \in (-1,0), \\ \dfrac{[2\gamma_E + \psi(a_{\alpha,0,\sigma_{\alpha,0}(z)}) + \psi(a_{\alpha,0,-\sigma_{\alpha,0}(z)})]\Gamma(1+\alpha)}{2^{-\alpha-1}\Gamma(a_{\alpha,0,\sigma_{\alpha,0}(z)})\Gamma(a_{\alpha,0,-\sigma_{\alpha,0}(z)})}, & \\ & \alpha \in [0,1), \end{cases}$$

$$
\widetilde{y}'_{2,\alpha,0,-1}(z,1) = \begin{cases} \dfrac{[2\gamma_E + \psi(a_{\alpha,0,\sigma_{\alpha,0}(z)}) + \psi(a_{\alpha,0,-\sigma_{\alpha,0}(z)})]\Gamma(1+\alpha)}{-2^{-\alpha-1}\Gamma(a_{\alpha,0,\sigma_{\alpha,0}(z)})\Gamma(a_{\alpha,0,-\sigma_{\alpha,0}(z)})}, \\ \qquad\qquad \alpha \in (-1,0), \\ \dfrac{[2\gamma_E + \psi((1+\sigma_{0,0}(z))/2) + \psi((1-\sigma_{0,0}(z))/2)]^2}{\Gamma((1+\sigma_{0,0}(z))/2)\Gamma((1-\sigma_{0,0}(z))/2)} \\ -\Gamma((1+\sigma_{0,0}(z))/2)\Gamma((1-\sigma_{0,0}(z))/2), \quad \alpha = 0, \\ \dfrac{-[2\gamma_E + \psi(a_{-\alpha,0,\sigma_{\alpha,0}(z)}) + \psi(a_{-\alpha,0,-\sigma_{\alpha,0}(z)})]\Gamma(-\alpha)}{\Gamma(a_{-\alpha,0,\sigma_{\alpha,0}(z)})\Gamma(a_{-\alpha,0,-\sigma_{\alpha,0}(z)})}, \\ \qquad\qquad \alpha \in (0,1), \end{cases}
$$

$$
\beta = 0. \tag{C.4}
$$

References

1. M. Abramowitz, I.A. Stegun, *Handbook of Mathematical Functions*, 9th printing (Dover, New York, 1972)
2. D. Alpay, J. Behrndt, Generalized Q-functions and Dirichlet-to-Neumann maps for elliptic differential operators. J. Funct. Anal. **257**, 1666–1694 (2009)
3. W.O. Amrein, D.B. Pearson, M operators: a generalization of Weyl–Titchmarsh theory. J. Comp. Appl. Math. **171**, 1–26 (2004)
4. P. Bailey, W. Everitt, A. Zettl, Algorithm 810: The SLEIGN2 Sturm-Liouville Code. ACM Trans. Math. Software **27**, 143–192 (2001)
5. J. Behrndt, M. Langer, Boundary value problems for elliptic partial differential operators on bounded domains. J. Funct. Anal. **243**, 536–565 (2007)
6. J. Behrndt, J. Rohleder, Titchmarsh–Weyl theory for Schrödinger operators on unbounded domains. J. Spectral Theory **6**, 67–87 (2016)
7. J. Behrndt, S. Hassi, H. De Snoo, *Boundary Value Problems, Weyl Functions, and Differential Operators*. Monographs in Math., vol. 108 (Birkhäuser, Springer, 2020)
8. S. Bochner, Über Sturm–Liouvillesche Polynomsysteme. Math. Z. **29**, 730–736 (1929; German)
9. N. Bonneux, Exceptional Jacobi polynomials. J. Approx. Theory **239**, 72–112 (2019)
10. J.F. Brasche, M. Malamud, H. Neidhardt, Weyl function and spectral properties of self-adjoint extensions. Integr. Eq. Oper. Th. **43**, 264–289 (2002)
11. B.M. Brown, G. Grubb, I.G. Wood, M-functions for closed extensions of adjoint pairs of operators with applications to elliptic boundary problems. Math. Nachr. **282**, 314–347 (2009)
12. J. Brüning, V. Geyler, K. Pankrashkin, Spectra of self-adjoint extensions and applications to solvable Schrödinger operators. Rev. Math. Phys. **20**, 1–70 (2008)
13. S. Clark, F. Gesztesy, R. Nichols, Principal solutions revisited, in *Stochastic and Infinite Dimensional Analysis*, ed. by C.C. Bernido, M.V. Carpio-Bernido, M. Grothaus, T. Kuna, M.J. Oliveira, J.L. da Silva. Trends in Mathematics (Birkhäuser, Springer, 2016), pp. 85–117
14. E.A. Coddington, N. Levinson, *Theory of Ordinary Differential Equations* (Krieger Publ., Malabar, 1985)
15. V.A. Derkach, M.M. Malamud, E.R. Tsekanovskii, Sectorial extensions of a positive operator, and the characteristic function. Sov. Math. Dokl. **37**, 106–110 (1988)

16. V. Derkach, S. Hassi, M. Malamud, H. de Snoo, Boundary relations and generalized resolvents of symmetric operators. Russian J. Math. Phys. **16**, 17–60 (2009)
17. W.F. Donoghue, On the perturbation of spectra. Commun. Pure Appl. Math. **18**, 559–579 (1965)
18. N. Dunford, J.T. Schwartz, *Linear Operators. Part II: Spectral Theory* (Wiley, Interscience, New York, 1988)
19. J. Eckhardt, F. Gesztesy, R. Nichols, G. Teschl, Weyl–Titchmarsh theory for Sturm–Liouville operators with distributional potentials. Opuscula Math. **33**, 467–563 (2013)
20. W.N. Everitt, A catalogue of Sturm–Liouville differential equations, in *Sturm-Liouville Theory: Past and Present*, ed. by W.O. Amrein, A.M. Hinz, D.B. Pearson (Birkhäuser, Basel, 2005), pp. 271–331
21. W.N. Everitt, K.H. Kwon, L.L. Littlejohn, R. Wellman, G.J. Yoon, Jacobi–Stirling numbers, Jacobi polynomials, and the left-definite analysis of the classical Jacobi differential expression. J. Comput. Appl. Math. **208**, 29–56 (2007)
22. D. Frymark, Boundary triples and Weyl m-functions for powers of the Jacobi differential operator. J. Diff. Eq. **269**, 7931–7974 (2020)
23. D. Frymark, C. Liaw, Properties and decompositions of domains for powers of the Jacobi differential operator. J. Math. Anal. Appl. **489**, 124–155 (2020)
24. G. Fucci, F. Gesztesy, K. Kirsten, L.L. Littlejohn, R. Nichols, J. Stanfill, The Krein–von Neumann extension revisited. Applicable Anal. 25 pp. (2021). https://doi.org/10.1080/00036811.2021.1938005
25. F. Gesztesy, L. Pittner, On the Friedrichs extension of ordinary differential operators with strongly singular potentials. Acta Phys. Austriaca **51**, 259–268 (1979)
26. F. Gesztesy, E. Tsekanovskii, On matrix-valued Herglotz functions. Math. Nachr. **218**, 61–138 (2000)
27. F. Gesztesy, R. Nichols, M. Zinchenko, *Sturm–Liouville Operators, Their Spectral Theory, and Some Applications*, (book manuscript in preparation)
28. F. Gesztesy, K.A. Makarov, E. Tsekanovskii, An addendum to Krein's formula. J. Math. Anal. Appl. **222**, 594–606 (1998)
29. F. Gesztesy, N.J. Kalton, K.A. Makarov, E. Tsekanovskii, Some applications of operator-valued Herglotz functions, in *Operator Theory, System Theory and Related Topics. The Moshe Livšic Anniversary Volume*, ed. by D. Alpay, V. Vinnikov. Operator Theory: Adv. Appl., vol. 123 (Birkhäuser, Basel, 2001), pp. 271–321
30. F. Gesztesy, R. Weikard, M. Zinchenko, On spectral theory for Schrödinger operators with operator-valued potentials. J. Diff. Eq. **255**, 1784–1827 (2013)
31. F. Gesztesy, S. Naboko, R. Weikard, M. Zinchenko, Donoghue-type m-functions for Schrödinger operators with operator-valued potentials. J. d'Analyse Math. **137**, 373–427 (2019)
32. F. Gesztesy, L.L. Littlejohn, R. Nichols, On self-adjoint boundary conditions for singular Sturm–Liouville operators bounded from below. J. Diff. Eq. **269**, 6448–6491 (2020)
33. F. Gesztesy, L.L. Littlejohn, R. Nichols, M. Piorkowski, J. Stanfill, Donoghue m-functions for singular Sturm-Liouville operators. Algebra i Analiz (St. Petersburg Math. J.) **35**(1), 134–183 (2023). http://arxiv.org/abs/2107.09832
34. U. Grünewald, Jacobische Differentialoperatoren. Math. Nachr. **63**, 239–253 (1974; German)
35. P. Hartman, *Ordinary Differential Equations* (SIAM, Philadelphia, 2002)
36. P. Hartman, A. Wintner, On the assignment of asymptotic values for the solutions of linear differential equations of second order. Am. J. Math. **77**, 475–483 (1955)
37. S. Hassi, M. Malamud, V. Mogilevskii, Unitary equivalence of proper extensions of a symmetric operator and the Weyl function. Integral Equ. Oper. Theory **77**, 449–487 (2013)
38. M. Ismail, *Classical and Quantum Orthogonal Polynomials in One Variable*. Encyclopedia of Mathematics and Its Applications, vol. 98 (Cambridge University Press, Cambridge, 2005)
39. K. Jörgens, F. Rellich, Eigenwerttheorie Gewöhnlicher Differentialgleichungen (Springer, Berlin, 1976; German)

40. H. Kalf, On the characterization of the Friedrichs extension of ordinary or elliptic differential operators with a strongly singular potential. J. Funct. Anal. **10**, 230–250 (1972)
41. H. Kalf, A characterization of the Friedrichs extension of Sturm–Liouville operators. J. Lond. Math. Soc. (2) **17**, 511–521 (1978)
42. H.G. Kaper, M.K. Kwong, A. Zettl, Characterizations of the Friedrichs extensions of singular Sturm–Liouville expressions. SIAM J. Math. Anal. **17**, 772–777 (1986)
43. T. Koornwinder, A. Kostenko, G. Teschl, Jacobi polynomials, Bernstein-type inequalities and dispersion estimates for the discrete Laguerre operator. Adv. Math. **333**, 796–821 (2018)
44. A. Krall, *Hilbert Space, Boundary Value Problems and Orthogonal Polynomials* (Birkhäuser-Verlag, Berlin, 2002)
45. M.G. Krein, Concerning the resolvents of an Hermitian operator with the deficiency-index (m, m). Comptes Rendue (Doklady) Acad. Sci. URSS (N.S.), **52**, 651–654 (1946; Russian)
46. M.G. Krein, H. Langer, Defect subspaces and generalized resolvents of an Hermitian operator in the space Π_κ. Funct. Anal. Appl. **5**, 136–146; 217–228 (1971)
47. M.G. Krein I.E. Ovčarenko, Inverse problems for Q-functions and resolvent matrices of positive Hermitian operators. Sov. Math. Dokl. **19**, 1131–1134 (1978)
48. A. Kuijlaars, A. Martinez-Finkelshtein, R. Orive, Orthogonality of Jacobi polynomials with general parameters. Electron. Trans. Numer. Anal. **19**, 1–17 (2005)
49. H. Langer, B. Textorius, On generalized resolvents and Q-functions of symmetric linear relations (subspaces) in Hilbert space. Pacific J. Math. **72**, 135–165 (1977)
50. W. Leighton, M. Morse, Singular quadratic functionals. Trans. Am. Math. Soc. **40**, 252–286 (1936)
51. M.M. Malamud, Certain classes of extensions of a lacunary Hermitian operator. Ukrain. Math. J. **44**, 190–204 (1992)
52. M. Malamud, H. Neidhardt, On the unitary equivalence of absolutely continuous parts of self-adjoint extensions. J. Funct. Anal. **260**, 613–638 (2011)
53. M. Malamud, H. Neidhardt, Sturm–Liouville boundary value problems with operator potentials and unitary equivalence. J. Diff. Eq. **252**, 5875–5922 (2012)
54. M. Marletta, Eigenvalue problems on exterior domains and Dirichlet to Neumann maps. J. Comp. Appl. Math. **171**, 367–391 (2004)
55. M. Marletta, A. Zettl, The Friedrichs extension of singular differential operators. J. Diff. Eq. **160**, 404–421 (2000)
56. V. Mogilevskii, Boundary triplets and Titchmarsh–Weyl functions of differential operators with arbitrary deficiency indices. Meth. Funct. Anal. Topol. **15**, 280–300 (2009)
57. S.N. Naboko, Boundary values of analytic operator functions with a positive imaginary part. J. Soviet Math. **44**, 786–795 (1989)
58. S.N. Naboko, The boundary behavior of $\mathfrak{S}_p$-valued functions analytic in the half-plane with nonnegative imaginary part, in *Functional Analysis and Operator Theory*. Banach Center Publications, vol. 30 (Institute of Mathematics, Polish Academy of Sciences, Warsaw, 1994), pp. 277–285
59. M.A. Naimark, *Linear Differential Operators. Part II: Linear Differential Operators in Hilbert Space*. Transl. by E.R. Dawson, Engl. translation edited by W.N. Everitt (Ungar Publishing, New York, 1968)
60. H.-D. Niessen, A. Zettl, Singular Sturm–Liouville problems: the Friedrichs extension and comparison of eigenvalues. Proc. Lond. Math. Soc. (3) **64**, 545–578 (1992)
61. F.W.J. Olver, D.W. Lozier, R.F. Boisvert, C.W. Clark (eds.), *NIST Handbook of Mathematical Functions* (National Institute of Standards and Technology (NIST), U.S. Dept. of Commerce, and Cambridge Univ. Press, 2010)
62. F.W.J. Olver et al., *NIST Handbook of Mathematical Functions*. http://dlmf.nist.gov/, Release 1.0.26 of 2020-03-15
63. K. Pankrashkin, An example of unitary equivalence between self-adjoint extensions and their parameters. J. Funct. Anal. **265**, 2910–2936 (2013)
64. D.B. Pearson, *Quantum Scattering and Spectral Theory* (Academic, London, 1988)

65. A. Posilicano, Boundary triples and Weyl functions for singular perturbations of self-adjoint operators. Meth. Funct. Anal. Topol. **10**, 57–63 (2004)
66. F. Rellich, Die zulässigen Randbedingungen bei den singulären Eigenwertproblemen der mathematischen Physik. (Gewöhnliche Differentialgleichungen zweiter Ordnung.). Math. Z. **49**, 702–723 (1943/1944; German)
67. F. Rellich, Halbbeschränkte gewöhnliche Differentialoperatoren zweiter Ordnung. Math. Ann. **122**, 343–368 (1951; German)
68. R. Rosenberger, Charakterisierungen der Friedrichsfortsetzung von halbbeschränkten Sturm–Liouville Operatoren. Ph.D. Thesis (Technical University of Darmstadt, 1984; German)
69. R. Rosenberger, A new characterization of the Friedrichs extension of semibounded Sturm–Liouville operators. J. Lond. Math. Soc. (2) **31**, 501–510 (1985)
70. V. Ryzhov, A general boundary value problem and its Weyl function. Opuscula Math. **27**, 305–331 (2007)
71. Sh.N. Saakjan, Theory of resolvents of a symmetric operator with infinite defect numbers. Akad. Nauk. Armjan. SSR Dokl. **41**, 193–198 (1965; Russian)
72. G. Szegő, *Orthogonal Polynomials*, 4th edn. Colloquium Publications, vol. 23 (American Mathematical Society, Providence, 1975)
73. G. Teschl, *Mathematical Methods in Quantum Mechanics. With Applications to Schrödinger Operators*, 2nd edn. Graduate Studies in Mathematics, vol. 157 (American Mathematical Society, Providence, 2014)
74. J. Weidmann, Linear Operators in Hilbert Spaces. Graduate Texts in Mathematics, vol. 68 (Springer, New York, 1980)
75. J. Weidmann, *Lineare Operatoren in Hilberträumen. Teil II: Anwendungen* (Teubner, Stuttgart, 2003; German)
76. S. Yao, J. Sun, A. Zettl, The Sturm–Liouville Friedrichs extension. Appl. Math. **60**, 299–320 (2015)
77. A. Zettl, *Sturm–Liouville Theory*. Mathematical Surveys and Monographs, vol. 121 (American Mathematical Society, Providence, 2005)

Entanglement Breaking Rank Via Complementary Channels and Multiplicative Domains

David W. Kribs, Jeremy Levick, Rajesh Pereira, and Mizanur Rahaman

Abstract Quantum entanglement can be studied through the theory of completely positive maps in a number of ways, including by making use of the Choi-Jamiolkowski isomorphism, which identifies separable states with entanglement breaking quantum channels, and optimal ensemble length with entanglement breaking rank. The multiplicative domain is an important operator structure in the theory of completely positive maps. We introduce a new technique to determine if a channel is entanglement breaking and to evaluate entanglement breaking rank, based on an analysis of multiplicative domains determined by complementary quantum channels. We give a full description of the class of entanglement breaking channels that have a projection as their Choi matrix, and we show the entanglement breaking rank and Choi rank of such channels are equal.

1 Introduction

Quantum entanglement remains one of the most important and deeply studied topics in modern quantum information theory. It is a broad topic that can be studied through a number of different perspectives, including via the theory of completely positive maps and the Choi-Jamiolkowski isomorphism, which identifies separable bipartite quantum states with the special class of quantum channels (completely positive, trace-preserving maps) called entanglement breaking [6, 9]. The identification also links the important measure of optimal ensemble cardinality of a separable state with the entanglement breaking rank of its associated channel.

Thus, the notion of entanglement breaking rank has received recent and growing interest. Another reason for this, is that when considering some special classes of

D. W. Kribs (✉) · J. Levick · R. Pereira
Department of Mathematics & Statistics, University of Guelph, Guelph, ON, Canada
e-mail: dkribs@uoguelph.ca; pereirar@uoguelph.ca

M. Rahaman
Univ Lyon, ENS Lyon, UCBL, CNRS, Inria, LIP, Lyon, France

Y. Choi et al. (eds.), *Operators, Semigroups, Algebras and Function Theory*,
Operator Theory: Advances and Applications 292,
https://doi.org/10.1007/978-3-031-38020-4_8

entanglement breaking channels, the entanglement breaking rank has a surprising connection with the geometry of the rank-one Kraus operators that define the channels. For example, as was shown in [15], the entanglement breaking rank of a particular quantum channel on $M_d(\mathbb{C})$ is d^2 if and only if there exist a set of d^2 equiangular lines in $\mathbb{C}^d$. Note that the latter existence statement is still an open problem in general and a popular conjecture of Zauner (see references in [15]).

In this paper, we first introduce a new technique to determine if a channel is entanglement breaking based on an analysis of complementary quantum channels [7–9]. In the theory of completely positive maps, the multiplicative domain of a map [2, 16] is an important operator structure that has arisen in a number of relatively recent investigations in quantum information theory [3, 10, 11]. We make use of multiplicative domains associated with complementary quantum channels to evaluate the entanglement breaking rank of entanglement breaking channels that have Choi matrix equal to a projection, showing that the entanglement breaking and Choi ranks of the channel are equal.

This paper is organized as follows. The rest of this section includes a presentation of the preliminary notions we require at the intersection of operator theory and quantum information. In Sect. 2 we recall the formulation of complementary quantum channels, and we present a description of when maps are entanglement breaking in terms of the complementary channel perspective. In Sect. 3, we begin by briefly recalling the multiplicative domain from the theory of completely positive maps, and then we derive our main result, equating the entangling breaking rank to the Choi rank for entanglement breaking channels with Choi matrix a projection, and we finish the section by giving a full description of what channels satisfy the result. We conclude in Sect. 4 with some remarks on related work and possible extensions of the approach.

1.1 Preliminaries

Let $M_n(\mathbb{C})$ be the set of $n \times n$ complex matrices for a fixed positive integer $n \geq 1$. A linear map $\Phi : M_n(\mathbb{C}) \to M_m(\mathbb{C})$ is called a (quantum) *channel* if it is completely positive and trace-preserving [8, 14]. By a *state* or *density matrix* we mean a positive semi-definite matrix with trace equal to one.

In order to be consistent with the interpretation of dual vectors as row vectors, we take inner products to be linear in the second element, so that if $v = \begin{pmatrix} v_1 \\ \vdots \\ v_n \end{pmatrix}$, we have $v^* = (\overline{v_1}, \cdots, \overline{v_n})$ and $v^* w = \langle v, w \rangle$. Hence, uv^* will denote the rank-one operator, whose action is defined in a way consistent with the notation; e.g., $(uv^*)w = \langle v, w \rangle u$.

Every channel $\Phi : M_n(\mathbb{C}) \to M_m(\mathbb{C})$ has an operator-sum decomposition (non-unique) given by,

$$\Phi(X) = \sum_{i=1}^{d} K_i X K_i^* \qquad \forall X \in M_n(\mathbb{C}), \tag{1.1}$$

with the *Kraus operators* $K_i \in M_{m\times n}(\mathbb{C})$ satisfying the trace-preserving condition $\sum_i K_i^* K_i = I_n$.

The minimal number d of Kraus operators in an operator-sum form Eq. (1.1) for Φ is called the *Choi rank* of Φ, which we will denote by Choi-rank(Φ). This rank is so-called, as it is equal to the rank of the *Choi matrix* $J(\Phi) \in M_n(\mathbb{C}) \otimes M_m(\mathbb{C})$, where

$$J(\Phi) = \sum_{i,j=1}^{n} E_{ij} \otimes \Phi(E_{ij}),$$

and with the 'matrix units' for $M_n(\mathbb{C})$ given by $E_{ij} = e_i e_j^*$ for a fixed orthonormal basis $\{e_1, \cdots, e_n\}$ for $\mathbb{C}^n$.

When viewed as a map, $J(\cdot)$ defines the *Choi-Jamiolkowski isomorphism* between completely positive maps from $M_n(\mathbb{C})$ to $M_m(\mathbb{C})$ and positive semi-definite operators in $M_n(\mathbb{C}) \otimes M_m(\mathbb{C})$. For a channel, the Choi matrix is a density matrix when scaled by the factor n^{-1}. The isomorphism can be used to study quantum entanglement; in particular because the channels that have a (normalised) separable Choi matrix are precisely the channels that break entanglement in the following sense.

Definition 1.1 A channel $\Phi : M_n(\mathbb{C}) \to M_m(\mathbb{C})$ is *entanglement breaking* if for all $k \geq 1$ and states $X \in M_k(\mathbb{C}) \otimes M_n(\mathbb{C})$, the output state $(\mathrm{id}_k \otimes \Phi)(X)$ is separable, where id_k is the identity map on $M_k(\mathbb{C})$.

This class of channels has been heavily investigated in recent years following the original works [6, 9] (for example, see recent related work of the authors and collaborators [1, 4, 12, 17, 18]). There are a number of equivalent characterisations of entanglement breaking channels beyond the definition; most importantly in our present setting is the rank-one form,

$$\Phi(X) = \sum_{i=1}^{r} (u_i v_i^*) X (u_i v_i^*)^*, \tag{1.2}$$

for some set of vectors $u_i \in \mathbb{C}^m$, which without loss of generality we take to be unit vectors, and a set of vectors $v_i \in \mathbb{C}^n$ that satisfy $\sum_i v_i v_i^* = I_n$ due to trace-preservation.

Our investigation focusses on the following notion for such maps [15].

Definition 1.2 Given an entanglement-breaking channel Φ, we call the minimum r such that Eq. (1.2) holds the *entanglement breaking rank* of Φ, or EB-rank(Φ) for short.

For an entanglement breaking channel Φ, it is not hard to see that the number EB-rank(Φ) is equal to the *length* or *optimal ensemble cardinality* of the corresponding separable state $J(\Phi)$ (i.e., the minimal number of pure states required to write the state as a convex linear combination of those pure states).

It is clear that the Choi rank is bounded above by the EB-rank, since it is equal to the minimal number of Kraus operators required in an operator-sum form for the map. It is quite hard to compute the EB-rank of an entanglement breaking channel as one has to search through all the rank-one Kraus decompositions of the map and find one with the least number of them. In the rest of the paper we identify a class of maps for which these ranks coincide.

2 Complementary Channels and a Characterisation of Entanglement Breaking Channels

We next discuss the complement of a quantum channel [7–9] and show how its behaviour can be used to detect when a map is entanglement breaking. We finish by focussing on the situation considered in the next section.

Definition 2.1 Given a channel $\Phi : M_n(\mathbb{C}) \rightarrow M_m(\mathbb{C})$ with Kraus operators $\{K_i\}_{i=1}^d$, the *complementary channel*, or *complement*, is the map $\Phi^C : M_n(\mathbb{C}) \rightarrow M_d(\mathbb{C})$ defined by

$$\Phi^C(X) := \sum_{i,j=1}^{d} \mathrm{tr}(K_i^* K_j X) E_{ji}. \tag{2.1}$$

There are a number of easily derived properties of complementary channels, including the following which we recall now as it will be used in what follows. We remind the reader of the (Hilbert-Schmidt) dual map $\Phi^\dagger$ for a map Φ, defined via the trace relation; $\mathrm{Tr}(\Phi^\dagger(X)Y) := \mathrm{Tr}(X\Phi(Y))$.

There is never a unique set of Kraus operators for a channel, but (as described in [14]) if $\{K_i\}_{i=1}^d$ is a minimal set (and so d is necessarily the Choi rank of Φ), and $\{L_j\}_{j=1}^r$ are any other set of Kraus operators, then there exists an isometry $W : \mathbb{C}^d \rightarrow \mathbb{C}^r$, such that if w_i are the columns of W^* with coordinates $w_i = (w_{ji})$, we have

$$L_i = \sum_{j=1}^{d} \overline{w_{ji}} K_j. \tag{2.2}$$

Note that changing Kraus operators corresponds to adjunction of the complement by the isometry W:

$$\begin{aligned}\sum_{i,j=1}^{r} \operatorname{tr}(L_j^* L_i X) E_{ij} &= \sum_{i,j=1}^{r} \sum_{k,l=1}^{d} \overline{w_{ki}} w_{jl} \operatorname{tr}(K_l^* K_k X) E_{ij} \\ &= \sum_{j=1}^{r} \overline{w_{ki}} E_{ik} \left(\sum_{k,l=1}^{d} \operatorname{tr}(K_l^* K_k X) E_{kl} \right) \sum_{i=1}^{r} w_{jq} E_{lj} \\ &= \sum_{i,p=1}^{r,d} \overline{w_{pi}} E_{ip} \left(\sum_{k,l=1}^{d} \operatorname{tr}(K_l^* K_k X) E_{kl} \right) \sum_{q,j=1}^{d,r} w_{jq} E_{qj} \\ &= W \left(\sum_{k,l=1}^{d} \operatorname{tr}(K_l^* K_k X) E_{kl} \right) W^*.\end{aligned}$$

For the rest of the paper, when we talk of 'the' complementary channel for a given channel Φ, we will be referring to the complement defined by a minimal set of Kraus operators for the map. By the above argument, such maps are equivalent up to adjunction by a unitary map.

As a useful observation in the calculations to come, note that Eq. (2.1) immediately implies that

$$\operatorname{tr}(\Phi^{C\dagger}(E_{ij})X) = \Phi^C(X)_{ji} = \operatorname{tr}(K_i^* K_j X),$$

and as this holds for all X, we have for all $1 \le i, j \le d$,

$$\Phi^{C\dagger}(E_{ij}) = K_i^* K_j. \tag{2.3}$$

The following result makes use of this fact to give a characterization of entanglement breaking channels in terms of complementary channels.

Proposition 2.2 *Let $\Phi : M_n(\mathbb{C}) \to M_m(\mathbb{C})$ be a channel of Choi rank d. Then Φ is entanglement breaking if and only if there exist vectors $\{w_i\}_{i=1}^r \subseteq \mathbb{C}^d$ such that*

$$\sum_{i=1}^{r} w_i w_i^* = I_d, \tag{2.4}$$

and there exist vectors $\{v_i\}_{i=1}^r \subseteq \mathbb{C}^n$ such that

$$\Phi^{C\dagger}(w_i w_i^*) = v_i v_i^* \quad \forall i. \tag{2.5}$$

Proof Suppose Φ is entanglement breaking; then as per Eq. (1.2) we may find a set of Kraus operators for Φ of the form $\{u_i v_i^*\}_{i=1}^r$. If $\{K_i\}_{i=1}^d$ are a minimal set of Kraus operators for Φ, then by Eq. (2.2) there must exist an isometry W, such that with w_i the columns of W^*,

$$u_i v_i^* = \sum_{j=1}^{d} \overline{w_{ji}} K_j. \tag{2.6}$$

Then multiply both sides of Eq. (2.6) by the respective adjoints (and recalling our default assumption that each u_i is a unit vector) to get

$$\begin{aligned} v_i v_i^* = \|u_i\|^2 v_i v_i^* &= \sum_{j,k=1}^{d} \overline{w_{ji}} w_{ki} K_k^* K_j \\ &= \sum_{j,k=1}^{d} w_{ki} \overline{w_{ji}} \Phi^{C\dagger}(E_{kj}) \\ &= \Phi^{C\dagger}\Big(\big(\sum_{k=1}^{d} w_{ki} e_k\big)\big(\sum_{j=1}^{d} \overline{w_{ji}} e_j^*\big)\Big) \\ &= \Phi^{C\dagger}(w_i w_i^*), \end{aligned}$$

and where we use Eq. (2.3) to replace $K_i^* K_j$ with $\Phi^{C\dagger}(E_{ij})$. Since W is an isometry,

$$W^* W = \sum_{i=1}^{r} w_i w_i^* = I_d.$$

For the converse, note that $\Phi^{C\dagger}(ww^*)$ for any (column) vector $w \in \mathbb{C}^d$ is $K(\overline{w})^* K(\overline{w})$ where $K : \mathbb{C}^d \to M_{m\times n}(\mathbb{C})$ is defined by

$$K(x) := \sum_{i=1}^{d} x_i K_i,$$

for any $x \in \mathbb{C}^d$ with coordinates x_i.

Thus, $\Phi^{C\dagger}(ww^*) = vv^*$ for some vector v if and only if $K(\overline{w}) \in M_{m\times n}(\mathbb{C})$ is a rank-one matrix. If there exist $\{w_i w_i^*\}$ each with $\Phi^{C\dagger}(w_i w_i^*) = v_i v_i^*$, then we may form the matrix W^* whose columns are w_i; $W^* W = \sum_{i=1}^r w_i w_i^*$. So W is an isometry if and only if Eq. (2.4) holds. If it does, then $\{L_i = K(\overline{w_i})\}_{i=1}^r$ is also a set of Kraus operators for Φ, and so together we have a set of rank-one Kraus operators for Φ. These must be of the form $L_i = u_i v_i^*$ for some vectors u_i, v_i, completing the proof. □

We can use this result to frame entanglement breaking rank as follows.

Corollary 2.3 *Let $\Phi : M_n(\mathbb{C}) \to M_m(\mathbb{C})$ be an entanglement breaking channel with Choi rank d. Then* EB-rank(Φ) *is the smallest r such that there exist r vectors $\{w_i\}_{i=1}^r \subseteq \mathbb{C}^d$ and $\{v_i\}_{i=1}^r \subseteq \mathbb{C}^n$ satisfying*

$$\sum_{i=1}^{r} w_i w_i^* = I_d; \quad \Phi^{C\dagger}(w_i w_i^*) = v_i v_i^* \quad \forall 1 \le i \le r.$$

Before proceeding to our main result in the next section, we link specific behaviour of the complement adjoint with structure of the Choi matrix.

Lemma 2.4 *Let $\Phi : M_n(\mathbb{C}) \to M_m(\mathbb{C})$ be a channel with* Choi-rank$(\Phi) = d$ *and complementary channel $\Phi^C : M_n(\mathbb{C}) \to M_d(\mathbb{C})$. The complement adjoint $\Phi^{C\dagger}$ is trace-preserving if and only if the Choi matrix $J(\Phi)$ is a projection.*

Proof Let $\{K_i\}_{i=1}^d$ be a set of Kraus operators in $M_{m\times n}(\mathbb{C})$ for Φ, and further assume $\mathrm{Tr}(\Phi^{C\dagger}(X)) = \mathrm{Tr}(X)$ for all $X \in M_d(\mathbb{C})$. As this holds for all X, we have $\Phi^C(I_n) = I_d$ and so,

$$\Phi^C(I_n) = \sum_{i,j=1}^{d} \mathrm{Tr}(K_j^* K_i) E_{ij} = I_d.$$

Hence, it follows that the operators K_i are orthogonal in the trace inner product; specifically, $\mathrm{Tr}(K_j^* K_i) = \delta_{ij}$.

The Kraus operators are related to the Choi matrix by defining vectors $k_i = (I_n \otimes K_i)\phi \in \mathbb{C}^n \otimes \mathbb{C}^m$ where $\phi = \sum_{i=1}^n e_i \otimes e_i$, so that $k_i = \mathrm{vec}(K_i)$, which is a block vector whose blocks are the columns of K_i in order. Then the Choi matrix is given by $J(\Phi) = \sum_{i=1}^d k_i k_i^*$.

It is clear that $\mathrm{Tr}(K_i^* K_j) = \langle k_i, k_j \rangle$, and so from the above calculation it follows that the vectors $\{k_i\}_{i=1}^d$ form an orthonormal set. Hence we have $J(\Phi) = \sum_{i=1}^d k_i k_i^* = P$, where P is a rank-d projection.

For the converse direction, the above argument can be reversed to show that $J(\Phi)$ being a projection implies Φ^C is unital and hence that its adjoint is trace-preserving. □

3 The Multiplicative Domain of a Completely Positive Map and Entanglement Breaking Rank

We now discuss the multiplicative domain of a unital completely positive map, and make use of its structure to identify entanglement breaking ranks.

Definition 3.1 Let $\Phi : M_n(\mathbb{C}) \to M_m(\mathbb{C})$ be a completely positive map. The *multiplicative domain* of Φ, $\mathcal{M}(\Phi)$, is the set

$$\mathcal{M}(\Phi) := \{A : \Phi(AX) = \Phi(A)\Phi(X) \;\&\; \Phi(XA) = \Phi(X)\Phi(A) \;\forall X\}. \tag{3.1}$$

The left- and right-multiplicative domains, $\mathcal{M}_L(\Phi)$, $\mathcal{M}_R(\Phi)$, are respectively defined by

$$\mathcal{M}_L(\Phi) = \{A : \Phi(AX) = \Phi(A)\Phi(X) \;\forall X\}, \tag{3.2}$$

$$\mathcal{M}_R(\Phi) = \{A : \Phi(XA) = \Phi(X)\Phi(A) \;\forall X\}. \tag{3.3}$$

The key internal description of the multiplicative domain, as derived originally by Choi, is given as follows. This is Theorem 3.1 of [2], and we include a proof that is slightly different from the original.

Theorem 3.2 *If Φ is a unital completely positive map, then $A \in \mathcal{M}_L(\Phi)$ if and only if*

$$\Phi(AA^*) = \Phi(A)\Phi(A^*) \tag{3.4}$$

and $A \in \mathcal{M}_R(\Phi)$ if and only if

$$\Phi(A^*A) = \Phi(A^*)\Phi(A). \tag{3.5}$$

Proof One direction is trivial, so we show the other; we only show the claim for the left-domain, as the proof of the other claim is basically identical. Let $V = \begin{bmatrix} I & A^* & X^* \end{bmatrix}$ so that $V^*V = \begin{bmatrix} I & A^* & X^* \\ A & AA^* & AX^* \\ X & XA^* & XX^* \end{bmatrix}$ is positive semidefinite.

Since Φ is completely positive, it is in particular 3-positive, so applying Φ to the blocks must yield a positive semidefinite matrix as well; as Φ is unital we get that

$$\begin{bmatrix} I & \Phi(A^*) & \Phi(X^*) \\ \Phi(A) & \Phi(AA^*) & \Phi(AX^*) \\ \Phi(X) & \Phi(XA^*) & \Phi(XX^*) \end{bmatrix}$$

is positive semidefinite. Take the Schur complement with respect to the top-left block to see that

$$\begin{bmatrix} \Phi(AA^*) - \Phi(A)\Phi(A^*) & \Phi(AX^*) - \Phi(A)\Phi(X^*) \\ \Phi(XA^*) - \Phi(X)\Phi(A^*) & \Phi(XX^*) - \Phi(X)\Phi(X^*) \end{bmatrix}$$

is positive semidefinite as well.

But if $\Phi(AA^*) = \Phi(A)\Phi(A^*)$, the top-left block of this positive semidefinite matrix is 0, so the blocks below and right of it must be zero as well. Hence we have $\Phi(AX^*) = \Phi(A)\Phi(X^*)$ for all X. □

The case of projections is of particular importance to us.

Corollary 3.3 *A projection P is in the multiplicative domain of a unital completely positive map Φ if and only if $\Phi(P)$ is also a projection.*

This follows as $\Phi(P^*P) = \Phi(PP^*) = \Phi(P) = \Phi(P)\Phi(P)^* = \Phi(P)^*\Phi(P)$, so that Eqs. (3.4), (3.5) are both satisfied simultaneously.

Thus, it follows that the multiplicative domain of a unital completely positive map Φ is a unital $*$-subalgebra (a finite-dimensional C*-algebra), and hence is unitarily equivalent to $\bigoplus_{k=1}^{M} I_{i_k} \otimes M_{j_k}(\mathbb{C})$ for some (unique) set of pairs of positive integers (i_k, j_k) [2].

We now apply our results from the previous section in this multiplicative domain context.

Lemma 3.4 *Suppose $\Phi : M_n(\mathbb{C}) \to M_m(\mathbb{C})$ is an entanglement breaking channel whose Choi matrix is a projection of rank d. Let $\{w_i\}_{i=1}^{r} \subseteq \mathbb{C}^d$ be the vectors guaranteed by Proposition 2.2 that satisfy*

$$\sum_{i=1}^{r} w_i w_i^* = I_d,$$

and for all $1 \leq i \leq r$,

$$\Phi^{C\dagger}(w_i w_i^*) = v_i v_i^*.$$

Then $w_i w_i^* \in \mathcal{M}(\Phi^{C\dagger})$ for all i.

Proof By Lemma 2.4, the map $\Phi^{C\dagger}$ is trace-preserving. As Φ is a channel, so is Φ^C and hence $\Phi^{C\dagger}$ is also unital. It follows that,

$$\|w_i\|^2 = \mathrm{Tr}(w_i w_i^*) = \mathrm{Tr}(\Phi^{C\dagger}(w_i w_i^*)) = \mathrm{Tr}(v_i v_i^*) = \|v_i\|^2,$$

and hence $||w_i|| = ||v_i||$. This means that $w_i w_i^*$ and $v_i v_i^*$ can be simultaneously re-scaled so they are both projections on their respective Hilbert spaces. That is, $P_i = \frac{w_i w_i^*}{\|w_i\|^2}$ and $Q_i = \frac{v_i v_i^*}{\|w_i\|^2} = \frac{v_i v_i^*}{\|v_i\|^2}$ are both projections, and they satisfy

$$\Phi^{C\dagger}(P_i) = Q_i,$$

for each i. Thus by Corollary 3.3, $w_i w_i^*$ are all in the multiplicative domain $\mathcal{M}(\Phi^{C\dagger})$ of $\Phi^{C\dagger}$, as claimed. □

We can now prove our main result.

Theorem 3.5 *Let Φ be a channel whose Choi matrix $J(\Phi)$ is a rank-d projection. Then Φ is entanglement breaking if and only if $\mathcal{M}(\Phi^{C\dagger})$ is unitarily equivalent to an algebra of the form $\bigoplus_{k=1}^{M} M_{j_k}(\mathbb{C})$. Moreover, in this case the entanglement breaking rank is the same as the Choi rank;*

$$\text{EB-rank}(\Phi) = \text{Choi-rank}(\Phi) = d.$$

Proof Suppose first that the multiplicative domain $\mathcal{M}(\Phi^{C\dagger})$ is unitarily equivalent to $\oplus_{k=1}^{M} M_{j_k}(\mathbb{C})$. In particular then, this algebra contains an algebra unitarily equivalent to the diagonal algebra, and hence contains rank-one projections unitarily equivalent to the diagonal matrix units, $E_{ii} = e_i e_i^*$. Thus, up to unitary equivalence, $e_i e_i^*$ are in the multiplicative domain of $\Phi^{C\dagger}$, which we note is unital following from the fact that Φ (and Φ^C) is trace-preserving. Hence by Corollary 3.3 we have that $P_i := \Phi^{C\dagger}(e_i e_i^*)$ is also a projection.

Moreover, from Lemma 2.4, we have that $J(\Phi)$ being a projection is equivalent to $\Phi^{C\dagger}$ being trace-preserving, as well as unital. Thus the rank of P_i is $\text{tr}(P_i) = \text{tr}(\Phi^{C\dagger}(e_i e_i^*)) = 1$, and so in fact P_i is a rank-one projection for each i; that is, there is some unit vector v_i such that $\Phi^{C\dagger}(e_i e_i^*) = v_i v_i^*$. It is also obvious that $\sum_i e_i e_i^* = I$, and so we satisfy the conditions of Proposition 2.2, and thus Φ is entanglement breaking.

For the converse direction, assume that Φ is entanglement breaking. We also have that $J(\Phi)$ is a projection, and so again by Lemma 2.4, it follows that $\Phi^{C\dagger}$ is both trace-preserving and unital. From Proposition 2.2, we know that Φ being entanglement breaking implies the existence of vectors $\{w_i\}$, $\{v_i\}$ satisfying $\sum_i w_i w_i^* = I$ and $\Phi^{C\dagger}(w_i w_i^*) = v_i v_i^*$. Further, from Lemma 3.4 we have that the vectors $\{w_i w_i^*\}$ are in the multiplicative domain of $\Phi^{C\dagger}$. Thus, $\mathcal{M}(\Phi^{C\dagger})$ contains rank-one projections summing to the identity.

From the structure theory for matrix algebras, we know that the multiplicative domain of $\Phi^{C\dagger}$ must have the form, up to unitary equivalence, $\oplus_{k=1}^{M} I_{i_k} \otimes M_{j_k}(\mathbb{C})$; we now show that if such an algebra contains a set of rank-one projections summing to the identity, that $i_k = 1$ for all k.

Suppose P_i are a set of projections in such an algebra summing to the identity; so it must be that each $P_i = \oplus_{k=1}^{M} I_{i_k} \otimes P_{ik}$ where each P_{ik} is a projection in $M_{j_k}(\mathbb{C})$, and furthermore we must have that $\sum_i P_{ik} = I_{j_k}$ for each k. Moreover, the rank of P_i is of course just its trace:

$$\text{tr}(P_i) = \sum_{k=1}^{m} i_k \text{tr}(P_{ik}) = \sum_k i_k \text{rank}(P_{ik}).$$

Thus, if P_i are rank-one projections summing to the identity, it must be that $1 = \sum_k i_k \text{rank}(P_{ik})$ and this is clearly only possible if $\text{rank}(P_{ik}) = 0$ for all but one k, in which case $\text{rank}(P_{ik}) = i_k = 1$. Hence for each i there is an index $1 \le q_i \le M$ such that $P_i = \oplus_{k=1}^{m} \delta_{k,q_i} P_{kq_i}$, and $i_{q_i} = 1$. So, we will be done if we can show that the map $i \mapsto q_i$ is a surjection onto $\{1, 2, \cdots, M\}$. Suppose not; then there is a direct summand, the qth one say, on which none of the P_i are supported; P_i restricted to the qth direct summand is always 0 for all i. But then, $\sum_i P_i$ restricted to that direct summand will also be 0, not I_{k_q}, as required for the projections to sum to the identity. So in fact, for each q, there is an i such that $q = q_i$, and thus $i_{q_i} = 1$. Hence the algebra has the form $\oplus_{k=1}^M I_1 \otimes M_{j_k} = \oplus_{k=1}^M M_{j_k}(\mathbb{C})$ as claimed.

Finally, we show that the EB-rank is the same as the Choi rank of Φ, which is d. We now know that for a channel with $J(\Phi)$ equal to a projection, being entanglement breaking is equivalent to $\Phi^{C\dagger}$ having a multiplicative domain (up to unitary equivalence) of the form $\oplus_{k=1}^M M_{j_k}(\mathbb{C})$. Moreover, we know that the matrix unit projections $E_{ii} = e_i e_i^*$ inside the algebra form a set of rank-one projections for which $\Phi^{C\dagger}(e_i e_i^*) = v_i v_i^*$, and so by Corollary 2.3, it is thus clear that the EB rank is no greater than d. For it to be smaller than d, we would have to find a set of fewer than d vectors such that $\sum_i w_i w_i^* = I_d$, but this is clearly impossible, so in fact the EB rank must be exactly d. □

A simple class of examples that Theorem 3.5 applies to arise from consideration of *Schur product channels*. These are channels $\Phi : M_d(\mathbb{C}) \to M_d(\mathbb{C})$ of the form

$$\Phi(X) = X \circ C,$$

where $\circ$ indicates the element-wise Schur product, and C is a correlation matrix; a positive semidefinite matrix all of whose diagonal entries are equal to one. Such channels arise in quantum information [5, 13]; for instance, any random unitary channel implemented by mutually commuting unitaries is also a Schur product channel in some basis.

As an observation we will use below, let us note how the double complement of a map may be identified with the map itself. Indeed, suppose we have a minimal set of Kraus operators for a map Φ, given by $\{K_i\}_{i=1}^n$; so we know that any other minimal set of Kraus operators gives a complement related to the first by $U\Phi^C(X)U^*$ for some unitary U. Then suppose $\{L_i\}_{i=1}^{n'}$ are a minimal set of Kraus operators for Φ^C; picking any other set of minimal Kraus operators for Φ will give instead the channel $U\Phi^C(X)U^*$ which clearly has Kraus operators $\{UL_i\}_{i=1}^{n'}$. Now, using the formula for the complement, for any U, we can write the complement of the complement:

$$(U\Phi^C(X)U^*)^C = \sum_{i,j=1}^{n'} \operatorname{tr}((UL_i)^*(UL_j)X)E_{ji} = \sum_{i,j} \operatorname{tr}(L_i^* L_j X)E_{ji};$$

that is, the presence of the unitary adjunction by U does not change the complement. In fact, even if U is an isometry rather than a unitary, this remains true, so that our choice of Kraus operators for Φ does not affect our formula for $(\Phi^C)^C$ at all; only a different choice of Kraus operators for Φ^C will change this. Thus, we can define $(\Phi^C)^C(X)$ up to adjunction to be $V \sum_{i,j} \operatorname{tr}(L_i^* L_j X)E_{ji} V^*$ for any unitary V, and some fixed minimal set of Kraus operators for Φ^C.

Returning to Schur product channels, note that any such channel is automatically unital, and so its complement has the property that $J(\Phi^C)$ is a projection by Lemma 2.4 (making use of the observation above). Further, it is not too hard to show that if the correlation matrix C is the Gram matrix for unit vectors $\{\overline{v_i}\}$, that is, $c_{ij} = \langle \overline{v_i}, \overline{v_j} \rangle$, then the complement has the form

$$\Phi^C(X) = \sum_{i=1}^{d} x_{ii} v_i v_i^* = \sum_{i=1}^{d} v_i e_i^* X e_i v_i^*,$$

and so is clearly entanglement-breaking, with rank-one Kraus operators $v_i e_i^*$ (and e_i the standard basis vectors for $\mathbb{C}^d$). In this case, $J(\Phi^C)$ is just the projection $P = \bigoplus_{i=1}^{d} v_i v_i^*$, and by Theorem 3.5 the EB-rank of Φ^C is equal to d.

In fact, we can show that such channels are the only ones that satisfy Theorem 3.5. By 'external' and 'internal' unitary equivalence to a map Φ, we mean that, respectively, there exist unitaries U and V and some map Φ' such that $\Phi'(X) = U\Phi(X)U^*$ and $\Phi'(X) = \Phi(VXV^*)$ for all X.

Theorem 3.6 *An entanglement breaking channel Φ has its Choi matrix equal to a projection if and only if Φ is internally or externally unitarily equivalent to the complement of a Schur product channel.*

Proof The only thing left to establish the forward direction is to prove the claim above that the complement of a Schur product channel has the desired properties. If $\Phi(X) = X \circ C$ where $c_{ij} = \langle v_i, v_j \rangle$ for some unit vectors $\{v_i\}_{i=1}^n \in \mathbb{C}^d$, then Φ has d Kraus operators $K_i = \sum_{j=1}^n \overline{v_{ji}} E_{jj}$;

$$\begin{aligned}
\Phi(X) &= \sum_{i=1}^{d} \sum_{j,k} \overline{v_{ji}} v_{ki} E_{jj} X E_{kk} \\
&= \sum_{j,k=1}^{n} x_{jk} E_{jk} \Big(\sum_{i=1}^{d} \overline{v_{ji}} v_{ki} \Big) \\
&= \sum_{j,k=1}^{n} x_{jk} E_{jk} \langle v_j, v_k \rangle \\
&= \sum_{j,k=1}^{n} c_{jk} x_{jk} E_{jk}.
\end{aligned}$$

Then Φ^C can be written as

$$\begin{aligned}\Phi^C(X) &= \sum_{i,j=1}^{d} \operatorname{tr}(K_i^* K_j X) E_{ji} \\ &= \sum_{i,j,k,l=1}^{d,n} v_{ki}\overline{v_{lj}} \operatorname{tr}(E_{kk} E_{ll} X) E_{ji} \\ &= \sum_{i,j,k=1}^{d,n} v_{ki}\overline{v_{kj}} x_{kk} E_{ji} \\ &= \sum_{k=1}^{n} x_{kk} \Big(\sum_{j,i=1}^{d} v_{ki}\overline{v_{kj}} E_{ji} \Big) \\ &= \sum_{k=1}^{n} x_{kk} \overline{v_k} v_k^T .\end{aligned}$$

This clearly has Kraus operators $\{\overline{v_k} e_k^*\}_{k=1}^n$ and so is obviously an entanglement breaking channel. Further, the Choi matrix of this is then $J(\Phi^C) = \sum_{i=1}^n E_{ii} \otimes v_i v_i^*$ and since each v_i is a unit vector, this is a projection. Notice also that the Choi rank of $J(\Phi^C)$ is n, as is the EB-rank, as witnessed by the n rank-one Kraus operators $\{\overline{v_k} e_k^*\}$.

Now we show the converse. By Theorem 3.5, Φ must have EB-rank equal to its Choi rank, which is also necessarily the dimension of the underlying space. This last fact follows from the fact that the rank of $J(\Phi)$, the Choi rank, is also the trace of $J(\Phi)$; but by trace-preservation we must have that $(\mathrm{id} \otimes \operatorname{tr})\big(J(\Phi)\big) = I_d$ and hence

$$\operatorname{tr}(J(\Phi)) = (\operatorname{tr} \otimes \operatorname{tr})\big(J(\Phi)\big) = \operatorname{tr}(I_d).$$

Thus, there exist d unit vectors $\{u_i\}$ and d vectors $\{v_i\}$ such that

$$\Phi(X) = \sum_{i=1}^{d} u_i v_i^* X v_i u_i^* . \tag{3.6}$$

Imposing trace-preservation on the Kraus operators requires

$$\sum_{i=1}^{d} v_i u_i^* u_i v_i^* = \sum_{i=1}^{d} v_i v_i^* = I_d,$$

and the only way to have d rank-ones adding up to the identity on $\mathbb{C}^d$ is for each to be a projection onto a subspace orthogonal to all the others; hence $\{v_i\}$ is an orthonormal basis for $\mathbb{C}^d$. Thus, if we form the channel $\Phi_V(X) = \Phi(VXV^*)$, where V is the unitary matrix defined by $Ve_i = v_i$, we see that

$$\Phi_V(X) = \sum_i u_i v_i^* V X V^* v_i u_i^* = \sum_{i=1}^{d} u_i e_i^* X e_i u_i^*.$$

Now, we calculate the complement of Φ_V^C:

$$\begin{aligned}\Phi_V^C(X) &= \sum_{i,j=1}^{d} \operatorname{tr}(e_i u_i^* u_j e_j^* X) E_{ji} \\ &= \sum_{i,j=1}^{d} \langle u_i, u_j \rangle \langle e_j, X e_i \rangle E_{ji} = \sum_{i,j=1}^{d} \langle u_i, u_j \rangle x_{ji} E_{ji},\end{aligned}$$

and it is clear that this is just X with the (j, i) entry multiplied by $\langle u_i, u_j \rangle = \overline{c_{ij}}$ where C is the Gram matrix for the unit vectors $\{u_i\}$. Thus, C is positive semidefinite with 1's down the diagonal, and so indeed we have that

$$\Phi_V^C(X) = \overline{C} \circ X$$

for the correlation matrix $\overline{C}$, and the claim is proved. □

4 Concluding Remarks

As noted above, computing entanglement breaking ranks is important in the theory of quantum entanglement in general, but quite challenging for specific classes of channels, or equivalently for separable states. Thus, even though Theorem 3.5 applies to a limited class of channels, as shown by Theorem 3.6, it still contributes to the short list of instances for which this rank can be computed explicitly.

The use of complementary channel properties in the study of entanglement breaking ranks is novel and could conceivably lead to further progress for other classes. Applications from an analysis of the structure of the multiplicative domain are also new in this setting, and are reminiscent of other applications in quantum information noted above of such operator structures. Further study of potential uses of multiplicative domain structures in entanglement theory is warranted.

All of this said, the possible extensions of the specific approach presented here to wider classes of entanglement breaking channels would appear to be limited. For instance, we have attempted to extend our results to channels with Choi matrix equal to a scalar multiple of a projection. Some of our results do extend to this more general setting. As an example, we can extend Lemma 2.4 to show the following, where we say that a map Φ is *trace-stabilizing* if for some fixed non-zero scalar α, we have $\mathrm{Tr}(\Phi(X)) = \alpha\,\mathrm{Tr}(X)$ for all X in the domain of Φ.

Lemma 4.1 *The complement adjoint $\Phi^{C\dagger}$ of a channel Φ is trace-stabilizing if and only if the Choi matrix $J(\Phi)$ is a scalar multiple of a projection.*

An important example of a channel that satisfies this condition, is the *completely depolarizing channel* [14] $\Phi_{\mathrm{CD}} : M_n(\mathbb{C}) \to M_n(\mathbb{C})$ defined by $\Phi_{\mathrm{CD}}(X) = \frac{\mathrm{Tr}(X)}{n} I_n$. Note this channel has Choi matrix equal to $\frac{1}{n} I_{n^2}$, and has Choi rank and EB rank equal to n^2.

However, Theorem 3.5 does not generalize to the case of a channel with Choi matrix equal to a scalar multiple of a projection. A counterexample can be given as follows. Consider the Werner-Holevo channel $\Phi : M_d(\mathbb{C}) \to M_d(\mathbb{C})$ defined by

$$\Phi(X) = \frac{\mathrm{Tr}(X) + X^t}{d+1},$$

where X^t is the transpose of X. This is known to be an entanglement breaking map, and its Choi matrix $J(\Phi)$ is the bipartite operator $\frac{I \otimes I + W}{d+1}$, where $W = \sum_{i,j} E_{i,j} \otimes E_{j,i}$. Now note that $P = (I \otimes I + W)/2$ is a projection in $M_d(\mathbb{C}) \otimes M_d(\mathbb{C})$ and hence

$$J(\Phi) = \frac{2}{d+1} P,$$

a multiple of a projection. It follows that the Choi rank of Φ is $\frac{d(d+1)}{2}$. But this channel has EB rank equal to d^2 as shown in [15]. In fact, separating the EB rank and the Choi rank is a non-trivial task and this is one of only a few examples known where these two ranks are different.

Acknowledgments We are grateful to the referee for helpful comments. D.W.K. was supported by NSERC Discovery Grant 400160. R.P. was supported by NSERC Discovery Grant 400550. M.R. is supported by the European Research Council (ERC Grant Agreement No. 851716). Part of this work was completed during IWOTA Lancaster UK 2021, and as such, the authors would like to express thanks to the conference organizers and acknowledge their funding via EPSRC grant EP/T007524/1.

References

1. J. Ahiable, D.W. Kribs, J. Levick, R. Pereira, M. Rahaman, Entanglement breaking channels, stochastic matrices, and primitivity. Linear Algebra Appl. **629**, 219–231 (2021)
2. M.-D. Choi, A Schwarz inequality for positive linear maps on C*-algebras. Ill. J. Math. **18**(4), 565–574 (1974)
3. M.-D. Choi, N. Johnston, D.W. Kribs, The multiplicative domain in quantum error correction. J. Phys. A Math. Theor. **42**(24), 245303 (2009)

4. M. Girard, D. Leung, J. Levick, C.-K. Li, V. Paulsen, Y.T. Poon, J. Watrous, On the mixed-unitary rank of quantum channels. Commun. Math. Phys. **394**, 919–951 (2022)
5. S.J. Harris, R.H. Levene, V.I. Paulsen, S. Plosker, M. Rahaman, Schur multipliers and mixed unitary maps. J. Math. Phys. **59**(11), 112201 (2018)
6. A.S. Holevo, Coding theorems for quantum channels. Russ. Math. Surv. **53**, 1295–1331 (1999)
7. A.S. Holevo, Complementary channels and the additivity problem. Theory Probab. Appl. **51**(1), 92–100 (2007)
8. A.S. Holevo, *Quantum Systems, Channels, Information: A Mathematical Introduction*, vol. 16 (Walter de Gruyter, Berlin, 2012)
9. M. Horodecki, P.W. Shor, M.B. Ruskai, Entanglement breaking channels. Rev. Math. Phys. **15**(06), 629–641 (2003)
10. N. Johnston, D.W. Kribs, Generalized multiplicative domains and quantum error correction. Proc. Am. Math. Soc. **139**(2), 627–639 (2011)
11. D.W Kribs, J. Levick, M.I. Nelson, R. Pereira, M. Rahaman, Quantum complementarity and operator structures. Quantum Inf. Comput. **19**, 67–83 (2019)
12. D.W. Kribs, J. Levick, K. Olfert, R. Pereira, M. Rahaman, Nullspaces of entanglement breaking channels and applications. J. Phys. A Math. Theor. **54**(10), 105303 (2021)
13. J. Levick, D.W. Kribs, R. Pereira, Quantum privacy and Schur product channels. Rep. Math. Phys. **80**(3), 333–347 (2017)
14. M.A. Nielson, I.L. Chuang, *Quantum Computation and Quantum Information* (Cambridge University Press, Cambridge, 2000)
15. S.K. Pandey, V.I. Paulsen, J. Prakash, M. Rahaman, Entanglement breaking rank and the existence of SIC POVMs. J. Math. Phys. **61**(4), 042203 (2020)
16. V. Paulsen, *Completely Bounded Maps and Operator Algebras*. Number 78 (Cambridge University Press, Cambridge, 2002)
17. M. Rahaman, A new bound on quantum Wielandt inequality. IEEE Trans. Inf. Theory **66**(1), 147–154 (2019)
18. M. Rahaman, S. Jaques, V.I. Paulsen, Eventually entanglement breaking maps. J. Math. Phys. **59**(6), 062201 (2018)

On the Bergman Projection and Kernel in Periodic Planar Domains

Jari Taskinen

Abstract We study Bergman kernels K_Π and projections P_Π in unbounded planar domains Π, which are periodic in one dimension. In the case Π is simply connected we write the kernel K_Π in terms of a Riemann mapping φ related to the bounded periodic cell ϖ of the domain Π. We also introduce and adapt to the Bergman space setting the Floquet transform technique, which is a standard tool for elliptic spectral problems in periodic domains. We investigate the boundedness properties of the Floquet transform operators in Bergman spaces and derive a general formula connecting P_Π to a projection on a bounded domain. We show how this theory can be used to reproduce the above kernel formula for K_Π. Finally, we consider weighted L^p-estimates for P_Π in periodic domains.

1 Introduction

In this paper we consider Bergman spaces $A^p(\Pi)$ and the Bergman projection P_Π on planar domains $\Pi \subset \mathbb{C}$, which have the special geometry of being periodic in one direction; the domain is obtained as the union of infinitely many translated copies of the bounded periodic cell $\varpi \subset \mathbb{C}$. We aim to work out the basic concepts of the Floquet transform theory in the context of operator theory in Bergman spaces of periodic domains. If the periodic domain Π is considered just as a domain in the two dimensional real Euclidean space, the Floquet transform, acting in appropriate Sobolev spaces, is a standard tool for example in spectral elliptic boundary value problems, see e.g. [15]. Here, we will consider the mapping properties of the Floquet transform in Bergman spaces. The main goal is to study the connection of the Bergman projection and kernel on Π with certain kernels related with the periodic cell, especially in the simply connected case.

J. Taskinen (✉)
Department of Mathematics and Statistics, University of Helsinki, Helsinki, Finland
e-mail: jari.taskinen@helsinki.fi

Y. Choi et al. (eds.), *Operators, Semigroups, Algebras and Function Theory*,
Operator Theory: Advances and Applications 292,
https://doi.org/10.1007/978-3-031-38020-4_9

Let us fix some notation. Given a domain Ω in the complex plane $\mathbb{C}$, we denote by $L^2(\Omega)$ the usual Lebesgue-Hilbert space with respect to the (real) area measure dA and by $A^2(\Omega)$ the corresponding Bergman space, which is the subspace consisting of analytic functions. It is a consequence of the Cauchy integral formula that the norm topology of $A^2(\Omega)$ is stronger than the topology of the uniform convergence on compact subsets, and this implies that the Bergman space is always a closed subspace of $L^2(\Omega)$, hence it is a Hilbert space and in particular complete. We denote by P_Ω the orthogonal projection from $L^2(\Omega)$ onto $A^2(\Omega)$. It can always be written with the help of the Bergman kernel $K_\Omega : \Omega \times \Omega \to \mathbb{C}$,

$$P_\Omega f(z) = \int_\Omega K_\Omega(z, w) f(w) dA(w) \tag{1.1}$$

and the kernel has the properties that $K_\Omega(z, \cdot) \in L^2(\Omega)$ for all z and $K(z, w) = \overline{K(w, z)}$ for all $z, w \in \Omega$. See e.g. [11] for a proof of these assertions.

We denote the periodic cell by ϖ and assume that $\varpi \subset (0, 1) \times (-M, M) \subset \mathbb{R}^2 \cong \mathbb{C}$ for some $M > 0$ and that its intersection with the axis $\{z : \operatorname{Re} z = \frac{1}{2} \pm \frac{1}{2}\}$ coincides with $\{\frac{1}{2} \pm \frac{1}{2}\} \times (a, b) =: J_\pm$ for some real numbers $b > a$. We denote the translates of ϖ by $\varpi_m = \varpi + m$, where $m \in \mathbb{Z} \subset \mathbb{C}$, and then define the periodic domain Π as the interior of the set

$$\bigcup_{m \in \mathbb{Z}} \mathrm{cl}(\varpi_m). \tag{1.2}$$

where $\mathrm{cl}(A)$ denotes the closure of the set A. Moreover, it is assumed that ϖ and Π are Lipschitz domains such that the boundaries $\partial\varpi$ and $\partial\Pi$ are in addition piecewise smooth: more precisely, $\partial\varpi$ consists of finitely many parametrized curves $\gamma : [0, 1] \to \mathbb{C}$ where γ is continuously differentiable on $[0, 1]$. Note that the Lipschitz assumption excludes cusps both in ϖ and Π. Consequently, $\partial\varpi$ is a Jordan curve. This implies that ϖ is a Carathéodory domain so that by an old result of Carleman, [3], polynomials form a dense subspace of the Bergman space $A^2(\varpi)$ (see also the introduction in the paper [2]). One more technical, geometric assumption (A) will be imposed in Sect. 2. Note also that Π is simply connected, if and only ϖ is; in the case where ϖ is multiply connected, the periodic domain is of course no longer finitely connected.

Our results are as follows. In Sect. 2 we consider simply connected Π, and in Proposition 2.3 we present a general formula for the kernel K_Π in terms of a Riemann mapping φ related to the periodic cell ϖ. This is actually a direct consequence of the fact that it is easy to write the Riemann mapping from Π onto the strip, where the kernel has an explicit formula. We consider cases of domains with polygonal boundaries with explicit Schwartz-Christoffel-type formulas for the Riemann mapping.

In Sect. 3 we present the general definition of the Floquet transform F for functions of real variables in periodic planar domains. Briefly, F transforms functions defined on the periodic domain Π into functions on ϖ depending on the so-called Floquet variable or quasimomentum $\eta \in [-\pi, \pi]$. We observe that the definition works nicely also in the case of analytic functions, and in particular in Theorem 3.6 we prove the isometry of F between the space $A^2(\Pi)$ and the corresponding parametrized Bergman-type space $L^2(-\pi, \pi; A^2_\eta(\varpi))$ on the periodic cell.

In Sect. 4 we apply the Floquet transform techniques to formulate a general formula involving the projection P_Π and η-dependent Bergman-type projections P_η related to the periodic cell ϖ. In contrast to Sect. 2, this consideration is made for general, not necessarily simply connected domains.

Section 5 contains some elementary preparations for Sect. 6, where we derive a presentation for the projections P_η. This formula is applied in Sect. 7 to reproduce the kernel formula for P_Π in the simply connected case by using Floquet transform techniques. We also give an application of the kernel formula to weighted L^p-estimates.

As for the notation used in this paper, we let $[x]$ be the integer part of the number $x \in \mathbb{R}$, i.e. the largest integer y with $y \le x$. We write $\mathbb{R}^+ = \{x + iy \in \mathbb{C} : x > 0\}$. If $a \in \mathbb{C}$ and $r > 0$, $B(a, r)$ denotes the open Euclidean disc with center a and radius r. Given a domain $\Omega \subset \mathbb{C}$ we denote by $\|f\|_\Omega$ the norm of $L^2(\Omega)$, which is the L^2-space with respect to the (real) area measure. If H is a Hilbert space, its inner product is denoted by $(f|g)_H$, where $f, g \in H$. We recall that if φ is a conformal mapping from the domain $\Omega \subset \mathbb{C}$ onto the domain Ω', then the $|\varphi'|^2$ is the Jacobian of the coordinate transform φ in the area integrals.

2 Bergman Kernels and Conformal Mappings on Periodic Domains

If Ω and Ω' are conformally equivalent domains in $\mathbb{C}$ and $\varphi : \Omega' \to \Omega$ is a conformal mapping and if $K(z, w)$ is the Bergman kernel of the domain Ω', then the Bergman kernel of the domain Ω can be got from the formula

$$K_\Omega(z, w) = K\big(\varphi(z), \varphi(w)\big)\varphi'(z)\overline{\varphi'(w)}, \tag{2.1}$$

see for example [1], formula (1), or [11], Proposition 2.7. We refer to the papers [4, 6, 7] for general constructions of conformal mappings onto periodic domains, but we will here construct one which leads to a useful formula of the Bergman kernel.

From now on we assume in this section that Π and ϖ are simply connected, denote $E(z) := e^{i2\pi z}$ and define the domain $\mathcal{D} := \{E(z) : z \in \varpi \cup J_+ \cup J_-\}$, which as a consequence of the geometric assumptions in Sect. 1 is doubly connected. Consequently, there exists a conformal mapping ϕ from $\mathcal{D}$ onto the annulus $\mathcal{A} =$

$\{z : 1/\rho < |z| < \rho\}$ for a uniquely determined $\rho > 1$ (see for example [7], [9], p. 362, the discussion in [23]). We also denote

$$\mathcal{D}_| = \{E(z) : z \in \varpi\} = \mathcal{D} \setminus \left(\mathcal{D} \cap \mathbb{R}^+\right) \text{ and } \mathcal{A}_| = \phi(\mathcal{D}_|). \tag{2.2}$$

We next observe that the function $z \mapsto (i2\pi)^{-1} \log(E(z))$ can be extended as an analytic function to the periodic domain Π. This is done by defining the branch cut of log on the annulus $\mathcal{A}$ on the curve $\Gamma := \phi(\mathcal{D} \cap \mathbb{R}^+) = \phi(E(J_+)) = \phi(E(J_-))$. In order to describe this in detail and for the sake of simplicity, we impose our final additional geometric assumption, which obviously could be made weaker.

(A) We assume that Γ *is contained, for some* $0 < \delta < 1$, *in the sector* $\{z \in \mathbb{C} : \delta < \arg z < \pi - \delta\}$.

Remark 2.1 It is plain that (A) is satisfied, if the mapping ϕ distorts the arguments of $z \in \mathcal{D} \cap \mathbb{R}^+$ at most by $\pm(\pi/2 - \delta)$ for some fixed $\delta > 0$. One then achieves (A) by applying a rotation of the annulus $\mathcal{A}$, and redefining ϕ accordingly. This holds for example, if Π is a small 1-periodic perturbation of a horizontal strip. However, we conjecture that assumption (A) is completely unnecessary for the results of this paper.

Let us denote $\varpi_+ = \varpi \cap \{z \in \mathbb{C} : \mathrm{Re} z \geq 1/2\}$ and $\varpi_- = \varpi \setminus \varpi_+$. If $\tilde{z} \in \Gamma$, then it has two preimages $z_\pm \in J_\pm$ with $\phi(E(z_\pm)) = \tilde{z}$. We consider a disc $B(\tilde{z}, r) \subset \mathcal{A}$, where $r = r(\tilde{z})$ is so small that both sets $B(\tilde{z}, r) \cap \phi(E(\varpi_\pm))$ are connected, and we define $\arg z \in [0, 3\pi]$ such that

$$\arg z < \pi \text{ for } z \in B(\tilde{z}, r) \cap \phi(E(\varpi_-)) \text{ and}$$
$$\arg z > 2\pi \text{ for } z \in B(\tilde{z}, r) \cap \phi(E(\varpi_+)).$$

It is clear that

$$\lim_{\substack{z \to \tilde{z} \\ z \in \phi(E(\varpi_+))}} \arg z = 2\pi + \lim_{\substack{z \to \tilde{z} \\ z \in \phi(E(\varpi_-))}} \arg z. \tag{2.3}$$

The logarithm is defined accordingly in $\phi(E(\varpi))$.

Lemma 2.2 *The mapping*

$$\varphi(z) = \frac{1}{i2\pi} \log\left(\phi(e^{i2\pi z})\right) + [\mathrm{Re}\, z] \tag{2.4}$$

is conformal from Π *onto the strip* $S = (-\infty, \infty) \times \left(-\frac{\log \rho}{2\pi}, \frac{\log \rho}{2\pi}\right)$.

Proof On $\varpi \ni z$ we have $\varphi(z) = \frac{1}{i2\pi}\log\big(\phi(e^{i2\pi z})\big)$, which is a conformal mapping, and we also have for every $m \in \mathbb{Z}$,

$$\varphi(\varpi_m) = \varphi(\varpi) + m \quad \text{with } \varphi(\varpi_m) \cap \varphi(\varpi_n) = \emptyset, \text{ if } m \neq n.$$

It remains to observe that φ is continuous on Π, by the explained branch cut and the defining formula (2.4), and that φ is a surjection, since $\phi \circ E$ is a surjection onto $\mathcal{A}$. □

Combining this lemma with formula (2.1) yields a kernel formula for the periodic domain Π. We denote $\operatorname{sech} z = 2/(e^z + e^{-z})$, which is the hyperbolic secant.

Proposition 2.3 *We have*

$$K_\Pi(z, w) = \widetilde{K}_\Pi(z, w)\frac{\pi^3}{4(\log\rho)^2}\operatorname{sech}^2\Big(\frac{\pi^2(\varphi(z) - \overline{\varphi(w)})}{2\log\rho}\Big), \tag{2.5}$$

where

$$\widetilde{K}_\Pi(z, w) = e^{i2\pi(z-\varphi(z)-\overline{w}+\overline{\varphi(w)})}\phi'(e^{i2\pi z})\overline{\phi'(e^{i2\pi w})},$$

Proof A conformal mapping from the strip $\Sigma = (-\infty, \infty) \times (-\pi, \pi)$ onto the upper half-plane is given by $z \mapsto ie^{z/2}$, and the kernel of the upper half-plane is $-1/(\pi(z - \overline{w})^2)$. By (2.1) one obtains

$$K_\Sigma(z, w) = \frac{1}{16\pi}\operatorname{sech}^2\big((z - \bar{w})/4\big) \tag{2.6}$$

Hence, (2.5) follows by applying (2.1) to the conformal mapping $z \mapsto 2\pi^2\varphi(z)/\log\rho$ from Π onto Σ and observing that $\phi(e^{i2\pi z}) = e^{i2\pi\varphi(z)}$, see (2.4). □

We will show later, among other things, how this formula also follows from the Floquet transform theory. We now mention a concrete example, where the boundary of Π consists of straight line segments: in this case, slightly more can be said about (2.5), namely, the inverse ψ of the mapping ϕ can be described up to some point, see [4, 6]. We assume that $\partial\Pi \cap R$, where $R = \{z : 0 < \operatorname{Re} z < 1\}$ is a strip, consists of two polygonal lines P_j, $j = 0, 1$ with $n_j \in \mathbb{N}$ edges and vertices at the points $z_k^{(j)}$, $k = 1, \ldots, n_j$, and we have $\operatorname{Re} z_1^{(0)} = \operatorname{Re} z_1^{(1)} = 1$, $\operatorname{Re} z_1^{(n_0)} = \operatorname{Re} z_1^{(n_1)} = 0$. We denote the interior angle at $z_k^{(j)}$ by $\alpha_k^{(j)}$, where

$$\alpha_k^{(j)} = \pi\big(\beta_k^{(j)} + 1\big) \quad \text{and} \quad \sum_{k=1}^{n_0}\beta_k^{(0)} = \sum_{k=1}^{n_1}\beta_k^{(0)} = 0\,.$$

According to [4], (8.14), the inverse $\psi = \phi^{-1} : \mathcal{A} \to \mathcal{D}$ of the conformal mapping ϕ can be defined by the formula

$$\psi(z) = A \int_1^z \prod_{j=0}^{1} \prod_{k=1}^{n_j} \left(P\Big(\frac{\zeta}{a_k^{(j)}}, \rho\Big)\right)^{\beta_k^{(j)}} \frac{d\zeta}{\zeta}$$

where $A \in \mathbb{C}$ is a constant, the points $a_k^{(j)} \in \partial\mathcal{A}$ are the images of the vertices $z_k^{(j)}$ and $P : \mathcal{A} \to \mathbb{C}$ is by (8.5) of [4] the mapping

$$P(\zeta, \rho) = (1 - \zeta) \prod_{k=1}^{\infty} \big(1 - \rho^{2k}\zeta\big)\big(1 - \rho^{2k}\zeta^{-1}\big).$$

However, as is well-known, calculating the points $a_k^{(j)}$ or the function ϕ explicitly is in general impossible.

3 Floquet Transform in Bergman Spaces

We next introduce the operator-theoretic tools. Since the origin of the Floquet transform, denoted here by F, is in the field of partial differential equations for functions of real variables, we put aside for a moment the complex variable, identify $\mathbb{C}$ with the plane $\mathbb{R}^2$, and denote the variable by (x_1, x_2) instead of z; the functions are still complex valued in the following. We refer to [12] for a general exposition and [13] for a concise treatment, and to [15, 16] for applications to general elliptic PDEs.

The Floquet transform with respect to the variable x_1 for functions $f \in L^2(\Pi)$ is given by

$$\mathsf{F}f(x_1, x_2, \eta) = \frac{1}{\sqrt{2\pi}} \sum_{m \in \mathbb{Z}} e^{-i\eta m} f(x_1 + m, x_2), \tag{3.1}$$

where $(x_1, x_2) \in \varpi$ and $\eta \in [-\pi, \pi)$ is the so-called Floquet parameter, or quasimomentum. As is well known (see Theorem 2.2.5 in [12] or Theorem 4.2 in [13]), the Floquet transform establishes an isometric isomorphism (isometric linear homeomorphism) between L^2-spaces,

$$\mathsf{F} : L^2(\Pi) \to L^2(-\pi, \pi; L^2(\varpi)), \tag{3.2}$$

where $L^2(-\pi, \pi; X)$ is the space of vector valued, Bochner-L^2 integrable functions on $[-\pi, \pi]$ with values in the Banach space X, and it is endowed with the norm

$$\|g\|_{L^2(0,2\pi;B)} = \left(\int_{-\pi}^{\pi} \|f(\eta)\|_X d\eta \right)^{1/2} .$$

In particular, the series (3.1) converges in the space $L^2(-\pi, \pi; L^2(\varpi))$ for every $f \in L^2(\Pi)$, and thus it also converges for example pointwise for almost every η and (x_1, x_2). The inverse transform is also explicitly given by the formula

$$\mathsf{F}^{-1}g(x_1, x_2) = \frac{1}{\sqrt{2\pi}} \int_{-\pi}^{\pi} e^{i[x_1]\eta} g(x_1 - [x_1], x_2, \eta) d\eta, \tag{3.3}$$

where $g \in L^2(-\pi, \pi; L^2(\varpi))$ and $[x]$ denotes the integer part of a real number x.

In the literature there is also another variation of the definition of the Floquet transform: if $f \in L^2(\Pi)$, one can also define it as

$$\mathcal{F}f(x_1, x_2, \eta) = \frac{1}{\sqrt{2\pi}} \sum_{m \in \mathbb{Z}} e^{-i\eta(x_1+m)} f(x_1 + m, x_2), \tag{3.4}$$

where $(x_1, x_2) \in \varpi$ and $\eta \in [-\pi, \pi]$. This differs from (3.4) only by a factor, which is a non-constant, unimodular, smooth function. In this case the inverse transform is defined on $L^2(-\pi, \pi; L^2(\varpi))$ by

$$\mathcal{F}^{-1}g(x_1, x_2) = \frac{1}{\sqrt{2\pi}} \int_{-\pi}^{\pi} e^{ix_1\eta} g(x_1 - [x_1], x_2, \eta) d\eta.$$

The isometry (3.2) holds also for $\mathcal{F}$.

Remark 3.1 We list some very elementary facts, some of which are consequences of the isomorphism relation (3.2) and will be needed throughout the paper.

(*i*) The space $L^2(-\pi, \pi; L^2(\varpi))$ can be identified in a canonical way with a L^2-space defined on a bounded Lipschitz domain of $\mathbb{R}^3$, namely the space $L^2\big([-\pi, \pi] \times \varpi\big)$, see [8], Prop. 1.2.24.

(*ii*) If a sequence $(f_n)_{n=1}^{\infty}$ converges in $L^2(\Pi)$, then, by definition, the sequence $(\mathsf{F}f_n)_{n=1}^{\infty}$ converges in $L^2(-\pi, \pi; L^2(\varpi))$, and consequently, there is a subsequence $\big(\mathsf{F}f_{n_k}(\cdot, \eta)\big)_{k=1}^{\infty}$, which converges in $L^2(\varpi)$ for almost all $\eta \in [-\pi, \pi]$. See [20], Theorem 3.12 and its vector valued version as explained after Definition 1.2.15. of [8]. Apparently, this does not need to happen for all η; a relevant case will be considered in Remark 3.3.

Although we will not need Sobolev-spaces in the sequel, we mention them briefly in order to explain the quasiperiodic boundary condition, which is essential later. For the next remark, see [13], p. 360. Given a domain $\Omega \subset \mathbb{R}^2$, let $H^1(\Omega)$ denote the standard Sobolev-Hilbert space of complex valued $L^2(\Omega)$-functions u which have weak partial derivatives of the first order belonging to the space $L^2(\Omega)$. According to the Sobolev embedding theorem, the space $H^1(\Omega)$ embeds into the space $C(\mathrm{cl}(\Omega))$ of continuous functions in the closure of Ω for example in the case where Ω is a bounded Lipschitz domain. In particular, the functions in $H^1(\Omega)$ have well-defined, continuous boundary values in $\partial\Omega$. The Floquet transform F is also an isomorphism from the Sobolev space $H^1(\Pi)$ onto $L^2(-\pi, \pi; H^1_\eta(\varpi))$, where $H^1_\eta(\varpi)$ consists of Sobolev functions u with the quasiperiodic boundary conditions

$$u(1, x_2) = e^{i\eta} u(0, x_2), \qquad a < x_2 < b \tag{3.5}$$

and $L^2(-\pi, \pi; H^1_\eta(\varpi))$ is defined as the closed subspace of $L^2(-\pi, \pi; H^1(\varpi))$ such that $f(\cdot, \eta) \in H^1_\eta(\varpi)$ for a.e. η. (If the other version $\mathcal{F}$ of the Floquet transform were used, one would replace $H^1_\eta(\varpi)$ by $H^1_{\mathrm{per}}(\varpi)$), where instead of (3.5) the periodic boundary conditions, i.e. (3.5) with $\eta = 0$, hold.)

The terminology for the Floquet transform varies from source to source: it may be called the Floquet-Bloch, Bloch, Zak or Gelfand transform. The last one is not to be confused with the Gelfand transform in Banach algebras. In the theory of spectral elliptic boundary problems, the transform F is used to convert a problem in the domain Π into another one in the cell ϖ, containing quasiperiodic boundary conditions, see, e.g. [5, 10, 14, 17]. Here, we instead wish to study the Bergman kernel in Π with the help of some kernels in ϖ.

We proceed to work in Bergman spaces of analytic functions. The transform (3.4), leading to periodic boundary conditions, is not useful as such for the study of analytic function spaces, since $\mathcal{F}f$ is not analytic even if f happens to be. Hence, we will use the transform (3.1), which is written using the complex variable, for $f \in A^2(\Pi)$,

$$\begin{aligned} \mathsf{F}f(z, \eta) &= \widehat{f}(z, \eta) \\ &= \frac{1}{\sqrt{2\pi}} \sum_{m \in \mathbb{Z}} e^{-i\eta m} f(z + m), \quad z \in \varpi, \ \eta \in [-\pi, \pi]. \end{aligned} \tag{3.6}$$

We note that F is a continuous (isometric) mapping from $A^2(\Pi)$ into $L^2(-\pi, \pi; L^2(\varpi))$. The series in (3.6) converges in $L^2(-\pi, \pi; L^2(\varpi))$, thus also pointwise for a.e. η, z, and also in $L^2(\varpi)$ for a.e. η. If $g \in L^2(-\pi, \pi; L^2(\varpi)$, we also denote

$$\mathsf{F}^{-1}g(z) = \frac{1}{\sqrt{2\pi}} \int_{-\pi}^{\pi} e^{i[\mathrm{Re} z]\eta} g(z - [\mathrm{Re}\, z], \eta) d\eta, \quad z \in \Pi. \tag{3.7}$$

The fact that formula (3.7) indeed gives the inverse of F will be established only in Theorem 3.6, since we first need to specify the proper function spaces. As for the motivation of the next proposition, we mention that the boundary condition (3.5) has to be taken into account in the subsequent considerations even though Bergman space functions in general do not have well defined boundary values.

Proposition 3.2

(i) If $f \in A^2(\Pi) \subset L^2(\Pi)$, then for almost all $\eta \in [-\pi, \pi]$, the function $z \mapsto \mathsf{F}f(z, \eta)$ is analytic in $\varpi \ni z$.
(ii) There is a dense subspace X of $A^2(\Pi)$, the elements f of which have the property that for all $\eta \in [-\pi, \pi]$, the function $\mathsf{F}f(\cdot, \eta)$ is analytic in ϖ and has an analytic extension to a neighborhood of the closure of ϖ in Π.

For an arbitrary $f \in A^2(\Pi)$, the analyticity of $\mathsf{F}f(\cdot, \eta)$ does not necessarily hold for *all* η, see Remark 3.3 below.

Proof

(*i*) Let $0 \neq f \in A^2(\Pi)$ and $\varepsilon > 0$, and denote

$$\varphi_\varepsilon(z) = e^{-\varepsilon z^2}. \tag{3.8}$$

Then, we have $\varphi_\varepsilon f \to f$ in $A^2(\Pi)$, as $\varepsilon \to 0$. (The proof of this can be obtained by choosing a large enough compact subset K such that the L^2-norm of the restriction of f to $\Pi \setminus K$ is small, and then observing that φ_ε converges to the constant function 1 uniformly on $K \cap \Pi$.)

We now consider the domain ϖ_U which is the interior of $\mathrm{cl}\,(\varpi \cup \varpi_{-1} \cup \varpi_1) \subset \Pi$. Let us fix $z_0 \in \varpi_U$ and a small enough open disc $B(z_0, \varrho)$, $\varrho > 0$, such that $\mathrm{cl}\,\big(B(z_0, 3\varrho)\big) \subset \varpi$. The Cauchy integral formula implies the estimate $|f(z)| \leq C\varrho^{-1}\|f\|_\varpi$ for all $z \in B(z_0, \varrho)$. Thus, applying a translation, or the same argument in all sets $\varpi_U + m$, $m \in \mathbb{Z}$, we get

$$\sup_{z \in \varpi_U} |f(z+m)| = \sup_{z \in \varpi_U + m} |f(z)| \leq \frac{C}{\varrho}\|f\|_{\varpi + m} \leq \frac{C}{\varrho}\|f\|_\Pi$$

for all $z \in B(z_0, \varrho)$. Moreover, for a fixed $\varepsilon < 1$ we have the estimate (see the choice of the domain ϖ for the number M)

$$|\varphi_\varepsilon(z)| = e^{-\varepsilon x^2 + \varepsilon y^2} \leq e^{M^2} e^{-\varepsilon x^2} \Rightarrow \sup_{z \in \varpi} |\varphi_\varepsilon(z+m)| \leq C e^{-\varepsilon |m|^2}$$

for all $m \in \mathbb{Z}$. As a consequence of these estimates, for a fixed ε the series in

$$F(\varphi_\varepsilon f)(z, \eta) = \frac{1}{\sqrt{2\pi}} \sum_{m \in \mathbb{Z}} e^{-i\eta m} \varphi_\varepsilon(z+m) f(z+m) \tag{3.9}$$

converges uniformly in the disc $B(z_0, \varrho)$ and thus defines an analytic function there. Since $z_0 \in \varpi_U$ was arbitrary, we infer that (3.9) thus defines an analytic function in ϖ_U.

Since the Floquet transform is an isometry from $L^2(\Pi)$ onto $L^2(-\pi, \pi; L^2(\varpi))$ and $\varphi_\varepsilon f \to f$ in $L^2(\Pi)$, we see that $\mathsf{F}(\varphi_\varepsilon f)$ converges to $\mathsf{F}f$ in $L^2(-\pi, \pi; L^2(\varpi))$ as $\varepsilon \to 0$. This implies, in view of the definition of the vector valued norm (see Remark 3.1(*ii*)), that there is a decreasing sequence $(\varepsilon_k)_{k=1}^\infty$ with $0 < \varepsilon_k \to 0$ and

$$\lim_{k\to\infty} \mathsf{F}(\varphi_{\varepsilon_k} f)(\cdot, \eta) = Ff(\cdot, \eta) \;\text{ in } L^2(\varpi) \tag{3.10}$$

for almost all $\eta \in [-\pi, \pi]$, and since $\mathsf{F}(\varphi_\varepsilon f)(\cdot, \eta)$ is analytic, by (3.9), the convergence in (3.10) happens in $A^2(\varpi)$ for almost all $\eta \in [-\pi, \pi]$. This completes the proof that $\mathsf{F}f$ is analytic in ϖ for almost all $\eta \in [-\pi, \pi]$.

(*ii*) The dense subspace X can be defined to consist of all functions $\varphi_\varepsilon f$, where $f \in A^2(\Pi)$; see just below (3.8). The result follows from the observations around (3.9). □

In (*i*) of the previous proposition, the analyticity does not need to hold for exactly all η, as shown in the next remark.

Remark 3.3 Consider the simple case that Π is the strip $(-\infty, \infty) \times (\frac{1}{4}, \frac{1}{2})$, corresponding to $\varpi = (0, 1) \times (\frac{1}{4}, \frac{1}{2})$. Then, the function $f(z) = 1/z$ is analytic on Π and belongs to $A^2(\Pi)$. Moreover, we have $|z| \le 1$ for all $z = x + iy \in \varpi$, so that the triangle inequality implies for $m \in \mathbb{Z}$, $m \neq 0, \pm 1$, and $z \in \varpi$,

$$\left| f(z+m) - \frac{1}{m} \right| = \left| \frac{x+iy}{(x+iy+m)m} \right| \le \frac{1}{(|m|-1)|m|} < \frac{1}{(|m|-1)^2}. \tag{3.11}$$

Thus for every $z \in \varpi$ and $\eta = 0$, the series

$$\sum_{m\in\mathbb{Z}} e^{-i\eta m} f(z+m) = \sum_{m\in\mathbb{Z}} \Big(\frac{1}{m} + \Big(f(z+m) - \frac{1}{m} \Big) \Big)$$

diverges by the triangle inequality and (3.11), since $\sum_m 1/m$ diverges. Thus $\mathsf{F}f$ is not a well defined analytic function for this particular value of η.

In order to treat the Bergman projection we will need the unitarity of certain operators, and this requires a selection of carefully defined function spaces. These are given in the next definition.

Definition 3.4

1°. Given the number $\eta \in [-\pi, \pi]$, we denote by $A^2_{\eta,\mathrm{ext}}(\varpi)$ the subspace of $A^2(\varpi)$ consisting of functions f which can be extended as analytic functions to a neighborhood in the domain Π of the set $\mathrm{cl}\,(\varpi) \cap \Pi$ and satisfy the boundary condition (3.5), i.e.

$$f(1+iy) = e^{i\eta} f(0+iy) \ \text{ for all } a < y < b. \tag{3.12}$$

We define the space $A^2_\eta(\varpi)$ as the closure of $A^2_{\eta,\mathrm{ext}}(\varpi)$ in $A(\varpi)$. We also denote by P_η the orthogonal projection from $L^2(\varpi)$ onto the subspace $A^2_\eta(\varpi)$. This definition will be motivated by the isomorphy property of F in Theorem 3.6; see also the remark before (3.5) and (ii) of Proposition 3.2. Formula (3.9) gives plenty of examples of functions belonging to the subspace $A^2_{E,\eta}(\varpi)$. In general, the functions belonging to $A^2(\varpi)$ may not have properly defined boundary values on $\partial\varpi$ so that the condition (3.12) cannot be posed directly.

2°. We denote by $L^2(-\pi,\pi;\, A^2_\eta(\varpi))$ the subspace of $L^2(-\pi,\pi;\, A^2(\varpi))$ consisting of functions f such that the function $z \mapsto f(z,\eta)$ belongs to $A^2_\eta(\varpi)$ for a.e. $\eta \in [-\pi,\pi]$. Since $A^2_\eta(\varpi)$ is by definition, for every η, a closed subspace of $A^2(\varpi)$ and thus of $L^2(\varpi)$, we obtain that $L^2(-\pi,\pi;\, A^2_\eta(\varpi))$ is a closed subspace of $L^2(-\pi,\pi;\, L^2(\varpi))$ (follows by using Remark 3.1(ii)).

3°. We define the subspace

$$\mathcal{H}_\eta := L^2\big(-\pi,\pi;\, A^2_{\eta,\mathrm{ext}}(\varpi)\big)$$

of $L^2(-\pi,\pi;\, A^2_\eta(\varpi))$ which consists of functions g such that the mapping $z \mapsto g(z,\eta)$ belongs to $A^2_{\eta,\mathrm{ext}}(\varpi)$ for a.e. η.

It is good to keep in mind that $L^2(-\pi,\pi;\, A^2_\eta(\varpi))$ and $\mathcal{H}_\eta$ are not vector valued L^2-spaces (since the space $A^2_\eta(\varpi)$ depends on η) but they only have the structure of a Banach vector bundle, see for example Section 1.3 of [12]. This fact complicates the proof of Theorem 3.6, since simple functions of $L^2\big(-\pi,\pi;\, A^2(\varpi)\big)$ are not contained in these subspaces. Thus, the following fact needs to be proven.

Lemma 3.5 *The subspace $\mathcal{H}_\eta$ is dense in $L^2\big(-\pi,\pi;\, A^2_\eta(\varpi)\big)$*

Proof Let us define the Bochner space $L^2\big(-\pi,\pi;\, A^2_{\mathrm{per}}(\varpi)\big)$, where $A^2_{\mathrm{per}}(\varpi)$ equals the space $A^2_\eta(\varpi)$ with $\eta = 0$, corresponding to the periodic boundary condition. We also denote by $A^2_{\mathrm{per,ext}}(\varpi) \subset A^2_{\mathrm{per}}(\varpi)$ the space $A^2_{\eta,\mathrm{per}}(\varpi)$ with $\eta = 0$. Finally, we define the subspace $L^2\big(-\pi,\pi;\, A^2_{\mathrm{per,ext}}(\varpi)\big)$ of $L^2\big(-\pi,\pi;\, A^2_{\mathrm{per}}(\varpi)\big)$ consisting of those functions g which have values in $A^2_{\mathrm{per,ext}}(\varpi)$ for almost every η.

By the theory of Bochner spaces, see Lemma 1.2.19 in [8], a dense subspace of $L^2\big(-\pi,\pi;\, A^2_{\mathrm{per}}(\varpi)\big)$ is formed by simple functions

$$\sum_j \chi_j(\eta) g_j(z), \tag{3.13}$$

where χ_j is the characteristic function of some measurable subset of $[-\pi, \pi]$, $g_j \in A^2_{\rm per}(\varpi)$ and the sum contains only finitely many terms. Moreover, since $A^2_{\rm per,ext}(\varpi)$ is dense in $A^2_{\rm per}(\varpi)$ by definition, a dense subspace of $L^2\big(-\pi, \pi; A^2_{\rm per}(\varpi)\big)$ is formed by functions (3.13), where every g_j belongs to $A^2_{\rm per,ext}(\varpi)$. This shows that $L^2\big(-\pi, \pi; A^2_{\rm per,ext}(\varpi)\big)$ is dense in $L^2\big(-\pi, \pi; A^2_{\rm per}(\varpi)\big)$.

The assertion of the lemma now follows from the facts that the mapping $f(z,\eta) \mapsto e^{i\eta z} f(z,\eta)$ is an isomorphism from $L^2\big(-\pi, \pi; A^2_{\rm per}(\varpi)\big)$ onto $L^2\big(-\pi, \pi; A^2_\eta(\varpi)\big)$ and from $L^2\big(-\pi, \pi; A^2_{\rm per,ext}(\varpi)\big)$ onto $\mathcal{H}_\eta$. The proofs of these isomorphisms are left to the reader. □

We are finally prepared to prove the main result of this section.

Theorem 3.6 *The Floquet transform* F *maps* $A^2(\Pi)$ *onto* $L^2(-\pi, \pi; A^2_\eta(\varpi))$. *Its inverse* $\mathsf{F}^{-1} : L^2(-\pi, \pi; A^2_\eta(\varpi)) \to A^2(\Pi)$ *is given by the formula* (3.7). *Moreover,* F *preserves the inner product and is thus a unitary operator.*

Proof We recall that F is an isometric isomorphism from $L^2(\Pi)$ onto $L^2(-\pi, \pi; L^2(\varpi))$ and first show that it maps the space $A^2(\Pi)$ onto $L^2(-\pi, \pi; A^2_\eta(\varpi))$. Let the dense subspace $X \subset L^2(\Pi)$ be as in Proposition 3.2(*ii*) and assume $f \in X$ so that for every η, $\mathsf{F}f(z,\eta)$ is an analytic function of z on ϖ and also has an analytic extension to a neighborhood of the closure of ϖ, see Definition 3.4. Thus $\mathsf{F}f(\cdot,\eta)$ also has well defined values on the lateral boundaries of ϖ. Moreover, given $\eta \in [-\pi, \pi]$, the function $\mathsf{F}f(\cdot,\eta)$ satisfies the quasiperiodic boundary condition (3.12), since

$$\mathsf{F}f(1+iy,\eta) = \frac{1}{\sqrt{2\pi}} \sum_{m\in\mathbb{Z}} e^{-i\eta(m-1)} f(1+iy+(m-1)) = e^{i\eta}\mathsf{F}f(iy,\eta).$$

Hence, F maps the dense subspace X into $L^2(-\pi, \pi; A^2_\eta(\varpi))$, and since F is an isometry and $L^2(-\pi, \pi; A^2_\eta(\varpi))$ is complete, it also maps $A^2(\Pi)$ into $L^2(-\pi, \pi; A^2_\eta(\varpi))$.

To see that F is surjective we first consider a function $g \in \mathcal{H}_\eta$. Clearly, for all $m \in \mathbb{Z}$ and a.e. $\eta \in [-\pi, \pi]$, the function

$$G_{\eta,m} : \varpi_m \to \mathbb{C}, \ \ G_\eta(z) = e^{im\eta} g(z-m, \eta), \tag{3.14}$$

is analytic on the subdomain ϖ_m, $m \in \mathbb{Z}$ (see (1.2)). Moreover, by Definition 3.4, it has an analytic extension, still denoted by $G_{\eta,m}$, to a neighborhood of the closure of

ϖ_m in Π. Since $g(\cdot, \eta)$ satisfies the boundary condition (3.12), we get for all $m \in \mathbb{Z}$, all $z = m + 1 + iy$ with $y \in [a, b]$,

$$
\begin{aligned}
G_{\eta,m}(z) &= \lim_{\varepsilon \to 0^+} G_{\eta,m}(m + 1 - \varepsilon + iy) = \lim_{\varepsilon \to 0^+} e^{i\eta m} g(1 + iy - \varepsilon, \eta) \\
&= e^{i\eta m} g(1 + iy, \eta) = e^{i\eta(m+1)} g(iy, \eta) = \lim_{\varepsilon \to 0^+} e^{i\eta(m+1)} g(iy + \varepsilon, \eta) \\
&= \lim_{\varepsilon \to 0^+} G_{\eta,m+1}(m + 1 + \varepsilon + iy) = G_{\eta,m+1}(m + 1 + iy) = G_{\eta,m+1}(z).
\end{aligned}
$$

In other words, $G_{\eta,m}$ and $G_{\eta,m+1}$ are two analytic functions with overlapping domains of definition and they coincide on the line segments $J_m = \{m+1\} \times [a, b]$. Thus, $G_{\eta,m}$ and $G_{\eta,m+1}$ coincide in their common domain of definition, and we can define the analytic function G_η on Π by setting $G_\eta(z) = G_{\eta,m}(z)$ for z belonging to the closure of ϖ_m in Π.

Comparing (3.14) with (3.7) we see that

$$
\mathsf{F}^{-1} g(z) = \frac{1}{\sqrt{2\pi}} \int_{-\pi}^{\pi} G_\eta(z) d\eta,
$$

which is an analytic function in Π, since G_η is. Hence, $\mathsf{F}^{-1} g \in A^2(\Pi)$. Then, we have $\mathsf{F}\mathsf{F}^{-1} g = g$ for all $g \in \mathcal{H}_\eta \subset L^2\big(-\pi, \pi; L^2(\varpi)\big)$, by the general Floquet inversion formula (3.4). We conclude that the image $\mathsf{F}\big(A^2(\Pi)\big)$ contains the subspace $\mathcal{H}_\eta$, which is dense in $L^2(-\pi, \pi; A^2_\eta(\varpi))$ by Lemma 3.5. This implies the surjectivity, since F is an isometry and thus bounded from below with respect to the L^2-norms.

It is well known that the Floquet transform preserves the inner product. This can also seen directly, since for $f, g \in L^2(\Pi)$ and $H := L^2(-\pi, \pi; L^2(\varpi))$ there holds

$$
\begin{aligned}
(\mathsf{F} f | \mathsf{F} g)_H &= \sum_{m,n \in \mathbb{Z}} \int_{\varpi} \int_{-\pi}^{\pi} e^{-im\eta} f(z + m) e^{in\eta} \overline{g(z + n)} dz d\eta \\
&= \sum_{m \in \mathbb{Z}} \int_{\varpi} f(z + m) \overline{g(z + m)} dz = \int_{\Pi} f(z) \overline{g(z)} dz = (f | g)_{L^2(\varpi)}.
\end{aligned}
$$

□

4 General Kernel Formula for the Periodic Domain

Our purpose is to show how the Bergman projection in Π can be presented with the help of an orthogonal projection in the periodicity cell ϖ by using the Floquet transform.

Definition 4.1 For all $\eta \in [-\pi, \pi]$ we denote by P_η the orthogonal projection from $L^2(\varpi)$ onto $A^2_\eta(\varpi)$; see Definition 3.41° for the notation.

The well-known proof for the existence of Bergman kernels, see e.g. [11], p. 1060, applies also in this case and implies that P_η can be written with the help of the integral kernel $K_\eta : \varpi \times \varpi \to \mathbb{C}$

$$P_\eta f(z) = \int_\varpi K_\eta(z, w) f(w) dA(w), \tag{4.1}$$

where $K_\eta(z, \cdot) \in L^2(\varpi)$ for all $z \in \varpi$. In Sect. 6 we will calculate the kernel K_η in some cases. We also define an operator $\mathcal{P}$ by denoting, for all $f \in L^2(-\pi, \pi; L^2(\varpi))$,

$$\mathcal{P} f(z, \eta) = \big(P_\eta f(\cdot, \eta)\big)(z), \quad z \in \varpi, \ \eta \in [-\pi, \pi]. \tag{4.2}$$

It follows directly from the definitions of the spaces and their norms that this is a bounded operator from $L^2(-\pi, \pi; L^2(\varpi))$ into $L^2(-\pi, \pi; A^2_\eta(\varpi))$. More precisely, the following holds.

Lemma 4.2 *The operator $\mathcal{P}$ is the orthogonal projection from $L^2(-\pi, \pi; L^2(\varpi))$ onto $L^2(-\pi, \pi; A^2_\eta(\varpi))$.*

Proof Since P_η is the orthogonal projection $L^2(\varpi)$ onto $A^2_\eta(\varpi)$, we have $\mathcal{P}^2 = \mathcal{P}$ and $\mathcal{P} f = f$ for every $f \in L^2(-\pi, \pi; A^2_\eta(\varpi))$. Thus, $\mathcal{P}$ is a projection operator between the given spaces, and it suffices to observe that it is also self-adjoint. With $H := L^2(-\pi, \pi; L^2(\varpi))$ and $f, g \in H$ we obtain, by the self-adjointness of P_η,

$$(\mathcal{P} f | g)_H = \int_\varpi \int_{-\pi}^{\pi} P_\eta f(\cdot, \eta) \overline{g(\cdot, \eta)} d\eta dA$$

$$= \int_\varpi \int_{-\pi}^{\pi} f(\cdot, \eta) \overline{P_\eta g(\cdot, \eta)} d\eta dA = (f | \mathcal{P} g)_H.$$

□

We can now describe the general relation of the orthogonal projections in the bounded domains and the full domain Π.

Theorem 4.3 *The Bergman projection P_Π from $L^2(\Pi)$ onto $A^2(\Pi)$ can be written as*

$$\mathsf{F}^{-1} \mathcal{P} \mathsf{F} f(z) = \frac{1}{\sqrt{2\pi}} \int_{-\pi}^{\pi} e^{i[\mathrm{Re} z]\eta} \big(P_\eta \widehat{f}(\cdot, \eta)\big)(z - [\mathrm{Re} z]) d\eta \tag{4.3}$$

$$= \frac{1}{2\pi} \int_\Pi \int_{-\pi}^{\pi} e^{i\eta([\mathrm{Re} z] - [\mathrm{Re} w])} K_\eta(z - [\mathrm{Re} z], w - [\mathrm{Re} w]) f(w) d\eta dA(w).$$

Proof The operator $\mathsf{F}^{-1}\mathcal{P}\mathsf{F}$ is self-adjoint, since F preserves the inner product (Theorem 3.6), and also clearly a projection, since $\mathcal{P}$ is. So, this operator must be the orthogonal projection from $L^2(\Pi)$ onto $A^2(\Pi)$, in view of Theorem 3.6. It remains to verify the rest of the formula (4.3).

The series (3.6) for $\mathsf{F}f$ converges in $L^2(\varpi)$ for almost all $\eta \in [-\pi, \pi]$, and since P_η is a bounded operator, we can write

$$\mathcal{P}\mathsf{F}f(z;\eta) = \frac{1}{\sqrt{2\pi}} \sum_{m\in\mathbb{Z}} e^{-i\eta m} \int_{\varpi} K_\eta(z,w) f(w+m) dA(w), \quad z \in \varpi, \tag{4.4}$$

for almost all η. On the other hand, the series (3.6) for $\mathsf{F}f$ converges also in $L^2(-\pi,\pi; L^2(\varpi))$, hence the series (4.4) converges in this space, too. Since F^{-1} is a bounded operator $L^2(-\pi,\pi; L^2(\varpi)) \to L^2(\Pi)$, the result follows from the defining formula for F^{-1} in (3.7),

$$\begin{aligned} &\mathsf{F}^{-1}\mathcal{P}\mathsf{F}f(z) \\ &= \frac{1}{2\pi} \sum_{m\in\mathbb{Z}} \int_{\varpi} \int_{-\pi}^{\pi} e^{i\eta([\mathrm{Re}z]-m)} K_\eta(z-[\mathrm{Re}z], w) f(w+m) d\eta dA(w) \\ &= \frac{1}{2\pi} \sum_{m\in\mathbb{Z}} \int_{\varpi_m} \int_{-\pi}^{\pi} e^{i\eta([\mathrm{Re}z]-[\mathrm{Re}w])} K_\eta(z-[\mathrm{Re}z], w-[\mathrm{Re}w]) f(w) d\eta dA(w) \\ &= \frac{1}{2\pi} \int_{\Pi} \int_{-\pi}^{\pi} e^{i\eta([\mathrm{Re}z]-[\mathrm{Re}w])} K_\eta(z-[\mathrm{Re}z], w-[\mathrm{Re}w]) f(w) d\eta dA(w). \end{aligned}$$

□

We will show in Sect. 6 that if the periodic cell ϖ is simply connected, the kernel K_η can be constructed by applying a certain canonical Riemann conformal mapping of doubly connected domains and derive again the kernel formula of Sect. 2 in this way.

5 Weights, Domains and Projections

In this section we collect some basic facts concerning the Bergman projection in domains and weighted spaces. These results are known at least to experts in this area, but we give some proofs for the sake of the completeness of the presentation.

Let $\Omega \subset \mathbb{C}$ be a domain. By a weight on Ω we mean a continuous function $V : \Omega \to \mathbb{R}^+ = (0, \infty)$, and we denote by $L^2_V(\Omega)$ the weighted L^2-space on Ω with norm and inner product

$$\|f\|^2_{\Omega,V} = \int_\Omega |f|^2 V\, dA \ , \quad (f|g)_{\Omega,V} = \int_\Omega f\overline{g} V\, dA, \tag{5.1}$$

where $f, g, \in L^2_V(\Omega)$. We also denote by $A^2_V(\Omega)$ the subspace of $L^2_V(\Omega)$ consisting of analytic functions. The proof of the unweighted case applies here, too, and shows that $A^2_V(\Omega)$ is a closed subspace. Our weight functions will mostly be of the form

$$V(z) = |v(z)|^2 = v(z)\overline{v(z)},$$

where $v : \Omega \to \mathbb{C} \setminus \{0\}$ is analytic.

We will use the following observations.

Lemma 5.1 *Let Ω and Ω' be conformally equivalent domains and let φ be a conformal mapping from Ω onto Ω'. Then, the composition operator $I : f \mapsto f \circ \varphi$ is a unitary isomorphism from the space $L^2_V(\Omega')$ onto $L^2(\Omega)$ and also from $A^2_V(\Omega')$ onto $A^2(\Omega)$, where the weight is $V(z) = |\psi'(z)|^2$ with $\psi = \varphi^{-1} : \Omega' \to \Omega$.*

If X is a closed subspace of $L^2_V(\Omega')$ and P_X is the orthogonal projection from $L^2_V(\Omega')$ onto X, then the orthogonal projection P from $L^2(\Omega)$ onto $I(X) \subset L^2(\Omega)$ is given by $P = IP_XI^{-1}$. If the function $K_X \in L^2(\Omega' \times \Omega')$ is the integral kernel of P_X, then the kernel of P is given by

$$K(z,w) = K_X\big(\varphi(z), \varphi(w)\big)|\varphi'(w)|^2, \quad z, w \in \Omega.$$

Proof The unitary isomorphism property of I is a direct consequence of the definitions and the fact that V is the Jacobian of the conformal transform ψ. Note that the inverse of the operator I is given by $I^{-1} : L^2(\Omega) \to L^2_V(\Omega'), f \mapsto f \circ \psi$.

Next, it is straightforward to see that the mapping $P = IP_XI^{-1} : L^2(\Omega) \to L^2(\Omega)$ is selfadjoint (since the adjoint of I is I^{-1} by the unitarity), a projection ($P^2 = P$), and maps $L^2(\Omega)$ onto $I(X)$. Hence, it is the claimed orthogonal projection. Finally, given $f \in L^2(\Omega)$ we write

$$Pf(z) = IP_XI^{-1}f(z) = \int_{\Omega'} K_X\big(\varphi(z), w\big) f\big(\varphi^{-1}(w)\big) dA(w)$$

$$= \int_\Omega K_X\big(\varphi(z), \varphi(w)\big) f(w) |\varphi'(w)|^2 dA(w).$$

□

6 Simply Connected Periodic Domain

In the following we construct the kernel for the operator P_η, assuming eventually that Π is simply connected and satisfies assumption (A) of Sect. 2. Due to the quasiperiodic boundary condition in the definition of the target space $A^2_\eta(\varpi)$, the Riemann mapping from ϖ onto $\mathbb{D}$ is not useful but instead we will use the doubly connected exponential image of ϖ and its Riemann map to an annulus. In the next section we will rewrite the formula of Theorem 4.3 of the Bergman projection P_Π with the help of the mentioned Riemann map of the bounded domain and show how formula (2.5) follows from these considerations.

We proceed to the construction of the kernel K_η of the orthogonal projection $P_\eta : L^2(\varpi) \to A^2_\eta(\varpi)$ for any $\eta \in [-\pi, \pi]$, see Definition 3.4, 2°. This will be accomplished in several steps, where we also employ certain classical methods in complex analysis. We will use the notation of Sect. 2 and recall that the exponential map $E : z \mapsto e^{i2\pi z}$ maps the set $\varpi \cup J_+ \cup J_-$ onto the domain $\mathcal{D}$, which is contained in an annulus so that there exist numbers $0 < \rho_0 < \rho_1$ with

$$\mathcal{D} \subset \{z \in \mathbb{C} : \rho_0 < |z| < \rho_1\}. \tag{6.1}$$

The domain $\mathcal{D}$ is doubly connected, if ϖ is simply connected. Let $\mathcal{D}_| = E(\varpi) \subset \mathcal{D}$ be as in (2.2). We will soon employ the branch cut of the logarithm as defined in Sect. 2, but we need to agree that when considered in $\mathcal{D}$, the branch cut of the logarithm happens on $\mathcal{D} \cap \mathbb{R}^+$ so that $\operatorname{Im} \log z \in (0, 2\pi)$ for $z \in \mathcal{D}_|$ and log is analytic on $\mathcal{D}_|$. Let us denote $L = E^{-1} : \mathcal{D}_| \to \varpi$, $L(z) = (i2\pi)^{-1} \log z$.

We define for all $\eta \in [-\pi, \pi]$ the weight function

$$W(z) = |L'(z)|^2 = \frac{1}{4\pi^2 |z|^2}, \quad z \in \mathcal{D}, \tag{6.2}$$

and consider the weighted Bergman space $A^2_W(\mathcal{D}_|)$ defined below (5.1) (restricting the weight W onto $D_|$, of course). The Bergman space $A^2_W(\mathcal{D})$ can be considered as a subspace $A^2_W(\mathcal{D}_|)$, which consists of functions that can be extended from $\mathcal{D}_|$ to the full domain $\mathcal{D}$ as analytic functions. Since $A^2_W(\mathcal{D})$ is complete, it is a closed subspace of $A^2_W(\mathcal{D}_|)$.

Lemma 6.1

(i) The operator $I_1 : f \mapsto f \circ L$ is a unitary isomorphism $A^2(\varpi) \to A^2_W(\mathcal{D}_|)$ and $L^2(\varpi) \to L^2_W(\mathcal{D})$.

(ii) The operator I_1 maps the space $A^2_\eta(\varpi)$ onto $A^2_{W,\eta}(\mathcal{D}_|)$, which is the closure of the subspace

$$A^2_{W,\eta,\text{ext}}(\mathcal{D}_|) = \left\{ z^{\eta/(2\pi)} g \,:\, g \in A^2_W(\mathcal{D}) \right\} \tag{6.3}$$

in $A^2_W(\mathcal{D}_|)$.

Proof The claim (i) follows by applying Lemma 5.1 to the inverse operator I_1^{-1} : $f \mapsto f \circ E$. As for the claim (ii), given an arbitrary $g \in A^2_W(\mathcal{D})$, the function

$$f := \left(z^{\eta/(2\pi)} g\right) \circ E =: h \circ E \in L^2(\varpi)$$

belongs to the subspace $A^2_{\eta,\mathrm{ext}}(\varpi)$ of $A_\eta(\varpi)^2$ (see Definition 3.4) and we have $I_1 f = h$, where h is an arbitrary element of $A^2_{W,\eta,\mathrm{ext}}$. Thus, the operator I_1 is a surjection onto $A^2_{W,\eta,\mathrm{ext}}$ and consequently also onto $A^2_{W,\eta}$, by (i) and the closedness of $A^2_\eta(\varpi)$ in $A^2(\varpi)$.

On the other hand, in view of Definition 3.4, we consider a function $f \in A^2_{\eta,\mathrm{ext}}(\varpi)$. This has well defined continuous boundary values satisfying (3.12). The function $f \circ L \in A^2_W(\mathcal{D}_|)$ has a discontinuity at the slit of $\mathcal{D}_|$, more precisely

$$\lim_{y \to 0^-} f \circ L(x + iy) = e^{i\eta} \lim_{y \to 0+} f \circ L(x + iy) \;\; \text{for all } x \in \mathcal{D} \cap \mathbb{R}^+.$$

Hence, the function $g = z^{-\eta/(2\pi)} f \circ L(z)$ has a continuous extension to $\mathcal{D}$, which makes it into an analytic function in $\mathcal{D}$. Moreover, since the multiplier $z^{-\eta/(2\pi)}$ is a bounded function, we also have $g \in L^2_W(\mathcal{D})$ and thus $g \in A^2_W(\mathcal{D})$, hence, we obtain $I_1 f = z^{\eta/(2\pi)} g \in A^2_{W,\eta,\mathrm{ext}}(\mathcal{D}_|)$.

Since, by definition, $A^2_\eta(\varpi)$ is the closure of $A^2_{\eta,\mathrm{ext}}(\varpi)$ in $A^2(\varpi)$ and I_1 is an isometry, we obtain that I_1 maps $A^2_\eta(\varpi)$ into $A^2_{W,\eta}(\mathcal{D}_|)$. □

From now on we will additionally assume that Π and ϖ are simply connected (thus $\mathcal{D}$ is doubly connected). We recall from Sect. 2 the conformal mapping

$$\phi : \mathcal{D} \to \mathcal{A} = \{z \in \mathbb{C} : 1/\rho < |z| < \rho\} \tag{6.4}$$

where the number $\rho > 1$ is uniquely determined by $\mathcal{D}$. Let us also agree that when considered on $\mathcal{A}$, the logarithm is defined as in Sect. 2. Note that our assumptions on the geometry of Π and ϖ, (1.2), do not imply smoothness of the boundary $\partial\mathcal{D}$ of $\mathcal{D}$. Hence, $\partial\mathcal{D}$ may include corners for example. This has the consequence that the derivative ϕ' does not need to be bounded or bounded away from zero. See for example [1, 19, 21, 22].

We define on $\mathcal{A}$ the weight

$$V(z) = W(\psi(z))|\psi'(z)|^2, \;\; z \in \mathcal{A},$$

where $\psi = \phi^{-1} : \mathcal{A} \to \mathcal{D}$ is the inverse map. According to (6.2),

$$V(z) = |(L \circ \psi)'(z)|^2 = \frac{|\psi'(z)|^2}{4\pi^2 |\psi(z)|^2} = v(z)\overline{v(z)}$$

$$\text{with } \; v(z) = \frac{1}{2\pi} \frac{\psi'(z)}{\psi(z)}, \tag{6.5}$$

where $z \in \mathcal{A}$. Moreover, we denote by $A^2_{V,\eta,\mathrm{ex}}(\mathcal{A}_|)$ the subspace of all functions of the form

$$z^{\eta/(2\pi)} g, \quad \text{where } g \in A^2_V(\mathcal{A}), \tag{6.6}$$

and by $A^2_{V,\eta}(\mathcal{A}_|)$ the closure of $A^2_{V,\eta,\mathrm{ex}}(\mathcal{A}_|)$ in $A^2_V(\mathcal{A}_|)$. We recall that the functions belonging to $A^2_{V,\eta}(\mathcal{A}_|)$ may be discontinuous on the curve $\Gamma = \phi(\mathcal{D} \cap \mathbb{R}^+)$, see Sect. 2. Finally, we denote by I_2 the composition operator $I_2 : f \mapsto f \circ \psi$.

Lemma 6.2

(i) The operator I_2 is a unitary isomorphism $L^2_W(\mathcal{D}) \to L^2_V(\mathcal{A})$.
(ii) The operator I_2 maps the space $A^2_{W,\eta}(\mathcal{D}_|)$ onto $A^2_{V,\eta}(\mathcal{A}_|)$ for every $\eta \in [-\pi, \pi]$.

As a consequence, the composition operator

$$I_2 I_1 : f \mapsto f \circ L \circ \psi \tag{6.7}$$

is a unitary isomorphism from $L^2(\varpi)$ onto $L^2_V(\mathcal{A})$ and from $A^2_\eta(\varpi)$ onto $A^2_{V,\eta}(\mathcal{A}_|)$.

Proof Here, the first assertion is just a variant of Lemma 5.1 with a similar proof. To prove (ii) we note that due to choice of the branch cut on $\mathcal{A}$ (see (2.3)) and (6.1), the function $F_\eta(z) = \big(\phi(z)/z\big)^{\eta/(2\pi)}$ is analytic, bounded and bounded away from zero on $\mathcal{D}$ for all $\eta \in [-\pi, \pi]$, and the same is thus true also for the function

$$\Big(\frac{\psi(z)}{z}\Big)^{\eta/(2\pi)} = \frac{1}{F_\eta \circ \psi(z)} \quad \text{on } \mathcal{A}.$$

Given $f = z^{\eta/(2\pi)} g \in A^2_{W,\eta,\mathrm{ext}}(\mathcal{D}_|)$ with $g \in A^2_W(\mathcal{D})$ as in (6.3) we thus have

$$I_2 f(z) = \psi(z)^{\eta/(2\pi)} g(\psi(z)) = z^{\eta/(2\pi)} \Big(\frac{\psi(z)}{z}\Big)^{\eta/(2\pi)} g(\psi(z)) =: z^{\eta/(2\pi)} h(z)$$

where

$$\|h\|^2_{\mathcal{D},V} = \int\limits_{\mathcal{A}} \Big|\frac{\psi(z)}{z}\Big|^{\eta/(2\pi)} |g(\psi(z))|^2 \frac{|\psi'(z)|^2}{4\pi^2 |\psi(z)|^2} dA(z)$$

$$\le C \int\limits_{\mathcal{A}} |g \circ \psi|^2 \frac{|\psi'|^2}{4\pi^2 |\psi|^2} dA = C \int\limits_{\mathcal{D}} |g|^2 W dA.$$

Hence, $h \in A^2_V(\mathcal{A})$, and I_2 maps $A^2_{W,\eta,\mathrm{ext}}(\mathcal{D}_|)$ into $A^2_{V,\eta,\mathrm{ext}}(\mathcal{A}_|)$. The converse relation

$$I_2^{-1}\big(A^2_{V,\eta,\mathrm{ext}}(\mathcal{A}_|)\big) \subset A^2_{W,\eta,\mathrm{ext}}(\mathcal{D}_|)$$

can be proved in the same way, starting by $f = z^{\eta/(2\pi)} g \in A^2_{V,\eta,\mathrm{ext}}(\mathcal{D}_|)$ with $g \in A^2_V(\mathcal{D})$ and using ϕ instead of ψ.

The proof is completed by observing that the operator $I_2 : A^2_{W,\eta}(\mathcal{D}_|) \to A^2_{V,\eta}(\mathcal{A}_|)$ as well as its inverse are bounded and by taking into account the densities of $A^2_{W,\eta,\mathrm{ext}}(\mathcal{D}_|)$ and $A^2_{V,\eta,\mathrm{ext}}(\mathcal{A}_|)$ in these spaces. □

Lemma 6.3 *An orthonormal basis in $A^2_{V,\eta}(\mathcal{A}_|)$ is formed by the functions*

$$f_{n,\eta}(z) = C_{n,\eta}\frac{z^{n+\eta/(2\pi)}}{v(z)}, \quad n \in \mathbb{Z}, \tag{6.8}$$

where $C_{n,\eta}$ are the normalization constants,

$$C_{n,\eta}^{-2} = \int\limits_{\mathcal{A}} \left|\frac{z^{n+\eta/(2\pi)}}{v(z)}\right|^2 V(z)\,dA(z) = \int\limits_0^{2\pi}\int\limits_{1/\rho}^{\rho} r^{2n+\eta/\pi+1}\,dr\,d\theta$$

$$= \frac{2\pi}{2(n+1)+\eta/\pi}\big(\rho^{2(n+1)+\eta/\pi} - \rho^{-2(n+1)-\eta/\pi}\big). \tag{6.9}$$

Proof The mutual orthogonality of the functions (6.8) follows from the usual orthogonality relation of the monomials z^n and the fact that the factor $1/v$ is cancelled by the weight in the inner product of $A^2_V(\mathcal{A})$. Indeed, given $n, m \in \mathbb{Z}$ and $\eta \in [-\pi, \pi]$ we have

$$\int\limits_{\mathcal{A}} f_{n,\eta}\overline{f_{m,\eta}}V\,dA = C_{n,\eta}C_{m,\eta}\int\limits_{\mathcal{A}} z^{n+\eta/(2\pi)}\overline{z^{m+\eta/(2\pi)}}\,dA \tag{6.10}$$

where $z^{\eta/(2\pi)} = (re^{i\theta})^{\eta/(2\pi)} = r^{\eta/(2\pi)}e^{i\theta(\eta/(2\pi)+\eta\ell(\theta))}$ for some $\ell(\theta) \in \mathbb{Z}$ hence, the integral in (6.10) reads in polar coordinates as

$$\int\limits_{1/\rho}^{\rho} r^{(\dots)}\,dr \int\limits_0^{2\pi} e^{i\theta(n-m+\eta/(2\pi)+\eta\ell(\theta)-\eta/(2\pi)-\eta\ell(\theta))}\,d\theta$$

which is null unless $n = m$.

As for the completeness of the orthonormal sequence (6.8), an arbitrary $h \in A^2(\mathcal{A})$ can be approximated in the Bergman space $A^2(\mathcal{A})$ by a linear combination of the functions z^n, $n \in \mathbb{Z}$, since these form an orthonormal basis of $A^2(\mathcal{A})$, after a proper normalization. Since the mapping $f \mapsto v^{-1}f$ is an isometry from $A^2(\mathcal{A})$ onto $A^2_V(\mathcal{A})$, we find that an arbitrary $g \in A^2_V(\mathcal{A})$ can be approximated in $A^2_V(\mathcal{A})$ by a linear combination of the functions $v^{-1}z^n$, $n \in \mathbb{Z}$. In view of (6.6), the linear combinations of the functions (6.8) are dense in $A^2_{V,\eta,\mathrm{ext}}(\mathcal{A})$ and thus the system (6.8) is complete in $A^2_{V,\eta}(\mathcal{A}_|)$. □

We now construct the kernel function $K_\eta(z, w)$, see (4.1). Due to Lemma 6.3, the kernel $K_{\eta,\mathcal{A}}$ of the orthogonal projection $f \mapsto \int_{\mathcal{A}} K_{\eta,\mathcal{A}}(\cdot, w) f(w) dA(w)$ from $L^2_V(\mathcal{A})$ onto $\mathcal{A}^2_{V,\eta}(\mathcal{A}_|)$ is

$$K_{\eta,\mathcal{A}}(z, w) = \sum_{n \in \mathbb{Z}} f_{n,\eta}(z) \overline{f_{n,\eta}(w)} V(w),$$

where the weight function V comes from the inner product of $L^2_V(\mathcal{A})$. We have, by (6.7), $I_2^{-1} I_1^{-1} f = f \circ \phi \circ E$ hence, Lemma 5.1 and (6.7) imply that the kernel K_η of the orthogonal projection from $L^2(\varpi)$ onto $A^2_\eta(\varpi)$ can be written as

$$\begin{aligned} K_\eta(z, w) &= K_{\eta,\mathcal{A}}\big(\phi \circ E(z), \phi \circ E(w)\big) \big|(\phi \circ E)'(w)\big|^2 \\ &= \sum_{n \in \mathbb{Z}} \frac{C^2_{n,\eta} \phi(e^{i2\pi z})^{n+\eta/(2\pi)} \overline{\phi(e^{i2\pi w})}^{n+\eta/(2\pi)}}{v(\phi(e^{i2\pi z})) \overline{v(\phi(e^{i2\pi w}))}}, \qquad z, w \in \varpi, \end{aligned}$$

where we used (6.5). Moreover, we have

$$v(\phi(e^{i2\pi z})) = \frac{1}{2\pi} \frac{e^{-i2\pi z}}{\phi'(e^{i2\pi z})}$$

so that denoting

$$\widetilde{K}(z, w) = 4\pi^2 e^{i2\pi(z - \bar{w})} \phi'(e^{i2\pi z}) \overline{\phi'(e^{i2\pi w})} \tag{6.11}$$

the kernel becomes

$$\begin{aligned} &K_\eta(z, w) \\ &= \widetilde{K}(z, w) \sum_{n \in \mathbb{Z}} C^2_{n,\eta} \phi(e^{i2\pi z})^{n+\eta/(2\pi)} \overline{\phi(e^{i2\pi \bar{w}})}^{n+\eta/(2\pi)}, \quad z, w \in \varpi. \end{aligned} \tag{6.12}$$

Next, we recall from Lemma 2.2 the conformal mapping

$$\varphi(z) = \frac{1}{i2\pi} \log\big(\phi(e^{i2\pi z})\big) + [\operatorname{Re} z] \quad \text{with } \phi(e^{i2\pi z}) = e^{i2\pi\varphi(z)}. \tag{6.13}$$

Taking into account (6.9) we now write (6.12) as

$$K_\eta(z,w) \tag{6.14}$$
$$= \widetilde{K}(z,w)\sum_{n\in\mathbb{Z}} \frac{2n+\eta/\pi}{2\pi(\rho^{2n+\eta/\pi}-\rho^{-2n-\eta/\pi})} e^{i(2\pi(n-1)+\eta)(\varphi(z)-\overline{\varphi(w)})}.$$

By (6.13), we have $|e^{i2\pi\varphi(z)}| = |\phi(e^{i2\pi z})|$ for all $z \in \varpi$, and since $\phi(e^{i2\pi z}) \in \mathcal{A}$, we have $1/\rho < \left|e^{i2\pi\varphi(z)}\right| < \rho$, which implies that the series (6.14) converges absolutely and uniformly on all compact subsets of ϖ. We arrive at the main result of this section.

Theorem 6.4 *Let the domain ϖ (equivalently, Π) be simply connected and satisfy assumption (A) of Sect. 2, and let $\eta \in [-\pi, \pi]$. The kernel K_η of the projection from $L^2(\varpi)$ onto $A^2_\eta(\varpi)$, see* (4.1)*, is given by formulas* (6.11), (6.14).

7 Applications

In order to demonstrate the potential applicability of Theorem 6.4, we derive the kernel formula (2.5) from it. To this end we combine the kernel formula (6.14) with (4.3) in Theorem 4.3,

$$K_\Pi(z,w) = \frac{\widetilde{K}(z,w)}{2\pi}\int_{-\pi}^{\pi}\sum_{n\in\mathbb{Z}} \frac{2n+\eta/\pi}{2\pi(\rho^{2n+\eta/\pi}-\rho^{-2n-\eta/\pi})}$$
$$\times\, e^{i\eta\big([\mathrm{Re}\, z]-[\mathrm{Re}\, w]\big)+i(2\pi(n-1)+\eta)\big(\varphi(z-[\mathrm{Re}\, z])-\varphi(\bar{w}-[\mathrm{Re}\, w])\big)}\,d\eta$$
$$= \frac{\widetilde{K}(z,w)}{4\pi^2} e^{-i2\pi(\varphi(z)-\overline{\varphi(w)})} 4\pi \int_{-\infty}^{\infty} \frac{t}{\rho^{2t}-\rho^{-2t}} e^{i2\pi t(\varphi(z)-\overline{\varphi(w)})}\,dt. \tag{7.1}$$

where we replaced summation and integration over $[-\pi, \pi]$ with integration over the real line (with $t = \eta/(2\pi)$) and used $\varphi\big(z - [\mathrm{Re}\, z]\big) = \varphi(z) - [\mathrm{Re}\, z]$ (see (6.13)) and similarly for the variable w. There holds the integral formula (Fourier transform)

$$\int_{-\infty}^{\infty} \frac{t}{2}\,\mathrm{csch}\,(at)e^{-ist}\,dt = \int_{-\infty}^{\infty} \frac{t}{e^{at}-e^{-at}} e^{-ist}\,dt$$
$$= \frac{1}{4a^2}\pi^2 \mathrm{sech}^2\Big(\frac{\pi s}{2a}\Big), \quad s \in \mathbb{R}, \tag{7.2}$$

where $a > 0$ is a parameter and csch denotes the hyperbolic cosecant; the identity (7.2) can be got from the Fourier cosine transform given in the table 1.9(18) of [18], since the transformed function is even. We apply (7.2) with $a = 2\log\rho$ and $s = 2\pi(\varphi(z) - \overline{\varphi(w)})$. The equations (6.11), (7.1) yield formula (2.5):

Corollary 7.1 *If the periodic domain* Π, (1.2), *is simply connected and satisfies the assumption* (A) *of Sect. 2, then its Bergman kernel equals*

$$K_\Pi(z, w) = \widetilde{K}_\Pi(z, w)\frac{\pi^3}{4(\log\rho)^2}\operatorname{sech}^2\Big(\frac{\pi^2(\varphi(z) - \overline{\varphi(w)})}{2\log\rho}\Big), \tag{7.3}$$

where

$$\widetilde{K}_\Pi(z, w) = e^{i2\pi(z-\varphi(z)-\bar{w}+\overline{\varphi(w)})}\phi'(e^{i2\pi z})\overline{\phi'(e^{i2\pi w})}, \tag{7.4}$$

$\phi : \mathcal{D} \to \mathcal{A}$ *is the Riemann mapping between doubly connected domains* $\mathcal{D} = \{e^{i2\pi z} : z \in \varpi\}$ *and the annulus* $\mathcal{A} = \{w : \rho^{-1} < w < \rho\}$, $\rho > 1$, *and* $\varphi : \Pi \to \mathbb{C}$ *is given in* (6.13).

Note that in the case where Π is the strip $\Sigma = (-\infty, \infty)\times(-\pi, \pi)$, see the proof of Proposition 2.3, the mapping ϕ is the identity, and obviously formula (7.3)–(7.4), with $\log\rho = 2\pi^2$, boils down to (2.6).

The invariance of the kernel under the mapping

$$(z, w) \mapsto (z + m, w + m), \quad m \in \mathbb{Z}, \tag{7.5}$$

follows from (6.13). We also have the following estimate for the kernel; estimates for arbitrary $(z, w) \in \Pi \times \Pi$ follow by applying the mapping (7.5).

Corollary 7.2 *Let* Π *be as in Corollary 7.1 and assume in addition that there exists a constant* $C > 0$ *such that*

$$\frac{1}{C} \le \sup_{w\in\mathcal{D}} |\phi'(w)| \le C. \tag{7.6}$$

There exist constants $0 < C_1 < C_2$ *such that*

$$C_1 e^{-\pi^2 n/(2\log\rho)} \le |K_\Pi(z, 0)| \le C_2 e^{-\pi^2 n/(2\log\rho)} \tag{7.7}$$

for all $n \in \mathbb{N}$ *and all* $z \in \Pi$ *with* $n - 1 \le |z| \le n$.

Proof We have for all $z \in \Pi$

$$\exp\Big(\frac{\pi^2\varphi(z)}{2\log\rho}\Big) = \exp\Big(\frac{\pi^2\varphi(z - [\operatorname{Re} z])}{2\log\rho}\Big)\exp\Big(\frac{\pi^2[\operatorname{Re} z]}{2\log\rho}\Big),$$

and on the right hand side the modulus of the first factor is bounded and bounded from below by a positive constant, since φ is a maps ϖ onto a bounded domain. The same holds for the factor $\widetilde{K}_\Pi$, by the assumption (7.6). The result follows from the properties of sech. □

Remark 7.3 Condition (7.6) holds, if the boundary of the doubly connected domain $\mathcal{D}$, equivalently, the boundary of Π, is for example C^2-smooth. There are non-smooth domains $\mathcal{D}$ such that for some boundary point $\zeta \in \partial\mathcal{D}$ we have $\lim_{z\to\zeta} |\phi'(\zeta)| = \infty$; see the references mentioned below (6.4). In this case the decay (7.7) of course fails, since (7.3), (7.4) imply $\sup_{z\in\varpi_m} |K_\Pi(z, 0)| = \infty$ for all $m \in \mathbb{Z}$.

The following decay estimate for the kernel does not depend on the geometry of the periodic cell ϖ.

Corollary 7.4 *Let Π be as in Corollary 7.1 and let $G \subset \varpi$ be a compact subset. Then, there exist constants $0 < C_1 < C_2$ such that*

$$C_1 e^{-\pi^2|n|/(2\log\rho)} \le |K(z, 0)| \le C_2 e^{-\pi^2|n|/(2\log\rho)}$$

for all $n \in \mathbb{Z}$ and all $z \in \Pi$ such that $z - n \in G$.

Remark 7.5 If the periodic cell ϖ were doubly connected, the entire domain Π would be infinitely connected, and in principle the above method yields a representation for the Bergman kernel K_Π in terms of Riemann mappings between finitely instead of infinitely connected domains. However, obtaining the concrete formula (7.3)–(7.4) required the use of a quite explicit, simple sequence of orthogonal functions on the annulus and an exact calculation of their norms. The author does not know a good enough example of such sequences for triply and higher connected domains.

We finally prove a simple boundedness result for the Bergman projection with respect to certain weighted L^p-norms. Let us consider continuous weights $W : \Pi \to \mathbb{R}^+$ which only depend on the real part of the variable $z \in \Pi$. We assume that there are constants $a, C > 0$ and $0 < b < 1$ such that for all $x \in \mathbb{R}$, $n \in \mathbb{Z}$

$$\frac{1}{C} W(x)e^{-a|n|^b} \le |W(x+n)| \le C W(x)e^{a|n|^b}. \tag{7.8}$$

Theorem 7.6 *Let Π be as in Corollary 7.2. Let $W : \Pi \to \mathbb{R}^+$ be a weight as above and $1 \le p < \infty$. Then, the projection operator $P_\Pi : L^p_W(\Pi) \to L^p_W(\Pi)$ is bounded.*

Proof We apply the Schur test, [24], Theorem 3.6, with the constant test function $h(z) \equiv 1$; note that we take $K_\Pi(z, w)W(w)^{-1}$ for the kernel K in the reference.

By also taking into account that $|\widetilde{K}_\Pi|$ is bounded by a constant $C > 0$ in $\Pi \times \Pi$, we obtain from (7.7), (7.8) for all $z = x + iy \in \Pi$

$$\begin{aligned}
&\int\limits_{\Pi} |K_\Pi(z,w)| W(w)^{-1} W(z) dA(z) \\
&\leq C \sum_{n \in \mathbb{N}} \int\limits_{n-1 \leq |x-\xi| \leq n} \exp\Big(-\frac{\pi^2 n}{2\log\rho}\Big) \exp\big(an^b\big) W(x)^{-1} W(x) dA(w) \\
&\leq C' \sum_{n \in \mathbb{N}} \exp\Big(-\frac{\pi^2 n}{2\log\rho} + an^b\Big) \\
&\leq C''.
\end{aligned}$$

The second integral $\int_\Pi |K_\Pi(z,w)| dA(w)$ in the Schur test is also bounded by a constant; the estimation is easier. □

Acknowledgments This research was partially supported by the Academy of Finland grant no. 324219 and a research grant from the Faculty of Science of the University of Helsinki.

References

1. D. Békollé, Projections sur des espaces de fonctions holomorphes dans des domaines plans. Can. J. Math. **XXXVIII**(1), 127–157 (1986)
2. J. Brennan, Approximation in the mean by polynomials on non- Carathéodory domains. Ark. Mat. **15**(1–2), 117–168 (1977)
3. T. Carleman, Über die approximation analytischer funktionen durch lineare aggregate von vorgegebenen potenzen. Ark. Mat. Astr. Fys. **17**, 1–30 (1923)
4. P. Baddoo, D. Crowdy, Periodic Schwarz-Christoffel mappings with multiple boundaries per period. Proc. A. Royal Soc. **475**(2228), 20190225 (2019)
5. G. Cardone, S.A. Nazarov, J. Taskinen, Spectra of open waveguides in periodic media. J. Funct. Anal. **269**(8), 2328–2364 (2015)
6. J.M. Floryan, Ch. Zemach, Schwarz-Christoffel methods for conformal mapping of regions with a periodic boundary. J. Comput. Appl. Math. **46**(1–2), 77–102 (1993)
7. G.M. Goluzin, *Geometric Theory of Functions of a Complex Variable* (American Mathematical Society, Providence, 1969)
8. T. Hytönen, J. van Neerven, J. Veraar, L. Weiss, *Analysis in Banach Spaces*, vol. 1. Ergebnisse der Mathematik und ihrer Grenzgebiete. 3. Folge/A Series of Modern Surveys in Mathematics (Springer International, 2016)
9. L.V. Kantorovich, V.I. Krylov, *Approximate Methods in Higher Analysis* (Noordhoff, Groningen, 1964)
10. V.Kozlov, J.Taskinen, Floquet problem and center manifold reduction for ordinary differential operators with periodic coefficients in Hilbert spaces. Algebra i Analiz **32**(3), 191–218 (2020)
11. S. Krantz, A tale of three kernels. Complex Variables Elliptic Equ. **53**(11), 1059–1082 (2008)
12. P. Kuchment, *Floquet Theory for Partial Differential Equations*. Operator Theory: Advances and Applications, vol. 60 (Birkhäuser Verlag, Basel, 1993)

13. P. Kuchment, An overview of periodic elliptic operators. Bull. Am. Math. Soc. (N.S.) **53**(3), 343–414 (2016)
14. P. Kuchment, J. Taskinen, Dispersion relations and spectra of periodically perforated structures. Pure Appl. Funct. Anal. **7**(2), 721–731 (2022)
15. S.A. Nazarov, Properties of spectra of boundary value problems in cylindrical and quasicylindrical domains, in *Sobolev Spaces in Mathematics*, vol. II, ed. by V. Maz'ya. International Mathematical Series, vol. 9 (Springer, New York, 2008), pp. 261–309
16. S.A. Nazarov, B.A. Plamenevski, *Elliptic Problems in Domains with Piecewise Smooth Boundaries* (Walter be Gruyter, Berlin, 1994)
17. S.A. Nazarov, K. Ruotsalainen, J. Taskinen, Essential spectrum of a periodic elastic waveguide may contain arbitrarily many gaps. Appl. Anal. **89**(1), 109–124 (2010)
18. M. Oberhettiger, F. Tricomi, *Tables of Integral Transforms* (McGraw-Hill, New York, 1954)
19. Ch. Pommerenke, *Boundary Behaviour of Conformal Maps*. Grundlehren der mathematischen Wissenschaften, vol. 299 (Springer, Berlin, 1992)
20. W. Rudin, *Real and Complex Analysis* (McGraw-Hill, New York, 1986)
21. A.A. Solov'ev, L^p estimates of integral operators associated with spaces of analytic and harmonic functions. Soviet Math. Dokl. **19**, 764–768 (1978)
22. A.A. Solov'ev, On the continuity of an integral operator with Bergman kernel in the L^p-space, Vest. Leningrad Universiteta **19**, 77–80 (1978; Russian)
23. G. Symm, Conformal mapping of doubly-connected domains. Numer. Math. **13**, 448–457 (1969)
24. K. Zhu, *Operator Theory in Function Spaces*, 2nd edn. AMS Mathematical Surveys and Monographs, vol. 138 (American Mathematical Society, Providence, 2007)

Brown Measure of R-diagonal Operators, Revisited

Ping Zhong

Abstract We use the perspective of subordination functions to reformulate Haagerup–Schultz's approach to the Brown measure of R-diagonal operators. This allows us to simplify the original argument and find a connection with the other approach due to Belinschi-Śniady-Speicher. The Brown measure formula can be rewritten in terms of subordination functions.

1 Introduction

The Brown measure of an operator in a finite von Neumann algebra with faithful, normal, tracial state is an extension of eigenvalue counting measure of finite dimensional square matrices. In light of Voiculescu's breakthrough work [25] on asymptotic freeness of Gaussian random matrices, the Brown measure of a free random variable is a natural candidate for the limit of empirical spectral distribution of suitable random matrix model. Let $(X_n)_{n=1}^{\infty}$ be a sequence of $n \times n$ matrices whose entries are i.i.d. copies of a complex random variable with mean zero and unit variance. The circular law [7] says that the limit empirical spectral distribution of $X_n/\sqrt{n}$ is the Brown measure of Voiculescu's circular operator. The limit distribution in the single ring theorem [12] is the Brown measure of R-diagonal operators, a large class of operators in the framework of free probability theory including circular operator.

The calculation of explicit Brown measure formula is technical in general and only some special classes of operators have explicit Brown measure formulas known up to now. The first nontrivial example was given by Haagerup–Larsen [14] in 2000, almost 20 years after L.G. Brown introduced his measure, where they calculated the Brown measure of all bounded R-diagonal operators. Haagerup–Schultz [16] extended the Brown measure formula for unbounded R-diagonal operators, and such

P. Zhong (✉)
Department of Mathematics and Statistics, University of Wyoming, Laramie, WY, USA
e-mail: pzhong@uwyo.edu

Y. Choi et al. (eds.), *Operators, Semigroups, Algebras and Function Theory*,
Operator Theory: Advances and Applications 292,
https://doi.org/10.1007/978-3-031-38020-4_10

work played an important role in their seminal work [17] on the invariant subspace problem in type II_1 factors (see also [9, 10] for some generalizations).

The proof in [14] uses Haagerup's work [13] on a different proof for the multiplicative property of Voiculescu's S-transforms and the fact that any R-diagonal operator T has the same distribution as a product ax of free self-adjoint symmetric distributed elements. The proof in [16] relies on a result of Nica–Speicher [22] which says that for two $*$-free R-diagonal operators S and T, the symmetrization of the distribution of $|S + T|$ is the free additive convolution of symmetrization distributions of $|S|$ and $|T|$ (see Proposition 2.3), and the following observation. Let U be a Haar unitary operator $*$-free from T. For any $\lambda \in \mathbb{C}$, then $|T - \lambda\mathbf{1}|$ has the same distribution as $|T + |\lambda| U|$. Hence, for any $\lambda \in \mathbb{C}\backslash\{0\}$ fixed, the distribution of $|T - \lambda\mathbf{1}|$ can be calculated as an addition of two R-diagonal operators. The problem is then reduced to understanding the free additive convolution of a symmetric probability measure and a Bernoulli distribution. The above observations are the key to find a formula for the Fuglede–Kadison determinant [11] of $T - \lambda\mathbf{1}$. The Brown measure of T is then obtained by taking the distributional Laplacian of the logarithm of the Fuglede–Kadison determinant. In a joint work with Bercovici [6], we extend some ideas in this paper to study the Brown measure of a sum of two free random variables, one of which is R-diagonal.

We observe that subordination functions can be used to make Haagerup–Schultz's approach more transparent and several technical arguments in their work can be simplified when it cooperates with subordination ideas. The reformulation of Fuglede–Kadison formulas in terms of subordination functions allows us to simplify the original proof for the Brown measure formula of R-diagonal operators. By using a well-known result from logarithmic potential theory and rotation invariant property of the Brown measure of R-diagonal operators, we obtain a new proof of the Brown measure formula. We can also express the Brown measure formula in terms of subordination functions.

The Hermitian reduction method is a very general and powerful technique to study the distributions of non-normal operators in free probability. It is known [23] that an operator T is R-diagonal if and only if the matrix-valued operator $\begin{bmatrix} 0 & T \\ T^* & 0 \end{bmatrix}$ and $M_2(\mathbb{C})$ are free with amalgamation over the algebra of diagonal matrices with a canonical conditional expectation. In [4], Belinschi–Śniady–Speicher used this characterization of R-diagonal operators and Hermitian reduction method to give a new proof of the Brown measure formula for R-diagonal operators. The subordination ideas found in [16] allow us to find the connection between the two methods, which seem to be quite different on a first impression.

The paper is organized as follows. In Sect. 2, we collect some preliminaries about Brown measure of R-diagonal operators to set up notations. In Sects. 3 and 4, we reformulate Haagerup–Schultz's work using subordination ideas. We present a new proof in Sect. 5 and discuss the connection with the Hermitian reduction method in Sect. 6. Finally, we show some applications in Sect. 7.

2 Preliminaries and Notations

Let $\mathcal{M}$ be a finite von Neumann algebra with a faithful, normal, tracial state τ. We let $\widetilde{\mathcal{M}}$ denote the set of closed, densely defined operators affiliated with $\mathcal{M}$. In 1983, L.G. Brown [8] introduced a spectral distribution for non-normal operators in $\widetilde{\mathcal{M}}$ which is a natural extension of eigenvalue counting measure of square matrices. The definition of Brown measure can be extended to a larger class $\mathcal{M}^{\Delta}$ of possibly unbounded operators in $\widetilde{\mathcal{M}}$ ([16] and [8, Appendix]).

Given $T \in \widetilde{\mathcal{M}}$, let $\mu_{|T|}$ be the spectral distribution of $|T| = (T^*T)^{1/2}$ with respect to τ. Let $\mathcal{M}^{\Delta}$ be the set of operators $T \in \widetilde{\mathcal{M}}$ satisfying the condition

$$\tau(\log^+ |T|) = \int_0^{\infty} \log^+(t)\, \mathrm{d}\mu_{|T|}(t) < \infty.$$

The *Fuglede–Kadison determinant* of T is defined by

$$\Delta(T) = \exp\Big(\int_0^{\infty} \log t\, \mathrm{d}\mu_{|T|}(t)\Big).$$

It is known [16] that the set $\mathcal{M}^{\Delta}$ is a subspace of $\widetilde{\mathcal{M}}$. In particular, for $T \in \mathcal{M}^{\Delta}$ and $\lambda \in \mathbb{C}$, $T - \lambda \mathbf{1} \in \mathcal{M}^{\Delta}$. If $T \in \mathcal{M}^{\Delta}$ and $\Delta(T) > 0$, then $\ker(T) = \{0\}$, its inverse $T^{-1} \in \mathcal{M}^{\Delta}$, and $\Delta(T^{-1}) = \frac{1}{\Delta(T)}$.

For $T \in \mathcal{M}^{\Delta}$, the function $\lambda \mapsto f(\lambda) := \log \Delta(T - \lambda \mathbf{1})$ taking values in $[-\infty, \infty)$ is subharmonic on $\mathbb{C}$. The *Riesz measure* associated with this subharmonic function is called the *Brown measure* of T and we denote it by μ_T. More precisely,

$$\mathrm{d}\mu_T = \frac{1}{2\pi} \nabla^2 f\, \mathrm{d}\lambda$$

taken in the distribution sense. The Brown measure μ_T is the unique probability measure on $\mathbb{C}$ satisfying

(i)

$$\int_{\mathbb{C}} \log^+ |z|\, \mathrm{d}\mu_T(z) < \infty,$$

(ii)

$$\forall \lambda \in \mathbb{C}: \quad \log \Delta(T - \lambda \mathbf{1}) = \int_{\mathbb{C}} \log |\lambda - z|\, \mathrm{d}\mu_T(z).$$

The family of R-diagonal operators was introduced by Nica–Speicher [21] which covers a large class of interesting operators in free probability theory. They have a number of remarkable symmetric properties. In [16], the class of R-diagonal operators was extended to the unbounded case. Recall that [16, 21] $T \in \widetilde{\mathcal{M}}$ is R-diagonal if there exist

- a von Neumann algebra $\mathcal{N}$, with a faithful, normal, tracial state,
- and $*$-free elements $U, H \in \mathcal{N}$ with U Haar unitary, $H \geq 0$,

such that T has the same $*$-distribution as UH.

Let $V|T|$ be the polar decomposition of T. If $T \in \widetilde{\mathcal{M}}$ is R-diagonal with $\ker(T) = 0$, then V is a Haar unitary which is $*$-free from $|T|$. If $T \in \widetilde{\mathcal{M}}$ is R-diagonal with $\ker(T) = 0$, then T has an inverse $T^{-1} \in \widetilde{\mathcal{M}}$, and T^{-1} is R-diagonal as well.

Observe that the Brown measure of a R-diagonal operator T is invariant under rotation. Hence, the Brown measure of T is completely determined by the distribution of $|T|$. In fact, the Brown measure of a R-diagonal operator T can be expressed in terms of some analytic transformation of the spectral probability measure μ_{T^*T}. For a probability measure μ on $\mathbb{R}_+ = [0, \infty)$, define $\psi_\mu : \mathbb{C}\backslash(0, \infty) \to \mathbb{C}$ by

$$\psi_\mu(z) = \int_0^\infty \frac{zt}{1 - zt}\, \mathrm{d}\mu(t), \qquad z \in \mathbb{C} \setminus (0, \infty).$$

One can check that: (i) $\psi_\mu'(x) > 0, x \in (-\infty, 0)$; (ii) $\psi_\mu(x) \to \delta - 1$ as $x \to -\infty$ where $\delta = \mu(\{0\})$; (iii) $\psi_\mu(x) \to 0$ as $x \to 0$ (see [5]). Hence, in a neighborhood $\mathcal{V}_\mu$ of $(\delta - 1, 0)$, the inverse χ_μ of ψ_μ exists: $\chi_\mu(z) = \psi_\mu^{\langle -1 \rangle}(z), z \in \mathcal{V}_\mu$. Then Voiculescu's S-transform is defined by

$$S_\mu(z) = \frac{z+1}{z} \chi_\mu(z), \quad z \in \mathcal{V}_\mu.$$

For two probability measures μ, ν on $\mathbb{R}_+$, the S-transform has the remarkable property that $S_{\mu \boxtimes \nu}(z) = S_\mu(z) S_\nu(z)$ in a region where all these functions are defined.

Let μ be a probability measure on $[0, \infty)$ which is not a Dirac measure, and put

$$\lambda_1(\mu) = \left(\int_0^\infty \frac{1}{u^2}\, \mathrm{d}\mu(u) \right)^{-\frac{1}{2}} \quad \text{and} \quad \lambda_2(\mu) = \left(\int_0^\infty u^2\, \mathrm{d}\mu(u) \right)^{\frac{1}{2}}, \tag{2.1}$$

with the convention that $\infty^{-\frac{1}{2}} = 0$. Then $0 \leq \lambda_1(\mu) < \lambda_2(\mu) \leq \infty$. Note that

$$\lambda_2(\mu_{|T|}) = \left(\int_0^\infty u^2\, \mathrm{d}\mu_{|T|}(u) \right)^{\frac{1}{2}} = \|T\|_2$$

and

$$\lambda_1(\mu_{|T|}) = \left(\int_0^\infty u^{-2}\,\mathrm{d}\mu_{|T|}(u)\right)^{-\frac{1}{2}} = \|T^{-1}\|_2^{-1},$$

where $\|T^{-1}\|_2 := +\infty$ in case $\ker(T) \neq 0$. Often, we denote

$$\lambda_1(T) = \lambda_2(\mu_{|T|}), \qquad \lambda_2(T) = \lambda_1(\mu_{|T|}).$$

The Brown measure of bounded R-diagonal operators was calculated by Haagerup-Larsen [14] and the result was further extended by Haaagerup-Schultz [16] to all R-diagonal operators in $\mathcal{M}^\Delta$.

Theorem 2.1 *Let T be an R-diagonal element in $\mathcal{M}^\Delta$ with Brown measure μ_T, and suppose μ_T is not a Dirac measure. The Brown measure μ_T is the unique rotationally invariant probability measure such that*

$$\mu_T\{\lambda \in \mathbb{C} : |\lambda| \le r\} = \begin{cases} 0, & \textit{for} \quad r < \lambda_1(\mu_{|T|}); \\ 1 + S^{\langle -1\rangle}_{\mu_{T^*T}}(r^{-2}), & \textit{for} \quad \lambda_1(\mu_{|T|}) < r < \lambda_2(\mu_{|T|}); \\ 1, & \textit{for} \quad r \ge \lambda_2(\mu_{|T|}). \end{cases}$$

When $\mu_{|T|}(\{0\}) = \delta > 0$, the Brown measure μ_T has a mass at the origin with the mass $\mu_T(\{0\}) = \delta$.

Proposition 2.2 ([15, Lemma 4]) *Let μ be a probability measure on $[0,\infty)$ with $\delta = \mu(\{0\}) < 1$. Then S_μ is decreasing on $(\delta - 1, 0)$ and $S_\mu((\delta - 1, 0)) = (b^{-1}, a^{-1})$, where*

$$a = \left(\int_0^\infty \frac{1}{u} d\mu(u)\right)^{-1}, \quad b = \int_0^\infty u\, d\mu(u).$$

Hence, the map $r \mapsto S^{\langle -1\rangle}_{\mu_{T^*T}}(r^{-2})$ is increasing on $(\lambda_1(\mu_{|T|}), \lambda_2(\mu_{|T|}))$ with range $(\delta - 1, 0)$, where $\delta = \mu(\{0\})$.

Given a probability measure σ on $\mathbb{R}$, let $\widetilde{\sigma}$ denote the *symmetrization* of μ. That is, $\widetilde{\sigma}$ is a symmetric probability measure given by

$$\widetilde{\sigma}(B) = \tfrac{1}{2}(\sigma(B) + \sigma(-B)),$$

where B is any Borel measurable set in $\mathbb{R}$. Let μ, ν be probability measures on $\mathbb{R}$, their free convolution is denoted by $\mu \boxplus \nu$. The following result is a powerful tool to calculate the distribution of the sum of R-diagonal operators.

Proposition 2.3 ([16, 22]) *Let $S, T \in \widetilde{\mathcal{M}}$ be $*$-free R-diagonal elements. Then $S+T \in \widetilde{\mathcal{M}}$ is R-diagonal and*

$$\widetilde{\mu}_{|S+T|} = \widetilde{\mu}_{|S|} \boxplus \widetilde{\mu}_{|T|}. \tag{2.2}$$

The calculation of free additive convolution of two probability measures μ, ν on $\mathbb{R}$ uses Voiculescu's R-transform. For a probability measure μ on $\mathbb{R}$, its Cauchy transform is defined as

$$G_\mu(z) = \int_{\mathbb{R}} \frac{1}{z-x} d\mu(x), \qquad z \in \mathbb{C}^+;$$

and the F-transform is its reciprocal, $F_\mu(z) = 1/G_\mu(z)$. The R-transform is defined by $R_\mu(z) = G_\mu^{\langle -1 \rangle}(z) - 1/z$ for z in a truncated Stolz angle $\{z \in \mathbb{C} : \Im z > \beta, |\Re z| < \alpha(\Im z)\}$ for some $\alpha, \beta > 0$. The R-transform has the remarkable property that $R_{\mu \boxplus \nu}(z) = R_\mu(z) + R_\nu(z)$ for z in some domain where all these functions are defined.

Another way to study free convolution is to use subordination functions, which are analytic functions $\omega_1, \omega_2 : \mathbb{C}^+ \to \mathbb{C}^+$ such that

$$F_\mu(\omega_1(z)) = F_\nu(\omega_2(z)) = F_{\mu \boxplus \nu}(z), \qquad z \in \mathbb{C}^+.$$

For a probability measure μ on $\mathbb{R}$, denote $H_\mu(z) = F_\mu(z) - z$. In [3], Belinschi and Bercovici showed that ω_1, ω_2 can be obtained from the following fixed point equations

$$\omega_1(z) = z + H_{\mu_2}(z + H_{\mu_1}(\omega_1(z))),$$
$$\omega_2(z) = z + H_{\mu_1}(z + H_{\mu_2}(\omega_2(z))).$$

In operator-valued free probability theory, all the above analytic transformations have operator-valued analogues. See [20, Chapter 9] and [26] for example.

3 The Subordinations and Haagerup–Schultz's Approach

Let $T \in \widetilde{\mathcal{M}}$ be a R-diagonal operator. The goal of this section is to understand the distribution of $T - \lambda\mathbf{1}$ and some technical functions introduced in [16], but from the subordination function perspective.

Let $U \in \mathcal{M}$ be a Haar unitary which is $*$-free from T. Then U^* is a Haar unitary which is also $*$-free from T. Consequently, the operator U^*T is R-diagonal and has the same $*$-distribution as T [21]. In addition, $-\lambda U$ has the same distribution as $|\lambda|U$. Hence, for $\lambda \in \mathbb{C}$, we have the following identification of distributions

$$|T - \lambda \mathbf{1}| \underset{\mathcal{D}}{\sim} |U^*T - \lambda \mathbf{1}| \underset{\mathcal{D}}{\sim} |T - \lambda U| \underset{\mathcal{D}}{\sim} |T + |\lambda| U|. \tag{3.1}$$

Throughout this section, $\mu_{|T|}$ denotes the spectral measure of the positive operator $|T|$ and $\mu = \widetilde{\mu}_{|T|}$ is its symmetrization, if we do not declare this assumption separately.

By Proposition 2.3 and (3.1), it follows that

$$\widetilde{\mu}_{|T-\lambda\mathbf{1}|} = \widetilde{\mu}_{|T|} \boxplus \widetilde{\mu}_{|\lambda|\mathbf{1}} = \mu \boxplus \nu,$$

where $\nu = \frac{1}{2}(\delta_{-|\lambda|} + \delta_{|\lambda|})$. We then observe that for any operator $x \in \mathcal{M}$ and $t > 0$, we have

$$\frac{1}{2}\frac{d}{dt}\big(\log \Delta(x^*x + t^2\mathbf{1})\big) = t\tau((x^*x + t^2\mathbf{1})^{-1}).$$

The above equation is connected with the Cauchy transform of $\widetilde{\mu}_{|x|}$ by

$$G_{\widetilde{\mu}_{|x|}}(it) = \int_{\mathbb{R}} \frac{1}{it - u} d\widetilde{\mu}_{|x|}(u) = \int_{\mathbb{R}} \frac{-it}{t^2 + u^2} d\widetilde{\mu}_{|x|}(u) = -it\tau((x^*x + t^2\mathbf{1})^{-1}), \tag{3.2}$$

where we used the symmetric property of $\widetilde{\mu}_{|x|}$.

Definition 3.1 Let $T \in \widetilde{\mathcal{M}}$ be an R-diagonal operator, and define

$$h(s) = s\,\tau\big((T^*T + s^2\mathbf{1})^{-1}\big), \qquad s > 0.$$

Moreover, for $\lambda \in \mathbb{C} \setminus \{0\}$, set

$$h_\lambda(t) = t\,\tau\big([(T - \lambda\mathbf{1})^*(T - \lambda\mathbf{1}) + t^2\mathbf{1}]^{-1}\big), \qquad t > 0.$$

Let ω_1, ω_2 be subordination functions such that $F_{\widetilde{\mu}_{|T-\lambda\mathbf{1}|}}(z) = F_{\widetilde{\mu}_{|T|}}(\omega_1(z)) = F_\nu(\omega_2(z))$. That is

$$F_{\mu\boxplus\nu}(z) = F_\mu(\omega_1(z)) = F_\nu(\omega_2(z)). \tag{3.3}$$

Note that ω_1, ω_2 are functions of z that depend on T and $|\lambda|$. As noted above, by symmetrization, the Cauchy transforms satisfy

$$G_{\widetilde{\mu}_{|T-\lambda\mathbf{1}|}}(it) = -ih_\lambda(t), \qquad G_{\widetilde{\mu}_{|T|}}(is) = -ih(s) \tag{3.4}$$

and the derivative of logarithm of Fuglede–Kadison determinants can be expressed as

$$\begin{aligned} \frac{1}{2}\frac{d}{dt}\big(\log \Delta((T-\lambda\mathbf{1})^*(T-\lambda\mathbf{1})+t^2\mathbf{1})\big) &= h_\lambda(t); \\ \frac{1}{2}\frac{d}{ds}\big(\log \Delta(T^*T+s^2\mathbf{1})\big) &= h(s). \end{aligned} \tag{3.5}$$

The subordination function builds a connection between $h_\lambda(t)$ and $h(s)$.

Proposition 3.2 *For $\lambda \in \mathbb{C}\backslash\{0\}$ and $t > 0$, the complex numbers $\omega_1(it), \omega_2(it)$ are pure imaginary. Set $\omega_1(it) = iW(t)$ for some $W(t) > t$. We have*

$$h_\lambda(t) = h(-i\omega_1(it)) = h(W(t)). \tag{3.6}$$

Proof In a domain of the form $\Gamma_{\alpha,\beta} = \{z = x+iy : y > \beta, |x/y| < \alpha\}$ for some $\alpha, \beta > 0$ (see [5]), we have

$$\omega_1(z) = F^{\langle -1\rangle}_{\widetilde{\mu}_{|T|}}\left(F_{\widetilde{\mu}_{|T-\lambda\mathbf{1}|}}(z)\right), \qquad z \in \Gamma_{\alpha,\beta}.$$

Hence, by (3.4), $\omega_1(it)$ is purely imaginary for $t > \beta$ and therefore $\omega_1(it)$ is purely imaginary for all $t > 0$ by analytic extension. The same argument works for ω_2. Note that we can express the subordination function as $\omega_1(z) = F_\sigma(z)$ for some probability measure σ on $\mathbb{R}$ (see [19, Definition 2.1] for example). Hence, by the asymptotics of the F-transform (see [5, Proposition 5.2] for example), we have $\Im\omega_1(z) > \Im(z)$. Set $W(t) = -i\omega_1(it)$. We have $W(t) = \Im\omega_1(it) > \Im(it) = t$. Finally,

$$h_\lambda(t) = -\Im G_{\widetilde{\mu}_{|T-\lambda\mathbf{1}|}}(it) = -\Im G_{\widetilde{\mu}_{|T|}}(\omega_1(it)) = h(W(t)),$$

which is exactly the subordination relation (3.6). □

Definition 3.3 For $\lambda \in \mathbb{C}\backslash\{0\}$ and $t > 0$, set $s(|\lambda|, t) = W(t) = \Im\omega_1(it)$.

We now outline the calculation of the Brown measure for T as four steps:

(1) The subordination function $s(|\lambda|, t)$ depends on λ. Study the analytic property of $\lim_{t\to 0} s(|\lambda|, t)$ as a function of λ. This is equivalent to studying the boundary value of the subordination function $\omega_1(it)$ as t goes to zero.
(2) Express $\Delta(|T-\lambda\mathbf{1}|^2 + t^2\mathbf{1})$ in terms of $\Delta(|T|^2 + s^2\mathbf{1})$ where s is some subordination function. To do so, we need to evaluate the following integrals

$$\int_{t_0}^{t} h_\lambda(t)dt, \qquad \text{and} \qquad \int_{s_0}^{s} h(s)ds.$$

The key in this step is to find a tractable formula for subordination functions.
(3) Find a formula for $\Delta(T-\lambda\mathbf{1})^2 = \lim_{t\to 0}\Delta(|T-\lambda\mathbf{1}|^2 + t^2)$.

(4) Finally, calculate the Brown measure of T.

Following [16], recall $\mu = \widetilde{\mu}_{|T|}$, the function h in Definition 3.1 reads

$$h(s) = \int_{\mathbb{R}} \frac{s}{s^2 + u^2} d\mu(u), \quad s > 0. \tag{3.7}$$

Note that the above definition is the same as the function in [16, Lemma 4.6] because

$$\int_{\mathbb{R}} \frac{s}{s^2 + u^2} d\mu(u) = \int_0^\infty \frac{s}{s^2 + u^2} d\mu_{|T|}(u).$$

Then put

$$k(s, t) = (s - t)\left(\frac{1}{h(s)} - s + t\right), \qquad s > 0,\ t \in \mathbb{R}. \tag{3.8}$$

In particular, we have

$$k(s, 0) = s\left(\frac{1}{h(s)} - s\right) = \frac{n(s)}{d(s)}$$

where

$$n(s) = \int_{\mathbb{R}} \frac{u^2}{s^2 + u^2} d\mu(u) \qquad \text{and} \qquad d(s) = \int_{\mathbb{R}} \frac{1}{s^2 + u^2} d\mu(u).$$

Recall that $\lambda_1(\mu), \lambda_2(\mu)$ were defined in (2.1).

Lemma 3.4 ([16, 22]) *The map $s \mapsto k(s, 0)$ is a strictly increasing bijection of $(0, \infty)$ onto $(\lambda_1(\mu)^2, \lambda_2(\mu)^2)$.*

Proof We have $k(s, 0) = \frac{s}{h(s)} - s^2$. One can verify that

$$\frac{d}{ds}\frac{s}{h(s)} = \frac{d}{ds}\frac{1}{\int_{\mathbb{R}} \frac{1}{s^2+u^2} d\mu(u)} = 2s \cdot \frac{\int_{\mathbb{R}} \frac{1}{(s^2+u^2)^2} d\mu(u)}{\left(\int_{\mathbb{R}} \frac{1}{s^2+u^2} d\mu(u)\right)^2}.$$

Since μ is non-trivial, the second factor in the right hand side of the above equation is strictly bigger than one, by the Cauchy-Schwartz inequality. Hence, $\frac{d}{ds}k(s, 0) > 0$.

We can verify directly

$$\lim_{s\to 0^+} k(s, 0) = \frac{1}{\int_{\mathbb{R}} \frac{1}{u^2} d\mu(u)} \qquad \text{and} \qquad \lim_{s\to\infty} k(s, 0) = \int_{\mathbb{R}} u^2 d\mu(u).$$

The result then follows. □

Lemma 3.5 ([16, Lemma 4.8]) *The function k is analytic on $(0,\infty)\times\mathbb{R}$. Moreover, for $t>0$ the map $s\mapsto k(s,t)$ is a strictly increasing bijection of (t,∞) onto $(0,\infty)$*

Proof The function $k(s,t)$ can be expressed as

$$k(s,t)=\left(1-\frac{t}{s}\right)s\left(\frac{1}{h(s)}-s\right)+(s-t)t=k(s,0)\left(1-\frac{t}{s}\right)+(s-t)t.$$

It follows from Lemma 3.4 that $s\mapsto k(s,t)$ is strictly increasing for $s>t$. One can check directly that $k(s,t)\to\infty$ as $s\to\infty$ for any $t>0$ fixed. □

The following result shows that our definition for $s(|\lambda|,t)=W(t)$ coincides with the function $s(|\lambda|,t)$ in [16, Definition 4.9]. In [16], Haagerup–Schultz studied various useful properties of the function s.

Proposition 3.6 *The analytic function ω_1 is the unique solution of the fixed point equation for $\omega\in\mathbb{C}^+$:*

$$\omega(z)=z+H_2(z+H_1(\omega(z))),\qquad z\in\mathbb{C}^+ \tag{3.9}$$

where $H_1(z)=F_\mu(z)-z$ and $H_2(z)=F_\nu(z)-z$. For $r>0$, the function $s(r,t)=W(t)=\Im\omega_1(it)$ is the unique solution $s\in(t,\infty)$ to the equation $k(s,t)=r^2$, which is

$$(s-t)^2-\frac{s-t}{h(s)}+r^2=0. \tag{3.10}$$

Proof It is known [3] that the subordination function (3.3) satisfies the fixed point equation (3.9). We next show that $s(r,t)$ is the unique solution to Eq. (3.10). Given $\lambda\in\mathbb{C}\backslash\{0\}$, recall that $\nu=\frac{1}{2}(\delta_{|\lambda|}+\delta_{-|\lambda|})$. We calculate that

$$G_\nu(z)=\frac{1}{2}\left(\frac{1}{z-|\lambda|}+\frac{1}{z+|\lambda|}\right)$$

and hence $H_2(z)=F_\nu(z)-z=\frac{-|\lambda|^2}{z}$. The fixed point equation (3.9) is expressed as

$$\omega(z)=z-\frac{|\lambda|^2}{z+H_1(\omega(z))}$$

which is equivalent to

$$-(\omega(z)-z)^2+\frac{\omega(z)-z}{G_\mu(\omega(z))}+|\lambda|^2=0. \tag{3.11}$$

When $z = it$, $\omega_1(it) = is(|\lambda|, t)$, using $F_\mu(is) = 1/G_\mu(is) = i/h(s)$, the above equation is further reduced to (3.10)

$$(s-t)^2 - \frac{s-t}{h(s)} + |\lambda|^2 = 0.$$

The uniqueness of the fixed point equation (3.9) implies that above equation has a unique solution. The uniqueness also follows from Lemma 3.5. □

Remark 3.7 For $\lambda \in \mathbb{C}\backslash\{0\}$, and set

$$K(\omega, z) = (\omega - z)\left(\omega - z - \frac{1}{G_\mu(\omega)}\right), \quad z, \omega \in \mathbb{C}^+.$$

For any $z \in \mathbb{C}^+$, the equivalence between (3.11) and the fixed point equation (3.9) as shown in the proof of Proposition 3.6 implies that there is a unique solution $\omega \in \mathbb{C}^+$ such that $K(\omega, z) = |\lambda|^2$. If $z = it$, then the solution $\omega = is$ for some $s > t$. Moreover, $K(is, it) = k(s, t)$.

Definition 3.8 For $r \in (\lambda_1(\mu), \lambda_2(\mu))$, let s_0 denote the unique solution $s \in (0, \infty)$ to the equation $k(s, 0) = r^2$ as in Lemma 3.5. For $r > 0$, define $s(r, 0)$ as follows.

$$s(r,0) = \begin{cases} 0, & \text{if} \quad 0 < r \leq \lambda_1(\mu); \\ s_0, & \text{if} \quad r \in (\lambda_1(\mu), \lambda_2(\mu)); \\ +\infty, & \text{if} \quad r \geq \lambda_2(\mu). \end{cases} \tag{3.12}$$

Proposition 3.9 *The function $r \mapsto s(r, 0)$ is increasing on $[0, \infty)$. It is strictly increasing on $(\lambda_1(\mu), \lambda_2(\mu))$ and a bijection of $(\lambda_1(\mu), \lambda_2(\mu))$ onto $(0, \infty)$.*

Proof It follows from the fact that the map $s \mapsto k(s, 0)$ is a strictly increasing bijection of $(0, \infty)$ onto $(\lambda_1(\mu)^2, \lambda_2(\mu)^2))$ and the definition that $s(r, 0)$ is the unique solution to the equation $k(s, 0) = r^2$ when $\lambda_1(\mu) < r < \lambda_2(\mu)$. □

It is known [2] that the subordination functions ω_1, ω_2 extend continuously to $\mathbb{R}$ with values in the extended complex plane $\overline{\mathbb{C}}$. The following result clarifies that the extension of ω_1, ω_2 at 0 is a nice function of λ.

Lemma 3.10 *The function $(r, t) \mapsto s(r, t)$ is analytic in $(0, \infty) \times (0, \infty)$. We have $s(r, t) > t$. Moreover,*

$$\lim_{t \to 0^+} s(r, t) = s(r, 0), \quad r > 0. \tag{3.13}$$

In addition, we have $\lim_{t \to \infty}(s(r, t) - t) = 0$ for all $r > 0$.

When $0 < r \leq \lambda_1(\mu)$, we have

$$\lim_{t \to 0^+} \frac{t}{s(r,t)} = \frac{\lambda_1(\mu)^2 - r^2}{\lambda_1(\mu)^2}.$$

When $r \geq \lambda_2(\mu)$, we have $\lim_{t \to 0^+} s(r, t)t = r^2 - \lambda_2(\mu)^2$.

Proof The analyticity of the function $(r, t) \mapsto s(r, t)$ was proved in [16, Lemma 4.10]. The limit $\lim_{t \to 0^+} s(r, t)$ was proved in [16, Lemma 4.10 and Remark 4.11] and this is equivalent to (3.13).

We now show that $\lim_{t\to\infty}(s(r, t) - t) = 0$. As we mentioned in the proof of Proposition 3.2, we can express subordination function as $\omega_1(z) = F_\sigma(z)$ for some probability measure σ on $\mathbb{R}$. Hence, by the asymptotic of F-transform (see [5, Proposition 5.2] for example), we have

$$\lim_{t\to\infty} (s(r, t) - t) = \lim_{t\to} F_\sigma(it) - it = 0.$$

By the monotone convergence theorem,

$$\lim_{s\to 0^+} \frac{h(s)}{s} = \lim_{s\to 0^+} \int_{\mathbb{R}} \frac{1}{s^2 + u^2} d\mu(u) = \frac{1}{\lambda_1(\mu)^2},$$

$$\lim_{s\to\infty} sh(s) = \lim_{s\to\infty} \int_0^\infty \frac{s^2}{s^2 + u^2} d\mu(u) = 1.$$

$$\lim_{s\to 0^+} s^2\Big(1 - sh(s)\Big) = \lim_{s\to 0^+} \int_0^\infty \frac{s^2u^2}{s^2 + u^2} d\mu(u) = \lambda_2(\mu)^2.$$

Hence, when $0 < r \le \lambda_1(\mu)$, by (3.10) and (3.13),

$$\begin{aligned} r^2 &= \lim_{t\to 0^+} \left(-(s(r, t) - t)^2 + \frac{s(r, t) - t}{h(s(r, t))} \right) \\ &= \lim_{s\to 0^+} \frac{s}{h(s)} \lim_{t\to 0^+} \left(1 - \frac{t}{s(r, t)}\right). \end{aligned}$$

Therefore, for $0 < r \le \lambda_1(\mu)$,

$$\lim_{t\to 0^+} \frac{t}{s(r, t)} = \frac{\lambda_1(\mu)^2 - r^2}{\lambda_1(\mu)^2}.$$

For the case $r \ge \lambda_2(\mu)$, we rewrite (3.10) as

$$\begin{aligned} r^2 &= -(s - t)^2 + \frac{s - t}{h(s)} \\ &= \frac{s^2\Big(1 - sh(s)\Big)}{sh(s)} + st\Big(2sh(s) - 1\Big) - t^2 sh(s). \end{aligned}$$

Hence, taking the limit of the above identity, when $r \geq \lambda_2(\mu)$, by (3.13),

$$\lim_{t\to 0^+} s(r,t)t = r^2 - \lim_{s\to\infty} \frac{s^2\Big(1 - sh(s)\Big)}{sh(s)}$$
$$= r^2 - \lambda_2(\mu)^2.$$

This finishes the proof. □

We summarize the discussions above. Let $T \in \widetilde{\mathcal{M}}$ be an R-diagonal element, let $\lambda \in \mathbb{C} \setminus \{0\}$. Let $\mu = \widetilde{\mu}_{|T|}$, and let $s(|\lambda|, t)$ be as in Definition 3.8. Then

$$h_\lambda(t) = h(s(|\lambda|, t)), \qquad t > 0. \tag{3.14}$$

Set $s = s(|\lambda|, t)$, then s satisfies

$$(s-t)^2 - \frac{s-t}{h(s)} + |\lambda|^2 = 0. \tag{3.15}$$

Hence

$$1 - 4|\lambda|^2 h(s)^2 = \Big(1 - 2h(s)(s-t)\Big)^2 \geq 0.$$

Moreover, by [16, Lemma 4.12], there exists a $t_\lambda > 0$ such that when $t > t_\lambda$ we have

$$s - t = \frac{1 - \sqrt{1 - 4|\lambda|^2 h(s)^2}}{2h(s)}. \tag{3.16}$$

Indeed, $\lim_{t\to\infty}(s-t) = 0$ and $\frac{1}{h(s)} \to 0$ as $s \to \infty$. Solve the quadratic equation (3.15) for $s - t$ to get the formula for $s - t$ when t is large. However, for small t, it is not clear if we need to change the sign in the quadratic formula from the right hand side of (3.16).

Lemma 3.11 *Let ω_2 be the subordination function such that $F_\nu(\omega_2(z)) = F_{\widetilde{\mu}_{|T-\lambda 1|}}(z)$. Then, for any $t > 0$,*

$$\omega_2(it) = i\frac{|\lambda|^2}{s(|\lambda|, t) - t} = \frac{|\lambda|^2}{it - i\omega_1(it)}.$$

Moreover,

$$\lim_{t\to 0^+} \omega_2(it) = \begin{cases} \infty, & \textit{if} \;\; 0 < |\lambda| \leq \lambda_1(\mu); \\ i\frac{|\lambda|^2}{s(|\lambda|,0)}, & \textit{if} \;\; |\lambda| \in (\lambda_1(\mu), \lambda_2(\mu)); \\ 0, & \textit{if} \;\; |\lambda| \geq \lambda_2(\mu). \end{cases}$$

When $0 < |\lambda| \leq \lambda_1(\mu)$, *we have*

$$\lim_{t \to 0^+} -it \cdot \omega_2(it) = \lambda_1(\mu)^2 - |\lambda|^2.$$

When $|\lambda| \geq \lambda_2(\mu)$, *we have*

$$\lim_{t \to 0^+} \frac{it}{\omega_2(it)} = \frac{|\lambda|^2 - \lambda_2(\mu)}{|\lambda|^2}.$$

Proof We have $\omega_1(z)+\omega_2(z) = z + F_{\widetilde{\mu}_{|T-\lambda 1|}}(z)$. Recall that $G_{\widetilde{\mu}_{|T-\lambda 1|}}(it) = -ih_\lambda(t)$ and $G_{\widetilde{\mu}_{|T-\lambda 1|}}(it) = G_{\widetilde{\mu}_{|T|}}(\omega_1(it)) = G_{\widetilde{\mu}_{|T|}}(is(|\lambda|, t)) = -ih(s(|\lambda|, t))$. Hence,

$$\omega_1(it) + \omega_2(it) = it + F_{\widetilde{\mu}_{|T-\lambda 1|}}(it)$$

is rewritten as

$$is(|\lambda|, t) + \omega_2(it) = it + \frac{1}{-ih(s(|\lambda|, t))}.$$

On the other hand, $s = s(|\lambda|, t)$ satisfies the identity

$$(s - t)^2 - \frac{s - t}{h(s)} + |\lambda|^2 = 0,$$

which implies $h(s) = \frac{s-t}{(s-t)^2+|\lambda|^2}$. Therefore,

$$\omega_2(it) = i\left[t - s + \frac{(s - t)^2 + |\lambda|^2}{s - t}\right] = i\frac{|\lambda|^2}{s(|\lambda|, t) - t} = \frac{|\lambda|^2}{it - \omega_1(it)}.$$

By taking the limit as $t \to 0^+$ and using Lemma 3.10, we obtain $\lim_{t\to 0^+} \omega_2(it)$.

When $0 < |\lambda| \leq \lambda_1(\mu)$, applying Lemma 3.10 and above formulas, we have

$$\begin{aligned}
\lim_{t \to 0^+} -it \cdot \omega_2(it) &= \lim_{t \to 0^+} \frac{t|\lambda|^2}{s(|\lambda|, t) - t} \\
&= \lim_{t \to 0^+} \frac{|\lambda|^2}{s(|\lambda|, t)/t - 1} \\
&= \lambda_1(\mu)^2 - |\lambda|^2.
\end{aligned}$$

When $|\lambda| \geq \lambda_2(\mu)$, we have

$$\lim_{t \to 0^+} \frac{it}{\omega_2(it)} = \lim_{t \to 0^+} \frac{s(|\lambda|, t)t - t^2}{|\lambda|^2} = \frac{|\lambda|^2 - \lambda_2(\mu)}{|\lambda|^2}.$$

□

We give a proof of the following lemma from [16] for the reader's convenience.

Lemma 3.12 ([16, Lemma 4.14]) *Let T be an unbounded R-diagonal element in $\mathcal{M}^{\Delta}$, let $\lambda \in \mathbb{C} \setminus \{0\}$, and let $t > 0$. With $\mu = \mu_{|T|}$ and $s(|\lambda|, t)$ as in Definition 3.3, we then have:*

$$\Delta\big((T - \lambda\mathbf{1})^*(T - \lambda\mathbf{1}) + t^2\mathbf{1}\big) = \frac{|\lambda|^2}{|\lambda|^2 + (s(|\lambda|, t) - t)^2}\Delta\big(T^*T + s(|\lambda|, t)^2\mathbf{1}\big). \tag{3.17}$$

Proof We use the relation between h, h_λ and the logarithm of the Fuglede–Kadison determinant (3.5). The function $s(|\lambda|, t)$ satisfies (3.10), which is

$$(s - t)^2 - \frac{s - t}{h(s)} + |\lambda|^2 = 0.$$

Fix $\lambda \in \mathbb{C}\backslash\{0\}$, and set $W(t) = s(|\lambda|, t)$. Hence, from Eqs. (3.6) and (3.10), for any $v > 0$, we have

$$h_\lambda(v) = (h \circ W)(v) = \frac{W(v) - v}{(W(v) - v)^2 + |\lambda|^2}.$$

Hence, for $0 < t_0 < t$, we have

$$\begin{aligned}
&\frac{1}{2}\left(\log\frac{\Delta((T - \lambda\mathbf{1})^*(T - \lambda\mathbf{1}) + t^2)}{\Delta((T - \lambda\mathbf{1})^*(T - \lambda\mathbf{1}) + t_0^2)}\right)\\
&= \int_{t_0}^{t} h_\lambda(v)dv\\
&= \int_{t_0}^{t} h(W(v))W'(v)dv - \int_{t_0}^{t}\frac{W(v) - v}{(W(v) - v)^2 + |\lambda|^2}(W'(v) - 1)dv\\
&= \int_{W(t_0)}^{W(t)} h(s)ds - \int_{W(t_0)-t_0}^{W(t)-t}\frac{x}{x^2 + |\lambda|^2}dx\\
&= \frac{1}{2}\log\frac{\Delta(T^*T + W(t)^2\mathbf{1})}{\Delta(T^*T + W(t_0)^2\mathbf{1})} - \frac{1}{2}\log\frac{(W(t) - t)^2 + |\lambda|^2}{(W(t_0) - t_0)^2 + |\lambda|^2},
\end{aligned}$$

where the integration above is interpreted as Riemann-Stieltjes integration. Hence there is a constant C such that

$$\begin{aligned}
&\log\Delta((T - \lambda\mathbf{1})^*(T - \lambda\mathbf{1}) + t^2\mathbf{1})\\
&= \log\Delta(T^*T + W(t)^2) - \log(|(W(t) - t)^2 + |\lambda|^2) + C.
\end{aligned}$$

Recall that $\lim_{t\to\infty} W(t) - t = \lim_{t\to\infty}(s(|\lambda|, t) - t) = 0$, we then figure out that $C = \log|\lambda|^2$. The result then follows. □

By Lemma 3.12 and the limit of $s(|\lambda|, t)$ as in (3.13), we obtain the formula for the Fuglede–Kadison determinant of $T - \lambda\mathbf{1}$.

Theorem 3.13 ([16, Theorem 4.15]) *Let $T \in \mathcal{M}^\Delta$ be R-diagonal, let $\mu = \mu_{|T|}$, and let $s(|\lambda|, 0)$ be as in Definition 3.8.*

(i) If $\lambda_1(\mu) < |\lambda| < \lambda_2(\mu)$, then

$$\Delta(T - \lambda\mathbf{1}) = \left(\frac{|\lambda|^2}{|\lambda|^2 + s(|\lambda|, 0)^2}\Delta(T^*T + s(|\lambda|, 0)^2\mathbf{1})\right)^{\frac{1}{2}}. \tag{3.18}$$

(ii) If $|\lambda| \le \lambda_1(\mu)$, then $\Delta(T - \lambda\mathbf{1}) = \Delta(T)$.
(iii) If $|\lambda| \ge \lambda_2(\mu)$, then $\Delta(T - \lambda\mathbf{1}) = |\lambda|$.

4 The Brown Measure of *R*-diagonal Operators

In this section, we calculate the gradient functions of the logarithmic Fuglede–Kadison determinant of $T - \lambda\mathbf{1}$ and provide a proof for the Brown measure formula of R-diagonal operators which simplifies Haagerup–Schultz's original arguments.

Lemma 4.1 *Let $T \in \mathcal{M}^\Delta$ be R-diagonal and $\lambda \in \mathbb{C}\backslash\{0\}$. Then*

$$\tau\Big((\lambda\mathbf{1} - T)[(\lambda\mathbf{1} - T)^*(\lambda\mathbf{1} - T) + t^2\mathbf{1}]^{-1}\Big) = \frac{\lambda}{|\lambda|^2}\frac{(s-t)^2}{|\lambda|^2 + (s-t)^2}$$

and

$$\tau\Big((\lambda\mathbf{1} - T)^*[(\lambda\mathbf{1} - T)^*(\lambda\mathbf{1} - T) + t^2\mathbf{1}]^{-1}\Big) = \frac{\overline{\lambda}}{|\lambda|^2}\frac{(s-t)^2}{|\lambda|^2 + (s-t)^2}$$

where $s = s(|\lambda|, t)$.

Proof Recall from Lemma 3.12 that

$$\Delta\big((\lambda\mathbf{1} - T)^*(\lambda\mathbf{1} - T) + t^2\mathbf{1}\big) = \frac{|\lambda|^2}{|\lambda|^2 + (s(|\lambda|, t) - t)^2}\Delta\big(T^*T + s(|\lambda|, t)^2\mathbf{1}\big).$$

Taking logarithms, we obtain

$$\begin{aligned}\log\Delta\big((\lambda\mathbf{1} - T)^*(\lambda\mathbf{1} - T) + t^2\mathbf{1}\big) &= \log\Delta\big(T^*T + s(|\lambda|, t)^2\mathbf{1}\big)\\ &\quad + \log\frac{|\lambda|^2}{|\lambda|^2 + (s(|\lambda|, t) - t)^2},\end{aligned}$$

which is

$$\tau\big(\log[(\lambda\mathbf{1}-T)^*(\lambda\mathbf{1}-T)+t^2\mathbf{1}]\big) = \tau\big(\log(T^*T+s^2\mathbf{1})\big)+\log|\lambda|^2-\log\big(|\lambda|^2+(s-t)^2\big).$$

The function $s(|\lambda|,t)$ is analytic for $(|\lambda|,t)$. Taking the derivative $\frac{\partial}{\partial\bar{\lambda}}$, it follows that

$$\begin{aligned}
&\tau\big((\lambda\mathbf{1}-T)[(\lambda\mathbf{1}-T)^*(\lambda\mathbf{1}-T)+t^2\mathbf{1}]^{-1}\big)\\
=&2s\cdot\tau\big((T^*T+s^2\mathbf{1})^{-1}\big)\cdot\frac{\partial s(|\lambda|,t)}{\partial\bar{\lambda}}+\frac{\lambda}{|\lambda|^2}\\
&-\frac{1}{|\lambda|^2+(s-t)^2}\left(\lambda+2(s-t)\cdot\frac{\partial s(|\lambda|,t)}{\partial\bar{\lambda}}\right)\\
=&2\left(h(s)-\frac{s-t}{|\lambda|^2+(s-t)^2}\right)\frac{\partial s(|\lambda|,t)}{\partial\bar{\lambda}}+\frac{\lambda}{|\lambda|^2}-\frac{\lambda}{|\lambda|^2+(s-t)^2}\\
=&\frac{\lambda}{|\lambda|^2}\frac{(s-t)^2}{|\lambda|^2+(s-t)^2},
\end{aligned}$$

where we used the identity (3.10) in the last step. The other formula can be obtained by taking the derivative with respect to λ. □

Remark 4.2 By applying the identity $x(x^*x+\varepsilon)^{-1}=(xx^*+\varepsilon)^{-1}x$ for $x\in\mathcal{M}$ and the tracial property, we can deduce

$$\tau\Big((\lambda\mathbf{1}-T)^*[(\lambda\mathbf{1}-T)^*(\lambda\mathbf{1}-T)+t^2\mathbf{1}]^{-1}\Big) = \tau\Big((\lambda\mathbf{1}-T)^*[(\lambda\mathbf{1}-T)(\lambda\mathbf{1}-T)^*+t^2\mathbf{1}]^{-1}\Big).$$

Lemma 4.3 *We have the partial derivative formulas for* $\log\Delta((T-\lambda\mathbf{1})^*(T-\lambda\mathbf{1}))$

$$\tau\Big((\lambda\mathbf{1}-T)[(\lambda\mathbf{1}-T)^*(\lambda\mathbf{1}-T)]^{-1}\Big) = \begin{cases}0, & \textit{for}\quad 0<|\lambda|\le\lambda_1(\mu_{|T|});\\ \frac{\lambda}{|\lambda|^2}\frac{s(|\lambda|,0)^2}{s(|\lambda|,0)^2+|\lambda|^2}, & \textit{for}\quad \lambda_1(\mu_{|T|})<|\lambda|<\lambda_2(\mu_{|T|});\\ \frac{\lambda}{|\lambda|^2}, & \textit{for}\quad |\lambda|\ge\lambda_2(\mu_{|T|})\end{cases}$$

and

$$\tau\Big((\lambda\mathbf{1}-T)^*[(\lambda\mathbf{1}-T)^*(\lambda\mathbf{1}-T)]^{-1}\Big)$$

$$=\begin{cases} 0, & \textit{for} \quad 0<|\lambda|\le\lambda_1(\mu_{|T|});\\ \frac{\overline{\lambda}}{|\lambda|^2}\frac{s(|\lambda|,0)^2}{s(|\lambda|,0)^2+|\lambda|^2}, & \textit{for} \quad \lambda_1(\mu_{|T|})<|\lambda|<\lambda_2(\mu_{|T|});\\ \frac{\overline{\lambda}}{|\lambda|^2}, & \textit{for} \quad |\lambda|\ge\lambda_2(\mu_{|T|}), \end{cases}$$

where $s(|\lambda|,0)$ is defined in Definition 3.8.

Proof If $\lambda_1(\mu_{|T|}) < |\lambda| < \lambda_2(\mu_{|T|})$, the result follows from Lemma 4.1 because $s(|\lambda|,0)\in(0,\infty)$ satisfies the same equation $k(s,0)=|\lambda|^2$ and is real analytic function of λ by Lemma 3.10.

If $0<|\lambda|\le\lambda_1(\mu_{|T|})$, by Theorem 3.13, we have $\Delta(T-\lambda\mathbf{1})=\Delta(T)$. Hence, the derivatives are equal to zero. Finally, when $|\lambda|\ge\lambda_2(\mu_{|T|})$, we have $\Delta(T-\lambda\mathbf{1})=|\lambda|$ which yields the other case of the formula. □

We can now give an alternative proof of Haagerup–Schultz's result on the Brown measure of R-diagonal operators inspired by the proof presented in [4].

Proof (Proof of Theorem 2.1) For λ such that $\lambda_1(\mu)<|\lambda|<\lambda_2(\mu)$, taking the limit $t\to 0^+$ in (3.10), we obtain

$$s(|\lambda|,0)^2+|\lambda|^2=\frac{s(|\lambda|,0)}{h(s(|\lambda|,0))},$$

which yields,

$$\begin{aligned}\tau\Big((\lambda\mathbf{1}-T)^*[(\lambda\mathbf{1}-T)^*(\lambda\mathbf{1}-T)]^{-1}\Big)&=\frac{\overline{\lambda}}{|\lambda|^2}\frac{s(|\lambda|,0)^2}{s(|\lambda|,0)^2+|\lambda|^2}\\ &=\frac{\overline{\lambda}}{|\lambda|^2}\frac{h(s(|\lambda|,0))}{s(|\lambda|,0)}s(|\lambda|,0)^2\\ &=\frac{\overline{\lambda}}{|\lambda|^2}\tau\Big([T^*T+s(|\lambda|,0)^2\mathbf{1}]^{-1}\Big)s(|\lambda|,0)^2\\ &=\frac{\overline{\lambda}}{|\lambda|^2}\int_0^\infty\frac{s(|\lambda|,0)^2}{s(|\lambda|,0)^2+u^2}d\mu_{|T|}(u)\\ &=\frac{\overline{\lambda}}{|\lambda|^2}\left(1+\psi_{\mu_{T^*T}}\left(-\frac{1}{s(|\lambda|,0)^2}\right)\right).\end{aligned}$$

We can check that

$$S^{\langle-1\rangle}_{\mu_{T^*T}}\left(\frac{1}{|\lambda|^2}\right)=\psi_{\mu_{T^*T}}\left(-\frac{1}{s(|\lambda|,0)^2}\right). \tag{4.1}$$

Indeed, recall that the S-transform satisfies $S_\mu(\psi_\mu(z)) = z \cdot \frac{\psi_\mu(z)+1}{\psi_\mu(z)}$. The above calculation shows that

$$\psi_{\mu_{T^*T}}\left(-\frac{1}{s(|\lambda|,0)^2}\right) = \frac{s(|\lambda|,0)^2}{s(|\lambda|,0)^2+|\lambda|^2} - 1 = -\frac{|\lambda|^2}{s(|\lambda|,0)^2+|\lambda|^2}.$$

Hence,

$$S_{\mu_{T^*T}}\left(\psi_{\mu_{T^*T}}\left(-\frac{1}{s(|\lambda|,0)^2}\right)\right) = -\frac{1}{s(|\lambda|,t)^2}\cdot -\frac{s(|\lambda|,t)^2}{|\lambda|^2} = \frac{1}{|\lambda|^2}$$

which yields (4.1). Therefore, we have

$$\begin{aligned}
&\tau\Big((\lambda\mathbf{1}-T)^*[(\lambda\mathbf{1}-T)^*(\lambda\mathbf{1}-T)]^{-1}\Big)\\
&=\begin{cases} 0, & \text{for } |\lambda|\le\lambda_1(\mu);\\ \frac{1}{\lambda}\left(1+\psi_{\mu_{T^*T}}\left(-\frac{1}{s(|\lambda|,0)^2}\right)\right), & \text{for } \lambda_1(\mu)<|\lambda|<\lambda_2(\mu);\\ \frac{1}{\lambda}, & \text{for } |\lambda|\ge\lambda_2(\mu).\end{cases}\\
&=\begin{cases} 0, & \text{for } |\lambda|\le\lambda_1(\mu);\\ \frac{1}{\lambda}\left(1+S^{\langle-1\rangle}_{\mu_{T^*T}}\left(\frac{1}{|\lambda|^2}\right)\right), & \text{for } \lambda_1(\mu)<|\lambda|<\lambda_2(\mu);\\ \frac{1}{\lambda}, & \text{for } |\lambda|\ge\lambda_2(\mu).\end{cases}
\end{aligned}$$

Recall that

$$\mu_T = \frac{2}{\pi}\frac{\partial}{\partial\bar\lambda}\frac{\partial}{\partial\lambda}\log\Delta(T-\lambda\mathbf{1})$$

and

$$\frac{\partial}{\partial\lambda}\log\Delta(T-\lambda\mathbf{1}) = \frac{1}{2}\tau\Big((\lambda\mathbf{1}-T)^*[(\lambda\mathbf{1}-T)^*(\lambda\mathbf{1}-T)]^{-1}\Big).$$

Hence, the Brown measure μ_T is supported in $\{\lambda\in\mathbb{C}:\lambda_1(\mu_{|T|})\le|\lambda|<\lambda_2(\mu_{|T|})\}$. When $\lambda_1(\mu_{|T|})<|\lambda|<\lambda_2(\mu_{|T|})$, the Brown measure μ_T has the density

$$-\frac{1}{2\pi|\lambda|^4}\left(S^{\langle-1\rangle}_{\mu_{T^*T}}\right)'\left(\frac{1}{|\lambda|^2}\right).$$

By Proposition 2.2, the map $r \mapsto S^{\langle -1\rangle}_{\mu_{T^*T}}(r^{-2})$ is increasing on $(\lambda_1(\mu_{|T|}), \lambda_2(\mu_{|T|}))$ with range $(\delta - 1, 0)$, where $\delta = \mu(\{0\})$. This allows us to conclude that the Brown measure $\mu_{|T|}$ is the rotationally invariant probability measure such that

$$\mu_T\{\lambda \in \mathbb{C} : |\lambda| \leq r\} = \begin{cases} 0, & \text{for} \quad r < \lambda_1(\mu_{|T|}); \\ 1 + S^{\langle -1\rangle}_{\mu_{T^*T}}(r^{-2}), & \text{for} \quad \lambda_1(\mu_{|T|}) < r < \lambda_2(\mu_{|T|}); \\ 1, & \text{for} \quad r \geq \lambda_2(\mu_{|T|}). \end{cases} \tag{4.2}$$

In particular, when $\delta > 0$, then $\lambda_1(\mu) = 0$ and the Brown measure μ_T has a mass at the origin with the mass $\mu_T(\{0\}) = \delta$. □

We can now restate the Brown measure of R-diagonal operators in terms of subordination functions as follows.

Theorem 4.4 *Let $T \in \mathcal{M}^\Delta$ be R-diagonal and $\lambda \in \mathbb{C}\backslash\{0\}$. The Brown measure μ_T is the rotationally invariant probability measure such that*

$$\mu_T\{\lambda \in \mathbb{C} : |\lambda| \leq r\} = \frac{s(r,0)^2}{s(r,0)^2 + r^2}, \tag{4.3}$$

where $s(|\lambda|, 0)$ is defined in Definition 3.8. When $\lambda_1(\mu_{|T|}) < r < \lambda_2(\mu_{|T|})$, the function $s(r, 0)$ is the unique positive solution s of the following equation

$$s^2 + r^2 - \frac{s}{h(s)} = 0.$$

Proof When $\lambda_1(\mu_{|T|}) < r < \lambda_2(\mu_{|T|})$, we recall that $s(r, 0)$ denotes the unique positive solution of the equation $k(s, 0) = r^2$ as shown in Definition 3.8. The equation $k(s, 0) = r^2$ can be rewritten as

$$s^2 + |\lambda|^2 - \frac{s}{h(s)} = 0,$$

or

$$sh(s) = \frac{s^2}{s^2 + r^2}.$$

Hence, from (4.1) in the above proof of Theorem 2.1, we have

$$\begin{aligned}1+S_{\mu_{T^*T}}^{\langle -1\rangle}\left(\frac{1}{r^2}\right) &= 1+\psi_{\mu_{T^*T}}\left(-\frac{1}{s(r,0)^2}\right)\\ &= \int_0^\infty \frac{s(r,0)^2}{s(r,0)^2+u^2}d\mu_{|T|}(u)\\ &= s(r,0)h(s(r,0))\\ &= \frac{s(r,0)^2}{s(r,0)^2+r^2}.\end{aligned}$$

When $0\le r\le \lambda_1(\mu_{|T|})$, $s(r,0)=0$; and when $r\ge\lambda_2(\mu_{|T|})$, $s(r,0)=\infty$. Hence, the formulation claimed follows from (4.2). □

Theorem 4.5 *Let $T\in\mathcal{M}^\Delta$ be R-diagonal and $\lambda\in\mathbb{C}\backslash\{0\}$, we have*

$$\tau\Big((\lambda\mathbf{1}-T)[(\lambda\mathbf{1}-T)^*(\lambda\mathbf{1}-T)]^{-1}\Big)$$

$$=\begin{cases}0, & \text{for} \quad 0<|\lambda|\le\lambda_1(\mu_{|T|});\\ \frac{\lambda}{|\lambda|^2}\mu_T\{\lambda\in\mathbb{C}:|\lambda|\le r\}, & \text{for} \quad \lambda_1(\mu_{|T|})<|\lambda|<\lambda_2(\mu_{|T|});\\ \frac{\lambda}{|\lambda|^2}, & \text{for} \quad |\lambda|\ge\lambda_2(\mu_{|T|})\end{cases}$$

and

$$\tau\Big((\lambda\mathbf{1}-T)^*[(\lambda\mathbf{1}-T)^*(\lambda\mathbf{1}-T)]^{-1}\Big)$$

$$=\begin{cases}0, & \text{for} \quad 0<|\lambda|\le\lambda_1(\mu_{|T|});\\ \frac{\overline{\lambda}}{|\lambda|^2}\mu_T\{\lambda\in\mathbb{C}:|\lambda|\le r\}, & \text{for} \quad \lambda_1(\mu_{|T|})<|\lambda|<\lambda_2(\mu_{|T|});\\ \frac{\overline{\lambda}}{|\lambda|^2}, & \text{for} \quad |\lambda|\ge\lambda_2(\mu_{|T|}),\end{cases}$$

where $s(|\lambda|,0)$ is defined in Definition 3.8.

Proof It follows directly from the partial derivative formula in Lemma 4.3 and Theorem 4.4. □

5 An Alternative Proof for the Brown Measure Formula

We present another proof for the Brown measure of R-diagonal operators in this section.

Lemma 5.1 ([24, Chapter II-Theorem 1.2]) *Let μ be a finite positive measure of compact support on the plane. Denote*

$$U^{\mu}(\lambda) = \int_{\mathbb{C}} \log \frac{1}{|\lambda - z|} d\mu(z), \qquad \lambda \in \mathbb{C},$$

and, for any z_0 and $r > 0$, denote the mean value

$$L(U^{\mu}; z_0, r) = \frac{1}{2\pi} \int_{-\pi}^{\pi} U^{\mu}(z_0 + re^{i\theta}) d\theta.$$

If r is a value for which $\frac{d}{dr} L(U^{\mu}; z_0, r)$ exists, then

$$\mu(\{\lambda : |\lambda - z_0| \leq r\}) = -r \frac{d}{dr} L(U^{\mu}; z_0, r).$$

Remark 5.2 If U^{μ} in Lemma 5.1 is a sufficiently smooth function of λ, for $z_0 = 0$ and $\lambda = x + iy$, we can prove Lemma 5.1 as follows:

$$\begin{aligned} -\frac{d}{dr} L(U^{\mu}; 0, r) &= -\frac{1}{2\pi} \frac{d}{dr} \int_{-\pi}^{\pi} U^{\mu}(re^{i\theta}) d\theta \\ &= -\frac{1}{2\pi r} \int_{|\lambda| = r} \left(\frac{\partial U^{\mu}}{\partial x} dy - \frac{\partial U^{\mu}}{\partial y} dx \right) \\ &= \frac{1}{2\pi r} \int_{|\lambda| = r} \nabla_{\lambda}^2 \Big(\int_{\mathbb{C}} \log |\lambda - z| d\mu(z) \Big) dx dy \\ &= \frac{1}{r} \mu(\{\lambda : |\lambda| \leq r\}). \end{aligned}$$

When applying Lemma 5.1 to calculate the Brown measure of a R-diagonal operator, the potential we deal with is $U^{\mu_T}(\lambda)$, which can be expressed as

$$U^{\mu_T}(\lambda) = -\log \Delta(T - \lambda \mathbf{1}).$$

As we know from Theorem 3.13 and regularity result of $s(|\lambda|, 0)$ in Lemma 3.10, the potential is a smooth function of λ within the domain $\lambda_1(\mu_{|T|}) < |\lambda| < \lambda_2(\mu_{|T|})$.

Proposition 5.3 *Let $r > 0$ and let $s(r, 0)$ be as in Definition 3.8 so that $s(r, 0)$ is the unique solution of equation* (3.10)

$$s^2 + r^2 - \frac{s}{h(s)} = 0,$$

where $h(s) = s\phi((T^*T + s^2\mathbf{1})^{-1})$*. Then, the Brown measure of* T *is invariant under rotation and it is given by*

$$\mu_T(\{\lambda : |\lambda| \le r\}) = \frac{s(r,0)^2}{s(r,0)^2 + r^2}, \tag{5.1}$$

for any $\lambda_1(\mu_{|T|}) < r < \lambda_2(\mu_{|T|})$*. If* $\mu_{|T|}(\{0\}) = \sigma > 0$*, then the Brown measure* μ_T *has a mass* σ *at the origin.*

Proof By Proposition 3.9, the function $r \to s(r, 0)$ is a real analytic function for $\lambda_1(\mu_{|T|}) < r < \lambda_2(\mu_{|T|})$. Recall the Fuglede–Kadison determinant formula (3.18)

$$\Delta(T - \lambda\mathbf{1}) = \left(\frac{r^2}{r^2 + s(r,0)^2}\Delta(T^*T + s(r,0)^2\mathbf{1})\right)^{\frac{1}{2}},$$

where $r = |\lambda|$. We then take the logarithmic derivative of $\Delta(T - \lambda\mathbf{1})$. Note that

$$s(r,0)\phi((T^*T + s(r,0)^2\mathbf{1})^{-1}) = h(s(r,0)).$$

We then obtain, for $\lambda_1(\mu_{|T|}) < r < \lambda_2(\mu_{|T|})$,

$$\begin{aligned}
\frac{\partial \log \Delta(T - \lambda\mathbf{1})}{\partial r} &= \frac{1}{r} - \frac{1}{r^2 + s(r,0)^2}\left(r + s(r,0)\frac{ds(r,0)}{dr}\right) + h(s(r,0))\frac{ds(r,0)}{dr} \\
&= \frac{1}{r} - \frac{r}{r^2 + s(r,0)^2} \\
&= \frac{1}{r}\frac{s(r,0)^2}{s(r,0)^2 + r^2},
\end{aligned} \tag{5.2}$$

where we used (3.10) to deduce the second identity. Now, recall that the Brown measure μ_T is the unique measure satisfying

$$\log \Delta(T - \lambda\mathbf{1}) = \int_{\mathbb{C}} \log|\lambda - z| d\mu_T(z), \qquad \lambda \in \mathbb{C}. \tag{5.3}$$

By adopting the notation as in Lemma 5.1, denote

$$U^{\mu_T}(\lambda) = -\int_{\mathbb{C}} \log|\lambda - z| d\mu_T(z).$$

Observe that T and $e^{i\theta}T$ have the same distribution. Hence, the Brown measure μ_T is invariant under the transformation $\lambda \mapsto \lambda e^{i\theta}$. The value of the function $\lambda \mapsto$

$U^{\mu_T}(\lambda)$ is also invariant if we replace λ by $\lambda e^{i\theta}$ for any $\theta \in [-\pi, \pi]$. The Brown measure μ_T is then determined by

$$\mu_T(\{\lambda : |\lambda - z_0| \leq r\}) = -r\frac{\partial}{\partial r}U^{\mu_T}(\lambda) = r\frac{\partial \log(\Delta(T - \lambda\mathbf{1}))}{\partial r} = \frac{s(r,0)^2}{s(r,0)^2 + r^2}.$$

This finishes the proof. □

We can then repeat the calculation in the proof of Theorem 2.1 to rewrite (5.1) in terms of the S-transform of T^*T, which recovers the Brown measure formula in Theorem 2.1.

6 Freeness with Amalgamation and R-diagonal Operators

We now describe the connection between the approach in the previous section with the Hermitian reduction and subordination method used in [4]. We denote by $\mathcal{D}$ the commutative C^*-algebra of diagonal matrices in $M_2(\mathbb{C})$, where we consider the same inclusion of $M_2(\mathbb{C})$ in $M_2(\mathcal{M})$. Let $T \in \mathcal{M}$. Recall from [23] that an element T in a tracial W^*-noncommutative probability space $(\mathcal{M}, \tau)$ is R-diagonal if and only if the matrix-valued operator $\begin{bmatrix} 0 & T \\ T^* & 0 \end{bmatrix}$ and $M_2(\mathbb{C})$ are free with amalgamation over $\mathcal{D}$ with respect to the expectation

$$\mathbb{E}_{\mathcal{D}}\colon M_2(\mathcal{M}) \to \mathcal{D}, \qquad \mathbb{E}_{\mathcal{D}}\begin{bmatrix} a_{11} & a_{12} \\ a_{21} & a_{22} \end{bmatrix} = \begin{bmatrix} \tau(a_{11}) & 0 \\ 0 & \tau(a_{22}) \end{bmatrix}.$$

Fix now a $\lambda \in \mathbb{C}$. The operators $\begin{bmatrix} 0 & \lambda \\ \overline{\lambda} & 0 \end{bmatrix}$ and $\begin{bmatrix} 0 & T \\ T^* & 0 \end{bmatrix}$ are free in the operator-valued noncommutative probability space $(M_2(\mathcal{M}), \mathbb{E}_{\mathcal{D}})$.

Let $X = \begin{bmatrix} 0 & \lambda \\ \overline{\lambda} & 0 \end{bmatrix}$, and denote by $\mathcal{D}\langle X\rangle$ the $*$-algebra generated by $\mathcal{D}$ and X. We now study the other operator-valued subordination function Ω_2 such that

$$\mathbb{E}_{\mathcal{D}\langle X\rangle}\left[\left(\begin{bmatrix} i\varepsilon & \lambda \\ \overline{\lambda} & i\varepsilon \end{bmatrix} - \begin{bmatrix} 0 & T \\ T^* & 0 \end{bmatrix}\right)^{-1}\right] = \left(\Omega_2\left(\begin{bmatrix} i\varepsilon & 0 \\ 0 & i\varepsilon \end{bmatrix}\right) - \begin{bmatrix} 0 & \lambda \\ \overline{\lambda} & 0 \end{bmatrix}\right)^{-1}. \tag{6.1}$$

Note that this algebra $\mathcal{D}\langle X\rangle = M_2(\mathbb{C})$. Hence, $\mathbb{E}_{\mathcal{D}\langle X\rangle}$ is actually the conditional expectation onto $M_2(\mathbb{C})$ given by evaluation of the trace τ on the entries of the matrix which is the unique conditional expectation onto $M_2(\mathbb{C})$ preserving the trace. Similar to Ω_1, the subordination function Ω_2 also takes values in $\mathcal{D}$. We then set

$$\begin{bmatrix} i\delta_1 & 0 \\ 0 & i\delta_2 \end{bmatrix} = \Omega_2\left(\begin{bmatrix} i\varepsilon & 0 \\ 0 & i\varepsilon \end{bmatrix}\right), \tag{6.2}$$

where $\delta_i = \delta_i(\lambda, \varepsilon)$(for $i = 1, 2$) are functions of λ and ε. Set $A_\lambda = \lambda\mathbf{1} - T$. The left hand side of (6.1) (see [4, formula (18)]) is given by

$$
\begin{aligned}
&\mathbb{E}_{\mathcal{D}\langle X\rangle}\left[\left(\begin{bmatrix} i\varepsilon & \lambda \\ \overline{\lambda} & i\varepsilon \end{bmatrix} - \begin{bmatrix} 0 & T \\ T^* & 0 \end{bmatrix}\right)^{-1}\right] \\
=&\mathbb{E}_{\mathcal{D}\langle X\rangle}\begin{bmatrix} -i\varepsilon([A_\lambda A_\lambda^* + \varepsilon^2\mathbf{1}]^{-1}) & A_\lambda([A_\lambda^* A_\lambda + \varepsilon^2\mathbf{1}]^{-1}) \\ A_\lambda^*([A_\lambda A_\lambda^* + \varepsilon^2\mathbf{1}]^{-1}) & -i\varepsilon([A_\lambda^* A_\lambda + \varepsilon^2\mathbf{1}]^{-1}) \end{bmatrix} \\
=&\begin{bmatrix} -i\varepsilon\tau([A_\lambda A_\lambda^* + \varepsilon^2\mathbf{1}]^{-1}) & \tau\left(A_\lambda([A_\lambda^* A_\lambda + \varepsilon^2\mathbf{1}]^{-1})\right) \\ \tau\left(A_\lambda^*([A_\lambda A_\lambda^* + \varepsilon^2\mathbf{1}]^{-1})\right) & -i\varepsilon\tau([A_\lambda^* A_\lambda + \varepsilon^2\mathbf{1}]^{-1}) \end{bmatrix}.
\end{aligned} \tag{6.3}
$$

Then the inverse matrix in the right hand side of (6.1) is given by

$$
\begin{bmatrix} i\delta_1 & \lambda \\ \overline{\lambda} & \delta_2 \end{bmatrix}^{-1} = \begin{bmatrix} -i\delta_2(|\lambda|^2 + \delta_1\delta_2)^{-1} & \lambda(|\lambda|^2 + \delta_1\delta_2)^{-1} \\ \overline{\lambda}(|\lambda|^2 + \delta_1\delta_2)^{-1} & -i\delta_1(|\lambda|^2 + \delta_1\delta_2)^{-1} \end{bmatrix}. \tag{6.4}
$$

Traciality of τ implies that the (1, 1) and (2, 2) entries of the above matrices (6.3) and (6.4) are equal, and thus $\delta_1(\lambda, \varepsilon) = \delta_2(\lambda, \varepsilon)$ and is denote by $\delta = \delta(\lambda, \varepsilon)$. Since the above matrices must be equal to each other due to (6.1), we then have

$$
\begin{aligned}
\frac{\delta}{\delta^2 + |\lambda|^2} &= \varepsilon\tau\Big([(\lambda\mathbf{1} - T)(\lambda\mathbf{1} - T)^* + \varepsilon^2\mathbf{1}]^{-1}\Big); \\
\frac{\lambda}{\delta^2 + |\lambda|^2} &= \tau\left((\lambda\mathbf{1} - T)\left[(\lambda\mathbf{1} - T)^*(\lambda\mathbf{1} - T) + \varepsilon^2\mathbf{1}\right]^{-1}\right); \\
\frac{\overline{\lambda}}{\delta^2 + |\lambda|^2} &= \tau\left((\lambda\mathbf{1} - T)^*\left[(\lambda\mathbf{1} - T)(\lambda\mathbf{1} - T)^* + \varepsilon^2\mathbf{1}\right]^{-1}\right).
\end{aligned} \tag{6.5}
$$

We now use the formulas in Lemma 4.1 to solve for δ in (6.5). We identify the function δ as the subordination studied in Sect. 3.

Proposition 6.1 *For any $\varepsilon > 0$, we have $\delta(\lambda, \varepsilon) = \omega_2(i\varepsilon)$ where ω_2 is the subordination function such that $G_{\widetilde{\mu}_{|T-\lambda\mathbf{1}|}}(z) = G_\nu(\omega_2(z))$ defined in Lemma 3.11. In particular, for $t > 0$, $\delta = \delta(\lambda, \varepsilon)$ is positive. Moreover,*

$$
\delta = \Im\omega_2(i\varepsilon) = \frac{|\lambda|^2}{s(|\lambda|, \varepsilon) - \varepsilon}
$$

where $s(|\lambda|, \varepsilon)$ is given in Definition 3.3.

Proof Recall that $\mu = \widetilde{\mu}_{|T|}$, $\nu = \frac{1}{2}\big(\delta_{|\lambda|} + \delta_{-|\lambda|}\big)$ and $\widetilde{\mu}_{|T-\lambda\mathbf{1}|} = \mu \boxplus \nu$. We have

$$
G_\nu(z) = \frac{-z}{-z^2 + |\lambda|^2}
$$

and by (3.2), we have

$$G_{\widetilde{\mu}_{|T-\lambda\mathbf{1}|}}(i\varepsilon) = -i\varepsilon\tau\Big([(\lambda\mathbf{1}-T)(\lambda\mathbf{1}-T)^*+\varepsilon^2\mathbf{1}]^{-1}\Big).$$

The subordination relation $G_{\widetilde{\mu}_{|T-\lambda\mathbf{1}|}}(i\varepsilon) = G_\nu(\omega_2(i\varepsilon))$ is equivalent to

$$\varepsilon\tau\Big([(\lambda\mathbf{1}-T)(\lambda\mathbf{1}-T)^*+\varepsilon^2\mathbf{1}]^{-1}\Big) = \frac{-\omega_2(i\varepsilon)}{-\omega_2(i\varepsilon)^2+|\lambda|^2}.$$

By comparing the above identity with (6.5), we deduce that $\omega_2(i\varepsilon) = i\delta$. Hence, by Lemma 3.11, it follows that $\delta = \delta(\lambda, \varepsilon) > 0$ and

$$\delta = \Im\omega_2(i\varepsilon) = \frac{|\lambda|^2}{s(|\lambda|, \varepsilon) - \varepsilon}.$$

This finishes the proof. □

We have another proof for the partial derivative formula in Lemma 4.1 by assuming that the R-diagonal operator T is bounded.

***Proof** (Second Proof for Lemma 4.1)* By (6.5) and Proposition 6.1, we have

$$\tau\left((\lambda\mathbf{1}-T)\Big[(\lambda\mathbf{1}-T)^*(\lambda\mathbf{1}-T)+t^2\mathbf{1}\Big]^{-1}\right) = \frac{\lambda}{\delta^2+|\lambda|^2} = \frac{\lambda}{|\lambda|}\frac{(s-t)^2}{|\lambda|^2+(s-t)^2}$$

where $s = s(\lambda, t)$. The other partial derivative formula follows in the same way by noticing Lemma 4.2. □

Remark 6.2 One can deduce Lemma 4.3 from Lemma 4.1 by letting $t \to 0$. Then apply the proof for Theorem 2.1 in Sect. 4 to recover Haagerup–Larsen–Schultz's density formula for Brown measure of R-diagonal operators. This is exactly the approach used in [4, Section 5].

7 Some Miscellaneous Applications

The properties of subordination functions studied in previous sections play an important role for proving convergence of empirical spectral distribution of random matrices in single ring theorem and its extension [1, 12]. The techniques from Sect. 3 could simplify various technical arguments proved in a recent work about local single ring theorem [1, Section 7, Supplement] (in particular, Lemmas 7.4, 7.5 and 7.6 in that paper).

Recall that $\mu = \widetilde{\mu}_{|T|}$, $\nu = \frac{1}{2}\big(\delta_{|\lambda|}+\delta_{-|\lambda|}\big)$ and $\lambda_1(\mu), \lambda_2(\mu)$ as in Sect. 3, and $s(|\lambda|, t) = \Im\omega_1(it)$.

Proposition 7.1 *For any λ_0 such that $\lambda_1(\mu) < |\lambda_0| < \lambda_2(\mu)$, there is a constant $C > 0$ depending on λ such that $\Im\omega_1(0) = s(|\lambda|, 0) > C$ in some neighborhood of λ_0.*

Proof It follows from the monotonicity of $k(s, t)$ in Lemmas 3.4 and 3.5. □

Proposition 7.2 *Let T be a bounded R-diagonal operator. For each λ such that $\lambda_1(\mu) < |\lambda| < \lambda_2(\mu)$, there are positive constants C_1, C_2 such that, for all $t \geq 0$, we have*

$$C_1 \min\left\{1, \frac{1}{t}\right\} \leq \Im\omega_1(it) - t = s(|\lambda|, t) - t \leq C_2 \min\left\{1, \frac{1}{t}\right\}.$$

Proof Recall from (3.16) that there is some $t_\lambda > 0$ so that for any $t > t_\lambda$, we have

$$s - t = \frac{1 - \sqrt{1 - 4|\lambda|^2 h(s)^2}}{2h(s)} = \frac{2|\lambda|^2 h(s)}{1 + \sqrt{1 - 4|\lambda|^2 h(s)^2}}.$$

Hence, for $t > t_\lambda$, we have

$$|\lambda|^2 h(s) < s - t < 2|\lambda|^2 h(s).$$

Since $h(s) = \int_{\mathbb{R}} \frac{s}{s^2+u^2} d\mu(u)$, for any $t > t_\lambda$, we then have

$$s - t < \frac{2|\lambda|^2}{s} \int_{\mathbb{R}} \frac{s^2}{s^2 + u^2} d\mu(u) < \frac{2|\lambda|^2}{s} < \frac{2|\lambda|^2}{t}.$$

As T is bounded, it follows that $\mu \boxplus \nu$ is compactly supported. Using the identity $h(s) = h_\lambda(t)$, we then have

$$s - t > |\lambda|^2 h(s) = |\lambda|^2 \int_{\mathbb{R}} \frac{t}{t^2 + u^2} d\mu \boxplus \nu(t) > \frac{|\lambda|^2 t}{t^2 + M}$$

for some $M > 0$. It is then clear that there are C_1, C_2 so the desired inequality holds. □

The negative moments of the resolvent of bounded R-diagonal operators were treated in [18]. Let us illustrate an example using subordination functions to calculate the first order negative moment for any $T \in \mathcal{M}^\Delta$. We expect that the approach can calculate higher order negative moments of unbounded R-diagonal operators.

Proposition 7.3 *Let $T \in \mathcal{M}^{\Delta}$ be R-diagonal and $\lambda \in \mathbb{C}\backslash\{0\}$. Then*

$$\tau\Big([(T-\lambda\mathbf{1})^*(T-\lambda\mathbf{1})]^{-1}\Big) = \lim_{t\to 0^+}\tau\Big([(T-\lambda\mathbf{1})^*(T-\lambda\mathbf{1})+t^2\mathbf{1}]^{-1}\Big)$$
$$= \begin{cases} \frac{1}{\lambda_1(\mu)^2-|\lambda|^2}, & for \quad 0<|\lambda|\le\lambda_1(\mu_{|T|}); \\ \infty, & for \quad \lambda_1(\mu_{|T|})<|\lambda|<\lambda_2(\mu_{|T|}); \\ \frac{1}{|\lambda|^2-\lambda_2(\mu)^2}, & for \quad |\lambda|\ge\lambda_2(\mu_{|T|}). \end{cases}$$

Proof Using Lemma 3.10, when $0<|\lambda|\le\lambda_1(\mu_{|T|})$ and $\lambda_1(\mu)>0$, we have

$$\begin{aligned}\tau\Big([(T-\lambda\mathbf{1})^*(T-\lambda\mathbf{1})]^{-1}\Big) &= \lim_{t\to 0^+}\tau\Big([(T-\lambda\mathbf{1})^*(T-\lambda\mathbf{1})+t^2\mathbf{1}]^{-1}\Big) \\ &= \lim_{t\to 0^+}\frac{h_\lambda(t)}{t} = \lim_{t\to 0^+}\frac{h(s(|\lambda|,t))}{t} \\ &= \lim_{s\to 0^+}\frac{h(s)}{s}\lim_{t\to 0^+}\frac{s(|\lambda|,t)}{t} \\ &= \int_0^\infty \frac{1}{u^2}d\widetilde{\mu}_{|T|}\frac{\lambda_1(\mu_{|T|})^2}{\lambda_1(\mu_{|T|})^2)-|\lambda|^2} \\ &= \frac{1}{\lambda_1(\mu_{|T|})^2-|\lambda|^2}.\end{aligned}$$

Similarly, when $|\lambda|\ge\lambda_2(\mu_{|T|})$, we have

$$\begin{aligned}\tau\Big([(T-\lambda\mathbf{1})^*(T-\lambda\mathbf{1})]^{-1}\Big) &= \lim_{t\to 0^+}\tau\Big([(T-\lambda\mathbf{1})^*(T-\lambda\mathbf{1})+t^2\mathbf{1}]^{-1}\Big) \\ &= \lim_{t\to 0^+}\frac{h_\lambda(t)}{t} \\ &= \lim_{s\to\infty} sh(s)\lim_{t\to 0^+}\frac{1}{s(|\lambda|,t)t} \\ &= \frac{1}{|\lambda|^2-\lambda_2(\mu_{|T|})^2)}.\end{aligned}$$

When $\lambda_1(\mu_{|T|})<|\lambda|<\lambda_2(\mu_{|T|})$, as $\lim_{t\to 0^+}s(|\lambda|,t)=s(|\lambda|,0)>0$, we have

$$\begin{aligned}\tau\Big([(T-\lambda\mathbf{1})^*(T-\lambda\mathbf{1})]^{-1}\Big) &= \lim_{t\to 0^+}\tau\Big([(T-\lambda\mathbf{1})^*(T-\lambda\mathbf{1})+t^2\mathbf{1}]^{-1}\Big) \\ &= \lim_{t\to 0^+}\frac{h(s(|\lambda|,t)}{s(|\lambda|,t)}\frac{s(|\lambda|,t)}{t} = \infty.\end{aligned}$$

This finishes the proof. □

Example 7.4 Let c_t be Voiculescu's circular operator with variance t, and g_t be a semicircular element of variance t, and let u be a Haar unitary operator that is $*$-free from $\{c_t, g_t\}$. In [27], the author calculated the Brown measure density of $g_t + u$. It is shown that the Brown measure of $g_t + u$ is the pushforward measure of $c_t + u$ under some natural pushforward map. The operator $c_t + u$ is R-diagonal, the outer radius of the support of the Brown measure of c_t+u is $\sqrt{t+1}$, and the inner radius is $\sqrt{(1-t)_+}$. Indeed, to calculate the inner radius $\lambda_1(c_t+u) = 1/\tau[(c_t+u)^*(c_t+u)]$, we observe that $|c_t + u| \underset{*\mathcal{D}}{\sim} |c_t + \mathbf{1}|$ and then apply Proposition 7.3.

Acknowledgments The author wants to thank Serban Belinschi, Hari Bercovici, Kamil Szpojankowski, Alexandru Nica, and Zhi Yin for helpful discussions and their interest in this work. The proof in Sect. 5 was found when the author was preparing his notes for an online learning seminar on Brown measures in summer 2021 organized by Professor Nica. He thanks the seminar participants for useful comments. He is grateful to an anonymous referee and to Yemon Choi for their careful reading and insightful suggestions. This project was partially supported by a grant from Simons Foundation and NSF grant No. 2316836.

References

1. Z. Bao, L. Erdös, K. Schnelli, Local single ring theorem on optimal scale. Ann. Probab. **47**(3), 1270–1334 (2019). https://doi.org/10.1214/18-AOP1284
2. S.T. Belinschi, The Lebesgue decomposition of the free additive convolution of two probability distributions. Probab. Theory Relat. Fields **142**(1–2), 125–150 (2008). https://doi.org/10.1007/s00440-007-0100-3
3. S.T. Belinschi, H. Bercovici, A new approach to subordination results in free probability. J. Anal. Math. **101**, 357–365 (2007). http://dx.doi.org/10.1007/s11854-007-0013-1
4. S.T. Belinschi, P. Śniady, R. Speicher, Eigenvalues of non-Hermitian random matrices and Brown measure of non-normal operators: Hermitian reduction and linearization method. Linear Algebra Appl. **537**, 48–83 (2018). https://doi.org/10.1016/j.laa.2017.09.024. http://www.sciencedirect.com/science/article/pii/S002437951730558X
5. H. Bercovici, D. Voiculescu, Free convolution of measures with unbounded support. Indiana Univ. Math. J. **42**(3), 733–773 (1993). http://dx.doi.org/10.1512/iumj.1993.42.42033
6. H. Bercovici, P. Zhong, The Brown measure of a sum of two free random variables, one of which is R-diagonal (2022). arXiv:2209.12379
7. C. Bordenave, D. Chafaï, Around the circular law. Probab. Surv. **9**, 1–89 (2012). https://doi.org/10.1214/11-PS183
8. L.G. Brown, Lidskii's theorem in the type II case, in *Geometric Methods in Operator Algebras (Kyoto, 1983)*. Pitman Res. Notes Math. Ser., vol. 123 (Longman Sci. Tech., Harlow, 1986), pp. 1–35
9. K. Dykema, F. Sukochev, D. Zanin, A decomposition theorem in II_1-factors. J. Reine Angew. Math. **708**, 97–114 (2015). https://doi.org/10.1515/crelle-2013-0084
10. K. Dykema, F. Sukochev, D. Zanin, An upper triangular decomposition theorem for some unbounded operators affiliated to II_1-factors. Israel J. Math. **222**(2), 645–709 (2017). https://doi.org/10.1007/s11856-017-1603-y
11. B. Fuglede, R.V. Kadison, Determinant theory in finite factors. Ann. Math. (2) **55**, 520–530 (1952). https://doi.org/10.2307/1969645
12. A. Guionnet, M. Krishnapur, O. Zeitouni, The single ring theorem. Ann. Math. (2) **174**(2), 1189–1217 (2011). https://doi.org/10.4007/annals.2011.174.2.10

13. U. Haagerup, On Voiculescu's R- and S-transforms for free non-commuting random variables, in *Free Probability Theory (Waterloo, ON, 1995)*. Fields Inst. Commun., vol. 12 (American Mathematical Society, Providence, 1997), pp. 127–148. https://doi.org/10.1215/s0012-7094-97-09004-9
14. U. Haagerup, F. Larsen, Brown's spectral distribution measure for R-diagonal elements in finite von Neumann algebras. J. Funct. Anal. **176**(2), 331–367 (2000). https://doi.org/10.1006/jfan.2000.3610. http://www.sciencedirect.com/science/article/pii/S0022123600936103
15. U. Haagerup, S. Möller, The law of large numbers for the free multiplicative convolution, in *Operator Algebra and Dynamics*. Springer Proc. Math. Stat., vol. 58 (Springer, Heidelberg, 2013), pp. 157–186. https://doi.org/10.1007/978-3-642-39459-1_8
16. U. Haagerup, H. Schultz, Brown measures of unbounded operators affiliated with a finite von Neumann algebra. Math. Scand. **100**(2), 209–263 (2007). https://doi.org/10.7146/math.scand.a-15023
17. U. Haagerup, H. Schultz, Invariant subspaces for operators in a general II_1-factor. Publ. Math. Inst. Hautes Études Sci. (109), 19–111 (2009). https://doi.org/10.1007/s10240-009-0018-7
18. U. Haagerup, T. Kemp, R. Speicher, Resolvents of R-diagonal operators. Trans. Am. Math. Soc. **362**(11), 6029–6064 (2010). https://doi.org/10.1090/S0002-9947-2010-05140-2
19. R. Lenczewski, Decompositions of the free additive convolution. J. Funct. Anal. **246**(2), 330–365 (2007). https://doi.org/10.1016/j.jfa.2007.01.010
20. J.A. Mingo, R. Speicher, *Free Probability and Random Matrices*. Fields Institute Monographs, vol. 35. (Springer, New York/Fields Institute for Research in Mathematical Sciences, Toronto, 2017). https://doi.org/10.1007/978-1-4939-6942-5
21. A. Nica, R. Speicher, R-diagonal pairs—a common approach to Haar unitaries and circular elements, in *Free Probability Theory (Waterloo, ON, 1995)*. Fields Inst. Commun., vol. 12 (American Mathematical Society, Providence, 1997), pp. 149–188
22. A. Nica, R. Speicher, Commutators of free random variables. Duke Math. J. **92**(3), 553–592 (1998). https://doi.org/10.1215/S0012-7094-98-09216-X
23. A. Nica, D. Shlyakhtenko, R. Speicher, R-diagonal elements and freeness with amalgamation. Canad. J. Math. **53**(2), 355–381 (2001). https://doi.org/10.4153/CJM-2001-015-6
24. E.B. Saff, V. Totik, *Logarithmic Potentials with External Fields*. Grundlehren der Mathematischen Wissenschaften [Fundamental Principles of Mathematical Sciences], vol. 316 (Springer, Berlin, 1997). https://doi.org/10.1007/978-3-662-03329-6. Appendix B by Thomas Bloom
25. D. Voiculescu, Limit laws for random matrices and free products. Invent. Math. **104**(1), 201–220 (1991). https://doi.org/10.1007/BF01245072
26. D. Voiculescu, The coalgebra of the free difference quotient and free probability. Int. Math. Res. Not. **2000**(2), 79–106 (2000). https://doi.org/10.1155/S1073792800000064
27. P. Zhong, Brown measure of the sum of an elliptic operator and a free random variable in a finite von Neumann algebra (2021). arXiv:2108.09844